5350
90 E

THE HISTORY OF
MODERN GEOGRAPHY

BIBLIOGRAPHIES OF THE HISTORY
OF SCIENCE AND TECHNOLOGY
(Vol. 9)

GARLAND REFERENCE LIBRARY
OF THE HUMANITIES
(Vol. 445)

Bibliographies of the History of Science and Technology

Editors

Robert Multhauf, Smithsonian Institution, Washington, D.C.
Ellen Wells, Smithsonian Institution, Washington, D.C.

1. *The History of Modern Astronomy and Astrophysics*
   A Selected, Annotated Bibliography
   by David DeVorkin
2. *The History of Science and Technology in the United States*
   A Critical, Selective Bibliography
   by Marc Rothenberg
3. *The History of the Earth Sciences*
   An Annotated Bibliography
   by Roy Sydney Porter
4. *The History of Modern Physics*
   An International Bibliography
   by Stephen Brush and Lanfranco Belloni
5. *The History of Chemical Technology*
   An Annotated Bibliography
   by Robert P. Multhauf
6. *The History of Mathematics from Antiquity to the Present*
   A Selective Bibliography
   by Joseph Dauben, editor
7. *The History of Meteorology and Geophysics*
   An Annotated Bibliography
   by Stephen G. Brush and Helmut E. Landsberg
8. *The History of Classical Physics*
   A Selected, Annotated Bibliography
   by R. W. Home, with assistance of Mark J. Gittins
9. *The History of Modern Geography*
   An Annotated Bibliography
   by Gary S. Dunbar
10. *The History of the Health Care Sciences and Health Care, 1700–1980*
    A Selective Annotated Bibliography
    by Jonathon Erlen

THE HISTORY OF MODERN GEOGRAPHY
An Annotated Bibliography of Selected Works

Gary S. Dunbar

GARLAND PUBLISHING, INC. • NEW YORK & LONDON
1985

© 1985 Gary S. Dunbar
All rights reserved

Library of Congress Cataloging in Publication Data
Dunbar, Gary S.
  The history of modern geography.

  (Bibliographies of the history of science and
technology ; v. 9) (Garland reference library of
the humanities ; v. 445)
  Includes indexes.
  1. Geography—History—Bibliography. I. Title.
II. Series. III. Series: Garland reference library
of the humanities ; v. 445.
Z6001.D86 1985 [G80] 016.9   83-48277
ISBN 0-8240-9066-7 (alk. paper)

Printed on acid-free, 250-year-life paper
Manufactured in the United States of America

This book is fondly dedicated to Elizabeth. As a librarian she can appreciate its usefulness, and as a wife she can overlook its shortcomings.

GENERAL INTRODUCTION

This bibliography is one of a series designed to guide the reader into the history of science and technology. Anyone interested in any of the components of this vast subject area is part of our intended audience, not only the student, but also the scientist interested in the history of his own field (or faced with the necessity of writing an "historical introduction") and the historian, amateur or professional. The latter will not find the bibliographies "exhaustive," although in some fields he may find them the only existing bibliographies. He will in any case not find one of those endless lists in which the important is lumped with the trivial, but rather a "critical" bibliography, largely annotated, and indexed to lead the reader quickly to the most important (or only existing) literature.

Inasmuch as everyone treasures bibliographies, it is surprising how few there are in this field. George Sarton's *Guide to the History of Science* (Waltham, Mass., 1952; 316 pp.), Eugene S. Ferguson's *Bibliography of the History of Technology* (Cambridge, Mass., 1968; 347 pp.), François Russo's *Histoire des Sciences et des Techniques. Bibliographie* (Paris, 2nd ed., 1969; 214 pp.) are justifiably treasured but they are, of necessity, limited in their coverage and need to be updated.

For various reasons, mostly bad, the average scholar prefers adding to the literature to sorting it out. The editors are indebted to the scholars represented in this series for their willingness to expend the time and effort required to pursue the latter objective. Our aim, and that of the publisher, has been to give the series enough uniformity to give some consistency to the series, but otherwise to leave the format and contents to the author/compiler. We have urged that introductions be used for essays on "the state of the field," and that selectivity be exercised to limit the length of each volume. Since the historical literature ranged from very large (e.g., medicine) to very small (e.g., chemical technology), some bibliographies will be limited to truly important writings while others will include modest "contributions" and even primary sources. The problem is intelligible guidance into a particular field—or subfield—and its solution is largely left to the author/compiler. In general, topical volumes (e.g., chemistry) will deal with the subject since about 1700, leaving earlier literature to area or chronological volumes (e.g., medieval science); but here, too, the volumes will vary according to the judgment of the author. The volumes are international (except for two, *Science and Technology in the United States* and *Science and Technology in Eastern Asia*) but the literature covered depends, of course, on the linguistic equipment of the author and his access to "exotic" literatures.

Robert Multhauf
Ellen Wells

Smithsonian Institution
Washington, D.C.

CONTENTS

Introduction ... xi

Part One—General and Topical

- I. General Bibliographical Works ... 3
- II. General Works on the History of Geography ... 10
- III. Discovery and Exploration ... 46
- IV. Cartography, Survey, and Navigation ... 60
- V. Physical Geography and Biogeography ... 82
- VI. Human Geography ... 96
- VII. Miscellaneous Subfields or Ancillary Disciplines ... 103
- VIII. Geographical Societies ... 105
- IX. Miscellaneous Organizations ... 129
- X. Geographical Periodicals ... 133
- XI. Geography in the Universities ... 137
- XII. Teaching (below University Level) ... 152

Part Two—Geography in Various Countries

- XIII. United States ... 155
- XIV. Great Britain ... 167
- XV. France ... 173
- XVI. Germany ... 181
- XVII. Italy ... 189
- XVIII. Other European Countries (including the USSR) ... 194
- XIX. Other Countries ... 206

Part Three—Biographical Works

- XX. United States ... 211
- XXI. Europe ... 235
- XXII. Other Countries ... 347

Author Index ... 353
Subject Index ... 369

INTRODUCTION

This book is intended to provide a guide to the literature in the history of geography from the mid-eighteenth century, when something like modern geography was beginning to take shape, to the present. A conscientious effort has been made to survey books and articles written in the major languages of western and central Europe. Although there is a fair body of interesting materials published in other languages—notably Russian, Polish, Hungarian, Finnish, and Japanese—the present work includes the vast majority of publications in the field. While this bibliography is designed for non-specialists, it is likely that specialists will find here numerous works that are new to them. It will also help them perceive trends and lacunae.

In order to keep the bibliography within manageable bounds, I necessarily had to limit the space devoted to some important fields, such as cartography and exploration, which deserve separate volumes. Indeed, one could easily devote a book to explorers' biographies alone. Instead, I have concentrated on academic geography, or the discipline of geography as it has emerged in the modern universities during the last two centuries. Geography in its earlier incarnations will be covered in the general volumes in this series on ancient, medieval, and Renaissance science. Since 1976, I have been making a special study of the professionalization and institutionalization of geography in Europe and North America, especially in the period 1870–1930, and this bibliography is obviously a useful adjunct to my other labors.

This is not a critical bibliography, because I did not include value judgments in my brief annotations. That may have been cowardly strategy, but I did not want to influence the reader. In fact, the state of this particular art is rather low. Most of the items in this bibliography would not measure up to the canons of scholarship in the history of science today. Only now are some historians of geography learning to write decent history by basing their work firmly on primary source materials and by placing their subjects in their general historical context. Many of the "histories" included herein are not histories at all but merely lists or simple narratives. In most cases, the authors were not proper historians or even proper geographers. Historians of science usually emphasize the physical and natural sciences, so that a field like geography is little tilled. Most geographers are concerned with their own tiny corner of reality and are unaware of their predecessors. Some of the histories of the field have been "whiggish" efforts highlighting certain individuals and events, with the author himself not mysteriously appearing in the center of the stage.

This volume is unique because there has never been a bibliography of this scope, but there are numerous bibliographical works that were useful to my compilation. I naturally made use of existing bibliographies, and I appropriated the references found in books and articles devoted to the history of geography and its subfields. In a sense, everything that has ever been published on any topic in geography can be useful to an historian of the field, but I am concentrating on those works that are *patently* historical. A scholar who wants to investigate academic geography in late 19th-century Germany would find the best sources in contemporary directories, catalogues, and unpublished university records, but such materials obviously cannot be listed here. There are scattered references to geography in publications devoted to other topics, but unless the references were lengthy or substantial, I did not include them in this bibliography. I have tried to emphasize utility and, to some degree, accessibility, although no single library, no matter how large, could be expected to contain all the references. A large percentage of these works can be found in any of several large North American libraries, which can reach out, with their interlibrary loan systems, almost to the ends of the earth. It has been my great good fortune to be situated at the University of California, Los Angeles (UCLA), with its marvelous Research Library only seventy-five paces from my office door. The UCLA libraries already possess most of the items in this bibliography, and what they did not have was usually

Introduction xiii

found by the interlibrary loan office in libraries near (e.g., Berkeley) and far (e.g., Biblioteca Nazionale Centrale, Rome). Many previously unseen titles were added in August 1982 during visits to several libraries in the eastern United States (Library of Congress, Johns Hopkins University, Princeton University, and Columbia University).

The following are perhaps the most useful general geographical bibliographies to an historian of geography:

1. *Bibliographie géographique internationale* (Paris). Annual 1891–1976; quarterly since 1976.
2. *Current Geographical Publications*, ten times a year since 1938. Issued by the American Geographical Society from New York, 1938–1978, and from Milwaukee since 1978.
3. *Research Catalogue of the American Geographical Society*, 15 vols. (Boston: G.K. Hall & Co., 1962). See especially Volume 2, which includes the history of geography.
4. *New Geographical Literature and Maps*, issued by the Royal Geographical Society, London, 1951–1981.
5. *Documentatio Geographica*, issued by the Bundesanstalt für Landeskunde und Raumforschung, Bad Godesberg, Germany, 1967–1973.
6. *Bibliotheca Geographica* (Gesellschaft für Erdkunde zu Berlin), vols. 1 (1891–1892) to 19 (1911–1912).
7. *The Bibliography of Cartography*, Library of Congress, Geography and Map Division, 5 vols. (Boston: G.K. Hall & Co., 1973), plus a 2-volume Supplement published in 1980.

Of considerable value for works published from 1750 to 1856 is *Bibliotheca Geographica,* edited by Wilhelm Engelmann (Leipzig: W. Engelmann, 1858) (reprinted Amsterdam: Meridian Publishing Company, 1956, in two volumes).

Among the general bibliographies, the most useful was Magda Whitrow, ed., *Isis Cumulative Bibliography: A Bibliography of the History of Science Formed from Isis Critical Bibliographies 1–90 (1913–1965)*, vol. 3: *Subjects* (London: Mansell, 1976). The remaining *Isis Critical Bibliographies* (post-1965) were also scanned.

Although there has never been a book-length bibliography of the history of geography, there have been some useful compilations of lesser scope. Of great value were the bibliographical essays published in the *Geographisches Jahrbuch* (Gotha, Germany) in the latter part of the 19th century and the early 20th century, but unfortunately they have not been continued. In the *Geographisches Jahrbuch,* as in many other sources, the history of geography is often confused with historical geography, which is now an altogether different field. The history of cartography and exploration often made up a large part of the essays that purportedly dealt with the history of geography.

Articles in the history of geography are published in a wide variety of journals. The only periodicals devoted solely to the subject are the *History of Geography Newsletter,* edited by Geoffrey J. Martin of Southern Connecticut State University (1981–), and an annual publication, *Geographers: Biobibliographical Studies,* edited by T.W. Freeman and published by Mansell of London (1977–). The ancillary fields of the history of cartography and exploration have annual publications, *Imago Mundi* (1935–) and *Terrae Incognitae* (1969–), respectively. Of immense value to the historian of geography is the *International List of Geographical Serials,* 3rd edition, edited by Chauncy D. Harris and Jerome D. Fellmann (University of Chicago, Department of Geography, Research Paper 193, 1980). Professor Harris is also the compiler of *Bibliography of Geography,* Part 1, *Introduction to General Aids* (Ibid., no. 179, 1976). Without these publications, my work would have been much more difficult.

A work of obvious value to the historian of geography, as it is to every other field of science, is the monumental *Dictionary of Scientific Biography,* edited by Charles C. Gillispie (16 vols.; New York: Charles Scribner's Sons, 1970–1980). Although the *Dictionary* naturally emphasizes the physical and natural sciences and mathematics, I nevertheless found biographical sketches of more than 130 individuals (geographers, cartographers, explorers, and workers in closely related fields) who have a place in the history of geography. Some major geographers, such as Alfred Hettner and Paul Vidal de la Blache, are missing, while some of their relatively less important contemporaries are included, but on the whole the *Dictionary* is a very useful work.

No bibliography can ever be complete, especially in an old discipline like geography that is in a constant state of change, but I hope that I have given interested readers some guidelines that

Introduction

they can use in charting their own courses. I did not usually supply information about editions other than those I used or about translations, but anyone who requires such data can find them in the various catalogues and national bibliographies to be found in any major library. Similarly, it might have been useful to include obituaries and book reviews, but such additions would have lengthened my task considerably. The reader can quite easily find such material in journals of the period immediately following the event (death or publication). I have assumed that it would be better to include a larger number of titles rather than heavy detail on a smaller number. The only unpublished materials included in this bibliography are several doctoral theses that I considered important.

In the course of their work, bibliographers become so deeply mired in debt that they can never hope to extricate themselves, but I should at least attempt a minimal listing of individuals who have rendered assistance: Ms. Janet Ziegler and Ms. Patrice Caravello, former and present heads of Interlibrary Services in the UCLA Research Library, and their staff; Professor Dr. Hanno Beck (HB) of the University of Bonn; Dr. Gerhard Engelmann of Potsdam; Professor Philippe Pinchemel (PP) of the University of Paris and his associates Marie-Claire Robic (MCR) and Gilles Palsky (GP): Dr. Numa Broc of the University of Perpignan; Professor Lucio Gambi (LG) of the University of Bologna; Professor Elio Migliorini and Dr. Claudia Mancini of the University of Rome; Mr. Merle Abbott of Birkbeck College, University of London; Professor J.A. van Ginkel (JAvG) of the University of Utrecht; Professor Geoffrey J. Martin of Southern Connecticut State University; Professor William A. Koelsch (WAK) of Clark University; Professor Clinton R. Edwards (CRE) of the University of Wisconsin-Milwaukee; Professor Lay J. Gibson of the University of Arizona; Dr. Anne Macpherson (AM) of El Cerrito, California; Dr. Susan J. Smith of Brunel University; and Dr. Jules Zentner of UCLA. Camera-ready copy was produced at the UCLA Central Word Processing Center, and my thanks go to Ms. Jane Bitar, Mrs. Irene Chow, and Mrs. Ethel Grigsby.

Annotations and translations are my own, except where initialed (see above for initials used), but I must acknowledge the aid of Mrs. Ursula Willenbrock Martin of the UCLA Research

Library in unraveling some particularly knotty German conundrums. All blame for errors of translation, transcription, or judgment should be laid at my door. Following current bibliographical practice, I have cited the last numbered page of each book, even though there might have been some printed matter on one or more succeeding pages.

The compilation of a bibliography is indeed a long-distance run, and the bibliographer plods his lonely course without experiencing the mysterious burst of strength that sometimes comes to runners. But bibliography is not entirely an act of masochism. The work has already been useful to me in my teaching and research, and if this book is similarly helpful to others, I shall feel that the marathon has been worth running.

G.S. Dunbar
University of California, Los Angeles
September 1984

The History of
Modern Geography

PART ONE

GENERAL AND TOPICAL

I. GENERAL BIBLIOGRAPHICAL WORKS IN THE HISTORY
   OF GEOGRAPHY

N.B.--One should also consult the general bibliographies listed in the Introduction. Those works contain many references to publications in the history of geography.

1. Altengarten, James S., and Molyneaux, Gary A. The History, Philosophy, and Methodology of Geography: A Bibliography Selected for Education and Research. Council of Planning Librarians (Monticello, Illinois), Exchange Bibliography 957 (January 1976). 56 p.

    A bibliography of 678 English-language works on the history, philosophy, and methodology of geography, including biographies and works on the history of exploration.

2. Broc, Numa. "De quelques bibliographies anciennes utiles à l'historien de la géographie (XVIe-XVIIIe siècles)," Revue d'histoire des sciences et de leurs applications, vol. 31, no. 2 (April 1978), 97-130.

    Examines 14 bibliographies published between 1545 and 1778. Only 2 of the bibliographies were published after 1750: Bibliothèque historique de la France by Jacques Le Long, edited by Charles-Marie Fevret de Fontette (5 vols., 1768-1778), and Méthode pour étudier la géographie by Abbé Nicolas Lenglet-Dufresnoy (1768 posthumous edition of work first published in 1716).

3. Geographisches Jahrbuch, published in Gotha, Germany, by Justus Perthes since 1866. Contains bibliographical

essays on various subfields of geography, including the following articles dealing with the history of geography (listed in chronological order):

4. Behm, E. "Einiges über die geographischen Reisen, Gesellschaften und Publikationen der Gegenwart," Geographisches Jahrbuch, vol. 1 (1866), 552-580.

   A resume of exploration, the work of geographical societies, and geographical publishing in recent decades (societies since 1821, publications since 1851).

5. Behm, E. "Die bedeutenderen geographischen Reisen in den Jahren 1866 und 1867, nebst Notizen über die geographischen Gesellschaften und Publikationen," Geographisches Jahrbuch, vol. 2 (1868), 419-479.

   Summary of recent travels, publications, and the activities of geographical societies.

6. Behm, E. "Die bedeutenderen geographischen Reisen in den Jahren 1868 und 1869, nebst Notizen über die geographischen Gesellschaften und Publikationen," Geographisches Jahrbuch, vol. 3 (1870), 482-572.

   Geographical activities of the last 2 years: travels, publications, and the work of geographical societies.

7. Spörer, J. "Zur historischen Erdkunde: Ein Streifzug durch das Gebiet der geographischen und historischen Literatur," Geographisches Jahrbuch, vol. 3 (1870), 326-420.

   This article does not concern historical geography as we know it today but is, rather, a bibliographical survey of the literature of geography, anthropology, and history. The great foundation-stones were Humboldt's Kosmos and Ritter's Erdkunde.

8. Spörer, J. "Zur historischen Erdkunde: Zweiter Streifzug durch das Gebiet der geographischen und historischen Literatur," Geographisches Jahrbuch, vol. 4 (1872), 184-272.

   Essay on the history of geography, with emphasis on German works in the field in the 19th century. See also pp. 542-552, "Nachtrag zu J. Sporer, 'Zur historischen Erdkunde.'"

Part One--General and Topical 5

9. Behm, E. "Die bedeutenderen geographischen Reisen in den Jahren 1870 und 1871, nebst Notizen über die geographischen Gesellschaften und Publikationen," Geographisches Jahrbuch, vol. 4 (1872), 374-450.

   Geographical activities of the last 2 years: travels, publications, and the work of geographical societies.

10. Behm, E. "Die bedeutenderen geographischen Reisen in den Jahren 1872 und 1873," Geographisches Jahrbuch, vol. 5 (1874), 190-322.

    Exploration and other travels conducted during the last 2 years.

11. Behm, E. "Geographische Gesellschaften und Zeitschriften," Geographisches Jahrbuch, vol. 5 (1874), 345-361.

    Journals and the activities of geographical societies in the past 2 years.

12. Behm, E. "Die bedeutenderen geographischen Reisen in den Jahren 1874 und 1875," Geographisches Jahrbuch, vol. 6 (1876), 434-544.

    Geographical travels during the last biennium.

13. Behm, E. "Geographische Gesellschaften und Zeitschriften," Geographisches Jahrbuch, vol. 6 (1876), 545-568.

    Concerns contemporary geographical societies and journals.

14. Behm, E. "Geographische Gesellschaften und Zeitschriften," Geographisches Jahrbuch, vol. 7 (1878), 636-660.

    Geographical societies and journals.

15. Wagner, Hermann. "Der gegenwärtige Standpunkt der Methodik der Erdkunde," Geographisches Jahrbuch, vol. 7 (1878), 550-636.

    Methodology of geography, instruction in geography in German-speaking universities, and a long section (pp. 565-598) on Ritter and Peschel and their legacy.

16. Wagner, Hermann. "Bericht über die Entwicklung der Methodik der Erdkunde," Geographisches Jahrbuch, vol. 8 (1880), 523-598.

   Concerns geographical methodology, the legacy of Ritter and Peschel, the character of geography in other European countries, and the establishment of professorships in European universities.

17. Wagner, Hermann, and Wichmann, H. "Geographische Gesellschaften, Kongresse und Zeitschriften," Geographisches Jahrbuch, vol. 8 (1880), 599-653.

   Geographical societies, journals, and congresses.

18. Wagner, Hermann. "Bericht über die Entwickelung des Studiums und der Methodik der Erdkunde," Geographisches Jahrbuch, vol. 9 (1882), 651-700.

   Scientific geography, its methodology and teaching.

19. Wichmann, H., and Wagner, Hermann. "Geographische Gesellschaften, Kongresse, Ausstellungen und Zeitschriften," Geographisches Jahrbuch, vol. 9 (1882), 701-714.

   Geographical societies, congresses, exhibitions, and journals.

20. Wagner, Hermann. "Bericht über die Entwickelung der Methodik und des Studiums der Erdkunde (1883-1885)," Geographisches Jahrbuch, vol. 10 (1884), 539-648.

   Methodology and teaching of scientific geography. Includes a name index for Wagner's 4 essays on geographical methodology in vols. 7-10.

21. Wichmann, H., and Wagner, Hermann. "Geographische Gesellschaften, Zeitschriften, Kongresse und Ausstellungen," Geographisches Jahrbuch, vol. 10 (1884), 651-674.

   Geographical societies, journals, congresses, and exhibits.

22. Wolkenhauer, W. "Geographische Nekrologie für die Jahre 1884, 1885, 1886 und 1887," Geographisches Jahrbuch, vol. 12 (1888), 349-408.

   Obituaries of geographers and people in related fields, 1884-1887. Name register (pp. 399-408) also includes names of people whose death was reported in Petermanns Geographische Mitteilungen, 1855-1884.

23. Wagner, Hermann. "Bericht über die Entwickelung der Methodik und des Studiums der Erdkunde (1885-1888)," Geographisches Jahrbuch, vol. 12 (1888), 409-460.

   Methodological writings since 1885.

24. Wichmann, H., and Wagner, Hermann. "Geographische Gesellschaften, Zeitschriften, Kongresse und Ausstellungen," Geographisches Jahrbuch, vol. 12 (1888), 461-474.

   Geographical societies, journals, and congresses.

25. Wagner, Hermann. "Bericht über die Entwickelung der Methodik und des Studiums der Erdkunde (1889-91)," Geographisches Jahrbuch, vol. 14 (1890/91), 371-462.

   Recent methodological writings. 2 long appendixes (pp. 412-462) on the geographical institutes in European universities and their instructional staffs.

26. Wichmann, H., and Wagner, Hermann. "Geographische Gesellschaften, Zeitschriften, Kongresse und Ausstellungen," Geographisches Jahrbuch, vol. 14 (1890/91), 463-484.

   Geographical societies, journals, and congresses.

27. Ruge, Sophus. "Die Litteratur zur Geschichte der Erdkunde in den letzten zehn Jahren (bis 1893) von Mittelalter an," Geographisches Jahrbuch, vol. 18 (1895), 1-60.

   Bibliographical essay on writings on the history of geography from the Middle Ages onward, published in the decade ending in 1893.

28. Wagner, Hermann. "Die Lehrstühle und Dozenten der Geographie an europäischen Hochschulen 1896," Geographisches Jahrbuch, vol. 19 (1896), 397-402.

   Professors and lecturers in European universities in 1896.

29. Kollm, Georg. "Geographische Gessellschaften, Zeitschriften, Kongresse und Ausstellungen," Geographisches Jahrbuch, vol. 19 (1896), 403-430.

Geographical societies, journals, congresses, and exhibits.

30. Ruge, Sophus. "Die Litteratur zur Geschichte der Erdkunde vom Mittelalter an (1894-96)," Geographisches Jahrbuch, vol. 20 (1897), 217-248.

    Bibliographical essay on recent writings in the history of geography from the Middle Ages onward.

31. Ruge, Sophus. "Die Litteratur zur Geschichte der Erdkunde vom Mittelalter an. 1897-1900," Geographisches Jahrbuch, vol. 23 (1900), 173-212.

    Recent works on the history of geography from the Middle Ages onward.

32. Kollm, Georg. "Geographische Gesellschaften, Zeitschriften und Kongresse," Geographisches Jahrbuch, vol. 24 (1901), 397-424.

    Geographical societies, journals, and congresses.

33. Ruge, Sophus. "Die Literatur zur Geschichte der Erdkunde vom Mittelalter an (1900-1903)," Geographisches Jahrbuch, vol. 26 (1903), 175-218.

    Recent writings on the history of geography since the beginning of the medieval period.

34. Ruge, Walther. "Die Literatur zur Geschichte der Erdkunde vom Mittelalter an (1903-07)," Geographisches Jahrbuch, vol. 30 (1907), 329-380.

    Bibliographical essay on the history of geography from the Middle Ages to the present. Sophus Ruge died in 1903, and his son Walther carried on his bibliographical work. Walther's citations were not always so complete as Sophus'.

35. Kollm, Georg. "Geographische Gesellschaften, Zeitschriften, Kongresse und Ausstellungen," Geographisches Jahrbuch, vol. 32 (1909), 409-438.

    Geographical societies, journals, and congresses.

36. Kretschmer, Konrad. "Die Literatur zur Geschichte der Erdkunde vom Mittelalter an (1907-25)," Geographisches Jahrbuch, vol. 41 (1926), 122-192.

Part One--General and Topical 9

Writings published in the period 1907-1925 on the history of geography from the Middle Ages to the present.

37. Herrmann, Albert. "Geschichte der Geographie (1926-39). I.: Bis zum Ausgange des Mittelalters," Geographisches Jahrbuch, vol. 55 (1940), 381-434.

 Writings published between 1926 and 1939 on the history of geography up to the end of the Middle Ages.

38. Herrmann, Albert. "Geschichte der Geographie (1926-39). II. Teil: Vom Zeitalter der Entdeckungen bis zum Ende des 19. Jahrhunderts," Geographisches Jahrbuch, vol. 56 (1941), 433-543.

 Writings (1926-39) on the history of geography from the Voyages of Discovery to the end of the 19th century.

39. Hervé, Roger; Lagarde, L.; and Grivot, F., comps. Bibliographie d'histoire de la géographie et de géographie historique 1976. (France, Ministère des universités, Comité des travaux historiques et scientifiques, Section de géographie.) Paris: Bibliothèque Nationale, 1978. 83 p.

 Annotated bibliography of recent publications (c. 1970-77) in the history of geography and exploration (pp. 5-34) and in historical geography (pp. 35-72). The 1977 issue was published in 1979 and contains 96 pages (pp. 5-40 cover the history of geography and exploration). The 1978 issue was published in 1980 and consists of 120 pages (pp. 5-45 treat the history of geography and exploration).

40. Ruge, Sophus; Ruge, Walther; and Kretschmer, Konrad. Die Litteratur zur Geschichte der Erdkunde vom Mittelalter an. Hildesheim, Germany, and New York: Georg Olms Verlag, 1980. 300 p.

 Reprint of 6 bibliographical essays (all cited above) that appeared in Geographisches Jahrbuch between 1895 (vol. 18) and 1926 (vol. 41).

II. GENERAL WORKS ON THE HISTORY OF GEOGRAPHY

41. Aay, Henry. "Textbook Chronicles: Disciplinary History and the Growth of Geographic Knowledge," pp. 291-301 in The Origin of Academic Geography in the United States, ed. Brian W. Blouet (Hamden, Connecticut: Archon Books, 1981).

    Essay seeks "to discover how the growth of geographic knowledge has been explored and presented in a number of large-scale English-language disciplinary histories" (Freeman's A Hundred Years of Geography, Dickinson's The Makers of Modern Geography, James' All Possible Worlds, Dickinson and Howarth's The Making of Geography, Warntz and Wolff's Breakthroughs in Geography, and Fischer, Campbell, and Miller's A Question of Place [all cited below]).

42. Ahmad, Kazi S. "Geography through the Ages," Pakistan Geographical Review, vol. 19, no. 1 (January 1964), 1-30.

    A general survey of geography from ancient times to modern, based on English-language sources.

43. Almagià, Roberto. "Concetto e indirizzi della geografia attraverso i tempi," pp. 7-51 in Introduzione allo studio della geografia (Milan: Marzorati, 1947).

    History of geography from Homer to the present. Reprinted in Almagià's Scritti geografici (1905-1957) (below).

44. Almagià, Roberto. Scritti geografici (1905-1957). Rome: Edizioni Cremonese, 1961. xi + 652 p.

    Contains several chapters on the history of geography and cartography, but mostly for earlier periods (i.e., Renaissance and earlier). Includes bibliography of Almagià's writings, 1902-1960.

45. Almagià, Roberto. "Storia della geografia," pp. 183-303 in Storia delle scienze, ed. Nicola Abbagnano, vol. 1 (Turin: Unione Tipografico-Editrice Torinese, 1962).

History of geography from ancient Greeks to the present, with considerable emphasis on cartography and exploration.

46. Antoniol, G.B. <u>Dello sviluppo storico del concetto scientifico della geografia</u>. Turin: Stamperia Reale di G.B. Paravia e Comp., 1903. 46p.

    Historical sketch of geography from ancient times to the present.

47. Auerbach, Bertrand. "L'évolution des conceptions et de la méthode en géographie," <u>Journal des savants</u>, n.s. vol. 6 (June 1908), 309-321.

    A review of trends in geography, beginning with Humboldt and Ritter. Particular emphasis on German philosophical debates about the nature of geography.

48. Babicz, Józef, ed. <u>Les écoles géographiques</u>. (Reprinted from <u>Organon</u>, no. 14.) Warsaw: PWN-- Polish Scientific Publishers, 1980. 94 p.

    Papers originally delivered in Leningrad in 1976 in a colloquium, "Les écoles de géographie, leurs caractéristiques et leurs liens internationaux," sponsored by the International Geographical Union's Commission on the History of Geographical Thought as part of the International Geographical Congress in Moscow. All of the papers are cited separately in this volume, with the exception of Manfred Büttner, "The Historical Conditions Affecting the Development of 'Geographia Generalis,'" which deals with an earlier period.

49. Babicz, Józef. "L'influence des sciences naturelles sur l'évolution de la géographie au XIXe siècle: postulats des recherches," <u>Actes du XIe Congrès International d'Histoire des Sciences</u> (Warsaw, etc., 1965), vol. 4 (Warsaw, etc.: Ossolineum, 1968), 307-310.

    Influence of natural sciences upon geography in the 19th century, with particular emphasis on Humboldt, Ritter, Moritz Wagner, and Ratzel.

50. Babicz, Józef, ed. <u>Studia z dziejów geografii i kartografii / Etudes d'histoire de la géographie et de la cartographie</u>. (Monografie z Dziejów Nauki i

Techniki, vol. 87.) Warsaw: Polish Academy of Science, 1973. 550 p.

Papers on the history of geography and cartography in Polish, French, Russian, German, and English. Several of the papers are cited separately elsewhere in this bibliography.

51. Baker, J.N.L. "Geography and Its History," The Advancement of Science, vol. 12, no. 46 (September 1955), 188-198.

    Presidential address to Section E of the British Association for the Advancement of Science. A rambling essay on geography from the 16th century onward, with emphasis on Great Britain. Reprinted in Baker's History of Geography (below).

52. Baker, J.N.L. The History of Geography. Oxford: Basil Blackwell, 1963. xxviii + 259 p.

    17 previously published essays on a wide range of topics, some of which do not concern the history of geography, despite the title.

53. Banse, Ewald. Entwicklung und Aufgabe der Geographie. Rückblicke und Ausblicke einer universalen Wissenschaft. ("Sammlung Die Universität," vol. 39.) Stuttgart and Vienna: Humboldt-Verlag, 1953. 239 p.

    Survey of geography from ancient times to modern, with heavy emphasis on German geographers.

54. Banse, Ewald. Lexikon der Geographie. 2 vols. Braunschweig and Hamburg: Georg Westermann, 1923. 786 + 785 p.

    A geographical dictionary or encyclopedia, with places, people, and topics arranged alphabetically. Article "Geographie" (I, 486-488) is an historical sketch of geography from ancient times to modern.

55. Beck, Hanno. "Entdeckungsgeschichte und geographische Disziplinhistorie," Erdkunde, vol. 9, no. 3 (July 1955), 197-204.

    (From English summary) "Until today history of geography has been understood as the history of discovery and exploration. For that reason an account of the development of geography as a branch of learning is still missing. The history of discovery and the history of geography are not identical but only related."

56. Beck, Hanno. Geographie: Europäische Entwicklung in Texten und Erläuterungen. ("Orbis Academicus," ed. Fritz Wagner.) Freiburg and Munich: Verlag Karl Alber, 1973. 510 p.

    Excerpts from geographical writings from ancient times to modern, amply supported by the editor's biographical and bibliographical notes. Emphasis on German geographers in the modern period.

57. Beck, Hanno. "Zu geographiegeschichtlichen Dissertationen 1961-1972," Erdkunde, vol. 28, no. 2 (June 1974), 125-129.

    Review article giving resumes of 10 German doctoral theses that concern the history of geography.

58. Beck, Hanno. "Krise der Geographie--Krise der Geschichte der Geographie? Geographiegeschichte und Wissenschaftstheorie," Sudhoffs Archiv: Zeitschrift für Wissenschaftsgeschichte, vol. 61, no. 1 (First Quarter 1977), 45-53.

    Methodological concerns in the field of geography in the 1960s and how they have affected the study of the history of geography. Now greater concern for scientific theory and for the history of science.

59. Beck, Hanno. "Methoden und Aufgabe der Geschichte der Geographie," Erdkunde, vol. 8, no. 1 (February 1954), 51-57.

    (From English summary) "From 1840 onwards a number of German geographers has paid more than passing attention to the history of their subject. At that time the essential ideas for an exact history of geography formulated by J.G. Lüdde and J. Löwenberg were overlooked because of the dominant interest in exploration. Since then various works on the history of geography have been devoted exclusively to the history of exploration, ignoring the fact that exploration is by no means the only or even a necessary concern of a geographer.... At present several German geographers are attempting to prepare monographs on the history of geography <u>stricto</u> <u>sensu</u> [sic]. It is suggested that in the first instance the history of present day controversies should be dealt with."

60. Beck, Hanno. "Das Problemfeld der Geschichte der Geographie. Erläuterung einer Strukturskizze," <u>Erdkunde</u>, vol. 31, no. 2 (June 1977), 81-85.

   The place of the history of geography ("geographische Disziplinhistorie") in relation to history and geography and to the history and philosophy of science, the history of cartography, and the history of exploration.

61. Beck, Hanno. "Zeittafel der Geographie von den Anfängen bis 1750," <u>Geographisches Taschenbuch 1962/63</u> (Wiesbaden:: Franz Steiner, 1962), pp. 1-20.

   Chronology of events in the history of geography and exploration beginning 1492/3 B.C. when the Pharaoh Hatshepsut sent a trading expedition to the Land of Punt.

62. Beck, Hanno. "Zeittafel der Geographie von 1905-1945," <u>Geographisches Taschenbuch 1964/65</u> (Wiesbaden: Franz Steiner, 1964), 1-18.

   Chronological table of major events in geography to the end of World War II. Emphasis on individuals, publications, and expeditions. Emphasis on Germans but refers also to British, French, and Americans.

63. Beck, Hanno. "Zeittafel der Geographie von 1859 bis 1905," <u>Geographisches Taschenbuch 1960/61</u> (Wiesbaden: Franz Steiner, 1960), pp. 1-14.

   Chronological table of major events in geography from 1859 (deaths of Humboldt and Ritter) to 1905 (deaths of Reclus, Richthofen, and Bastian). Divided into 2 parts: 1859-1870 (Prelude to modern geography) and 1870-1905 (Geography under the influence of Richthofen and Ratzel).

64. Beck, Hanno. "Zeittafel der präklassischen und klassischen Geographie," <u>Geographisches Taschenbuch 1958/59</u> (Wiesbaden: Franz Steiner Verlag GMBH, 1958), pp. 29-48.

Part One--General and Topical   15

> A chronological table of facts and events in the history of geography, 1750-1859. "Preclassical" period was from 1750 to 1798; "classical," 1799-1859.

65. Berdoulay, Vincent. "The Contextual Approach," pp. 8-16 in Geography, Ideology and Social Concern, ed. D.R. Stoddart (Oxford: Basil Blackwell, 1981).

    Argues for a greater concern for the study of historical contexts or intellectual climates in the history of geography.

66. Berdoulay, Vincent. "Professionnalisation et institutionnalisation de la géographie," pp. 15-22 in Les écoles géographiques, ed. Józef Babicz (Warsaw: PWN--Polish Scientific Publishers, 1980).

    Describes the various ways in which geography was institutionalized and professionalized in Europe and the United States. Treats the role of the geographical societies and especially the universities in the process of professionalization.

67. Berry, Brian J.L. "Geographical Theories of Social Change," pp. 18-35 in The Nature of Change in Geographical Ideas ("Perspectives in Geography," 3), ed. B.J.L. Berry (DeKalb: Northern Illinois University Press, 1978).

    Traces what the author considers "the larger paradigm changes that have characterized the scientific progress of the field" from ancient times onward. Emphasis on the paradigms of environmental determinism and cultural diffusion in the 19th and 20th centuries.

68. Bertacchi, Cosimo. Nuovo dizionario geografico universale. Vol. 1. Turin: Unione Tipografico-Editrice, 1904. civ + 783 + 190 (supplement) p.

    An historical sketch of geography from ancient times to the end of the 19th century is given in the "Methodological and Historical Introduction" (pp. xiii-c).

69. Bobek, Hans. "Die Entwicklung der Geographie--Kontinuität oder Umbruch?," Mitteilungen der Osterreichischen Geographischen Gesellschaft, vol. 114, nos. 1-2 (1972), 3-18.

Survey of post-Hettnerian methodological developments. Counters some points made in recent articles by Dieter Bartels. Although recognizing that quantification and model-building are important, Bobek would like to retain the "synthesizing perspective" of the older style of geography. It is not necessary to set the Classical and New geographies up in opposition to each other.

70. Bonetti, Eliseo. "Attraverso la storia della geografia. I precursori della moderna geografia," Geopolitica (Milan), vol. 3, nos. 8-9 (1941), 3-11.

    A survey of geographical thought from Peter Apian (1524) to Humboldt and Ritter.

71. Bonetti, Eliseo. "L'evoluzione del pensiero geografico dall'antichità agli inizi del XIX secolo," Notiziario dell'Istituto de geografia dell'Università di Trieste, no. 1 (March 1950), 4-10.

    Brief sketch of the history of geography from ancient times to the beginning of the 19th century. [LG]

72. Bonifacio, Antoine. "La géographie," pp. 1131-1162 in Histoire de la science, ed. Maurice Daumas ("Encyclopédie de la Pléiade," vol. 5) (Paris: Librairie Gallimard, 1957).

    An historical sketch of geography from ancient times to the present.

73. Bowen, Margarita. Empiricism and Geographical Thought from Francis Bacon to Alexander von Humboldt. Cambridge: Cambridge University Press, 1981. xv + 351 p.

    (From Introduction, p. 1) "As part of a widespread reassessment of the positivist movement in science, this study considers its impact on geography during the period from 1600 to 1860, commencing with the sense-empiricism of Francis Bacon and concluding with the work of Alexander von Humboldt."

74. Bowman, Isaiah. Geography in Relation to the Social Sciences. (Also includes Geography in the Schools of Europe by Rose B. Clark.) (Report of the American Historical Association Commission on the Social

Part One--General and Topical   17

Studies, part 5.) New York: Charles Scribner's Sons,
1934. xxii + viii (Introduction to Clark's section)
+ 382 p.

Bowman's work is a survey of human geography. It
is not a history of geography but does mention some
historical matters, especially in the first chapter,
"By Way of Definition" (pp. 1-39), and in a later
discussion of regional systems (pp. 154-163). Bowman's
work ends on p. 227 and Clark's section is continuously
paged (viii, 229-366). Clark's work is on the teaching
of geography in European schools and is not historical.

75. Broc, Numa. "Histoire et historiens de la géographie.
Notes bio-bibliographiques (milieu du XVIIIe siècle--
1914)," France, Comité des travaux historiques et
scientifiques, Bulletin de la section de géographie,
vol. 84 (1979--published 1981), 72-116.

Biobibliographical essay on historians of geography
and their publications, from the middle of the 18th
century to 1914. Emphasis on classical, medieval, and
Renaissance geography and on the history of
exploration.

76. Brusa, A. "Cento anni di geografia," Quaderno di
Studi e Ricerche di Geografia Economica e Regionale
(Genoa), no. 4 (1966/1969), 89-130.

A review article based on T.W. Freeman's A Hundred
Years of Geography (1962). [LG]

77. Büttner, Manfred. "Geographie und Theologie im 18.
Jahrhundert," Verhandlungen des deutschen
Geographentages, Bochum 1965 (Wiesbaden, 1966), pp.
352-356, followed by discussion on pp. 356-359.

The relations between geography and theology in the
18th century. Explains why so many Protestant
ministers were geographers. [HB]

78. Büttner, Manfred, ed. Wandlungen im geographischen
Denken von Aristoteles bis Kant. ("Abhandlungen und
Quellen zur Geschichte der Geographie und
Kosmologie," 1.) Paderborn, etc.: Ferdinand
Schöningh, 1979. 276 p.

The first 4 sections deal with geography and
cosmography from the ancient Greeks through the 17th

century. The fifth section, "The Beginnings of Modern Geographical Thought in the 18th Century," contains 4 chapters--on Christian Wolff, Eberhard David Hauber, Johann Michael Franz, and Immanuel Kant ("Immanuel Kant and the Conception of Geography at the End of the 18th Century" by Karl Hoheisel, pp. 263-276).

79. Buttimer, Anne. "On People, Paradigms, and 'Progress' in Geography," pp. 81-98 in Geography, Ideology and Social Concern, ed. D.R. Stoddart (Oxford: Basil Blackwell, 1981).

    Essay on the appropriateness of the concept of paradigm to the history of geography. Looks for insights from oral history.

80. Buttimer, Anne. The Practice of Geography. London and New York: Longman, 1983. xiii + 298 p.

    Illustrates the practice of geography by 12 autobiographical essays by contemporary senior geographers (2 Americans and 10 Europeans), transcripts of 3 group discussions ("The Environment of Graduate School" [U.S.], "French Geography in the Forties," and "American Geography in the Fifties"), and an appendix on "Highlights of the Decades (1900-80) in Nine Countries."

81. Campbell, J.A., and Livingstone, D.N. "Neo-Lamarckism and the Development of Geography in the United States and Great Britain," Transactions of the Institute of British Geographers, n.s. vol. 8, no. 3 (1983), 267-294.

    Comte and Darwin have received considerable attention, but "very rarely ... has recognition been given to the impact on geography, both directly and indirectly, of another equally significant strand in late nineteenth century naturalism, namely, the selective revival of Lamarckian doctrines which, when used in evolutionary theory to supplement, rather than to supplant, Darwinian natural selection, came to be known as Neo-Lamarck(ian)ism" (p. 267).

82. Capel, Horacio. Filosofía y ciencia en la geografía contemporánea: Una introducción a la geografía. Barcelona: Barcanova, 1981. xiii + 509 p.

Part One--General and Topical   19

Book consists of 3 parts: Part 1, "The Putative
Fathers of Contemporary Geography" (Humboldt and
Ritter), pp. 3-76; Part 2, "The Institutionalization of
Geography in the 19th Century," pp. 77-241; and Part 3,
"The Course of Scientific Ideas" (positivism,
historicism, neopositivism, Marxism, etc.), pp. 243-
456.

83. Capel, Horacio. "Institutionalization of Geography and
    Strategies of Change," pp. 37-69 in Geography,
    Ideology and Social Concern, ed. D.R. Stoddart
    (Oxford: Basil Blackwell, 1981).

    (Page 38) "Today's geography has its origins in the
    process of institutionalization that from the mid
    nineteenth century, and after a period of decline,
    leads to the appearance of the scientific community of
    geographers, extending without interruption to the
    present time."

84. Claval, Paul. "La brève histoire de la nouvelle
    géographie," Rivista Geografica Italiana, vol. 83,
    no. 4 (December 1976), 395-424.

    The development of a "new" geography from the 1950s.
    The new geography is quantitative and theoretical
    (spatial analysis). It brings geography at last into
    the social sciences. It is mostly an Anglo-Saxon
    development but includes Torsten Hägerstrand of Sweden.
    Claval cites mostly English-language sources.

85. Claval, Paul. "Epistemology and the History of
    Geographical Thought," pp. 227-239 in Geography,
    Ideology and Social Concern, ed. D.R. Stoddart
    (Oxford: Basil Blackwell, 1981).

    Consideration of the applicability of the ideas of
    the French philosopher Michel Foucault to the analysis
    of the development of geographical thought. Also
    published in Progress in Human Geography, vol. 4, no. 3
    (September 1980), 371-384.

86. Claval, Paul. "L'influence de la géographie physique
    et de la géographie naturelle sur les concepts et les
    méthodes de la géographie humaine," Revue
    géographique des Pyrénées et du Sud-Ouest, vol. 41,
    no. 2 (April 1970), 113-122.

The influence of physical geography and biogeography on human geography has not been great, but it has usually been indirect and has tended to retard the growth of human geography. Gives 19th- and 20th-century examples.

87. Claval, Paul. Les mythes fondateurs des sciences sociales. Paris: Presses Universitaires de France, 1980. 261 p.

    A geographer's attempt to synthesize the trends in all the social sciences since the 17th century.

88. Claval, Paul. La pensée géographique: introduction à son histoire. ("Publications de la Sorbonne, Série N.-S. Recherches," 2.) Paris: Société d'Edition d'enseignement supérieur, 1972. 116 p.

    Discusses the general context of studies in the history of geography, with particular emphasis on the 19th and 20th centuries.

89. Claval, Paul. "Le rôle des écoles géographiques dans le développement de la discipline," pp. 23-28 in Les écoles géographiques, ed. Józef Babicz (Warsaw: PWN--Polish Scientific Publishers, 1980).

    Discusses the nature of "schools" of geography. The term "school" may be best applied to the schools of thought that developed in universities in the period from 1875 to 1950, after which changes came so quickly that "mandarins" around whom schools could form could not emerge.

90. Clozier, René. Histoire de la géographie. 5th edition. ("Que sais-je?," 65.) Paris: Presses Universitaires de France, 1972. 128 p..

    A survey of geography from ancient times to the present, with emphasis on exploration. First published as Les étapes de la géographie in 1942.

91. Corley, Nora T. "Geographical Literature," Encyclopedia of Library and Information Science, vol. 9 (New York: Marcel Dekker, Inc., 1973), 266-282.

    An essay on the history of geography, ancient times to modern, followed by bibliography of works on the

Part One--General and Topical   21

   history and methodology of geography and a list of 30
   leading geographical serials.

92. Crone, Gerald R. Background to Geography. London:
    Museum Press, Ltd., 1964. 224 p.

    A collection of disparate essays on geography, the
    first 5 of which (60 pages) deal with the history of
    geography from the Middle Ages to the beginning of the
    20th century.

93. Crone, G.R. Modern Geographers: An Outline of Progress
    since AD 1800. London: The Royal Geographical
    Society, 1970. 55 p. (First published 1951, revised
    1960, expanded and revised 1970.)

    Brief essays on the growth of geography in Europe and
    North America, focusing on the work of certain
    influential men: Joseph Banks, Alexander von Humboldt,
    Carl Ritter, Paul Vidal de la Blache, Ferdinand von
    Richthofen, Friedrich Ratzel, Halford Mackinder, and
    Isaiah Bowman. Originated as a series of 7 articles,
    "The Men behind Modern Geography," published in the
    Geographical Magazine between August 1949 and April
    1951.

94. Dalla Vedova, Giuseppe. "I progressi della geografia
    nel secolo XIX," Bollettino della Società Geografica
    Italiana, vol. 11, no. 7 (July 1901), 615-636.

    A review of geography in Europe from the 1760s
    onward, including European exploration overseas.

95. Dardel, Eric. L'homme et la terre: nature de la
    réalité géographique. ("Nouvelle encyclopédie
    philosophique.") Paris: Presses Universitaires de
    France, 1952. 133 p.

    Conceptions of space in various epochs and among
    various cultures, culminating in the truly scientific
    geography that appears toward the end of the 18th
    century.

96. Darmstaedter, Ludwig, and Du Bois-Reymond, René. 4000
    Jahre Pionier-Arbeit in den exakten Wissenschaften.
    Berlin: J.A. Stargardt, 1904. v + 389 p.

Chronological list of facts of interest to the historian of science. Includes items of interest to historians of geography and exploration.

97. Davis, William Morris. "A Retrospect of Geography," Annals of the Association of American Geographers, vol. 22, no. 4 (December 1932), 211-230.

    Signs of progress during the preceding 40 years.

98. Denucé, Jean. "L'histoire de la géographie," Revue de l'Université de Bruxelles, vol. 19, no. 4 (January 1914), 281-292.

    An introductory lecture in the course on the history of geography for students preparing for the doctorate in philosophy and letters at the Free University of Brussels. Deals mostly with the history of exploration and cartography in the Renaissance and earlier.

99. Dickinson, Robert E. The Makers of Modern Geography. New York: Frederick A. Praeger, Publishers, 1969. xiv + 305 p.

    "The purpose of this book is to trace the development of modern geography as an organised body of knowledge in the light of the works of its foremost German and French contributors." After an introductory chapter, "From Strabo to Kant," the author devotes attention to Humboldt and Ritter and the 3 "generations" that followed. Complemented by a later book on British and American geographers (below).

100. Dickinson, Robert E. Regional Concept: The Anglo-American Leaders. London: Routledge & Kegan Paul Ltd., 1976. xxi + 408 p.

    A survey of geography in Britain and North America over the last century, beginning with Patrick Geddes and Halford Mackinder. Does not deal solely with regional geography, as the title might lead one to believe. Complements The Makers of Modern Geography (above).

101. Dickinson, Robert E., and Howarth, O.J.R. The Making of Geography. Oxford: Clarendon Press, 1933. 164 p.

    A textbook on the history of geography from ancient times to modern. The first 15 chapters cover the subject more or less chronologically from the Greeks to the time of Humboldt and Ritter, but the next 4 chapters treat the recent development of some of the major subfields: physical, human, bio-, and regional geography.

102. Dörflinger, Johannes. Die Geographie in der "Encyclopedie." Eine wissenschaftsgeschichtliche Studie. (Osterreichische Akademie der Wissenschaften, Philosophisch-Historische Klasse, Sitzungsberichte, vol. 304, part 1.) (Veröffentlichungen der Kommission für Geschichte der Mathematik, Naturwissenschaften und Medizin, no. 17.) Vienna, 1976. 116 p.

    Analyzes articles on geography in Diderot's Encyclopédie (1751-1765).

103. Donazzolo, Pietro. Storia della geografia. Feltre, Italy: Premiata Tipografia Panfilo Castaldi, 1902. 236 p.

    History of geography from ancient times to modern, with emphasis on exploration. Last 90 pages deal with the period 1750-1900. Last chapter, "Scientific Geography in the Nineteenth Century" (pp. 229-234), begins with Herder and Kant and lists leading personalities, societies, and journals.

104. Downes, Alan. "The Bibliographic Dinosaurs of Georgian Geography (1714-1830)," Geographical Journal, vol. 137, part 3 (September 1971), 379-387.

    An exploration of the "dark age in the history of academic geography" between Varenius' General Geography (1650) and Humboldt's Kosmos (1845-). "This article intends to bridge the gap by suggesting how, particularly in France and Scotland, eighteenth century philosophy led to a redirection of geographical styles and interpretations which were passed on to their nineteenth century descendants, among whom were the relatively unproclaimed French geographers, Malte Brun and Balbi. It demonstrates that modern geography grew not solely out of the work of Humboldt and Ritter," but also out of the "compendium" geographies produced in France and Great Britain in the Georgian or Hanoverian period. J.K. Wright has termed these geographies veritable dinosaurs not only because of their size but also because they represent an extinct species of geographical literature.

105. Dunbar, Gary S., ed. The History of Geography: Translations of Some French and German Essays. Malibu, California: Undena Publications, 1983. 121 p.

English translations of 3 French (Emmanuel de Martonne, Philippe Pinchemel, and Paul Claval) and 3 German (Hermann Wagner, Alfred Hettner, and Hanno Beck) essays on the history of geography and a 19-page appendix containing biobibliographical notes.

106. Ferro, Gaetano, and Caraci, Ilaria. Ai confini dell'orizzonte: storia della esplorazioni e della geografia. ("Viaggi, esplorazioni e scoperte," 8.) Milan: U. Mursia editore, 1979. 207 p.

Part 1, pp. 5-99, by Ferro, is a history of exploration. Part 2, pp. 101-185, by Caraci, treats the history of geography from ancient times to the end of the 19th century, with a very brief concluding chapter on trends in geography at the beginning of the 20th century.

107. Fèvre, Joseph. Petite histoire de la géographie (Voyageurs et géographes). (Collection "Tout pur tous.") Paris: J. de Gigord, Editeur, 1947. 127 p.

Primarily a history of exploration, but most of the chapters end with a resume of the state of geographical science in that period (e.g., pp. 116-126, "Geographical Science in the 19th and 20th Centuries," emphasizing French contributions).

108. Fischer, Eric; Campbell, Robert D.; and Miller, Eldon S., eds. A Question of Place: The Development of Geographic Thought. 2nd ed. Arlington, Virginia: Beatty, 1969. vii + 446 p. (1st ed. 1967.)

Excerpts from the writings of geographers from ancient times onward, with emphasis on the modern period beginning with Immanuel Kant. The selections were chosen to illustrate the changing currents of geographical thought.

109. Fitzgerald, Walter. "Progress in Geographical Method," Nature, vol. 153, no. 3886 (22 April 1944), 481-483.

(Page 481) "This article is ... concerned ... with those trends, in both method and concept, which have marked the development of the subject abroad, and particularly in Germany, from the late eighteenth century onwards."

110. Fochler-Hauke, Gustav(o). Introducción a la historia de la geografía. Universidad Nacional de Tucumán, Argentina, Facultad de Filosofía y Letras, Instituto de Estudios Geográficos, Serie didáctica, 5 (1953). 123 p.

History of geography from ancient times to modern, with considerable emphasis on exploration.

111. Freeman, Thomas Walter. "Forty Years of Geography," Irish Geography, vol. 5, no. 5 (1968), 355-371.

A review of geography since World War I, with emphasis on developments in Great Britain.

112. Freeman, T.W. The Geographer's Craft. Manchester: Manchester University Press; New York: Barnes & Noble, Inc., 1967. xi + 204 p.

After an introductory chapter, "On the Work of Geographers," the book consists of biographical sketches of several geographers: Francis Galton, Paul Vidal de la Blache, Jovan Cvijić, Ellsworth Huntington, Sten de Geer, Percy Roxby, and Alan Ogilvie.

113. Freeman, T.W. A Hundred Years of Geography. Chicago: Aldine Publishing Company, 1962. 334 p.

An historical survey of geography from about the middle of the 19th century onward. Several chapters treat many of the major subfields: physical, regional, economic, social, and political geography, as well as cartography. Valuable notes and biographical appendix.

114. Fulvi, Fulvio. Le ragioni della geografia. ("Tangenti," 93.) Messina and Florence: Casa Editrice G. D'Anna, 1981. 208 p.

History and methodology of geography, from ancient times to the present. The largest section of the book, "Confronti" (pp. 87-201), contains statements about geography from numerous writers, mostly 20th-century geographers.

115. Fuson, Robert H. A Geography of Geography: Origins and Development of the Discipline. Dubuque, Iowa: William C. Brown Company Publishers, 1969. xi + 127 p.

   A brief textbook on the history of geography from ancient times to the present. It is not a geography of geography, despite the title.

116. Gallois, Lucien. "L'évolution de la géographie," Congrès national des Sociétés françaises de géographie, 21st session, Paris, 20-24 August 1900, Comptes rendus (published by the Société de géographie), 1901, pp. 108-119.

   Brief treatment of the development of geography in its fundamental relations with the progress of science since the time of the ancient Greeks. Since geography is concerned with a "very complex world, [it] does not have the simplicity of the ordinary sciences." Human geography completes physical geography; it studies the human phenomena that depend on the physical world. [MCR]

117. [Anonymous]. "Les géographes. Esquisse d'une histoire de la géographie," Le magasin pittoresque (Paris), vol. 44 (1876), 146-148, 198, 230, 259-260, 339-340; vol. 45 (1877), 161-162, 349-351; vol. 46 (1878), 282-283; vol. 49 (1881), 146-148, 206-207, 250-251.

   Chronology of geography and exploration, with emphasis on the latter, from ancient times onward.

118. Glacken, Clarence J. Traces on the Rhodian Shore: Nature and Culture in Western Thought from Ancient Times to the End of the Eighteenth Century. Berkeley and Los Angeles: University of California Press, 1967. xxviii + 763 p.

   (From Preface, pp. vii-viii) "The main theme of this work is that, in Western thought until the end of the eighteenth century, concepts of the relationship of human culture to the natural environment were dominated ... by ... three ideas"--"the idea of a designed earth; the idea of environmental influence; and the idea of man as a geographic agent."

Part One--General and Topical   27

119. Granö, Olavi. "External Influence and Internal Change in the Development of Geography," pp. 17-36 in <u>Geography, Ideology and Social Concern,</u> ed. D.R. Stoddart (Oxford: Basil Blackwell, 1981).

   (Page 17) "The selection and moulding of knowledge to the domain of a given social institution is a decisive event in the development of geography. This paper considers the background and general character of this process of change, namely geography's development into a formal academic discipline, without slavishly adhering to any one country's tradition."

120. Günther, Siegmund. <u>Entdeckungsgeschichte und Fortschritte der wissenschaftlichen Geographie im neunzehnten Jahrhundert.</u> ("Am Ende des Jahrhunderts: Rückschau auf 100 Jahre geistiger Entwickelung," vol. 23.) Berlin: Verlag Siegfried Cronbach, 1902. 231 p.

   The progress of geography and exploration from about 1750 to 1900, with heavy emphasis on the work of Germans.

121. Günther, Siegmund. <u>Geschichte der Erdkunde.</u> ("Die Erdkunde," ed. Maximilian Klar, part 1.) Leipzig and Vienna: Franz Deuticke, 1904. xi + 343 p.

   History of geography from ancient times through the 19th century.

122. Hahn, Friedrich Gustav. "Die Klassiker der Erdkunde und ihre Bedeutung für die geographische Forschung der Gegenwart," <u>Königsberger Studien,</u> vol. 1 (1887), 213-242.

   The classics of geography and their significance for geographical work in the present day. Emphasis on the works of Varenius, Humboldt, Ritter, and Peschel.

123. Hard, Gerhard. "Die Diffusion der 'Idee der Landschaft': Präliminarien zu einer Geschichte der Landschaftsgeographie," <u>Erdkunde,</u> vol. 23, no. 4 (December 1969), 249-264.

   (From English summary) "... The professional interest of the geographer in the landscape is revealed as a branch of a general interest in the landscape (and, above all, the German landscape) which is characteristic of German literature since about 1900-

1910 and has no parallel in other language areas. This interest found its peak at first in the literature of the aesthetic, in literary essays, in the literature of the humanities, then first in geography around 1930-1940. While the general interest, however, has been declining for about two decades, especially in literature of rank, it has lived on since 1950 in the literature of institutionalised 'landscape architecture and landscape conservation' and, in unchanged form, in academic geography." Several graphs show incidence of word "landschaft" in titles of German works from about 1886 to 1965.

124. Hartshorne, Richard. "The Concept of Geography as a Science of Space, from Kant and Humboldt to Hettner," Annals of the Association of American Geographers, vol. 48, no. 2 (June 1958), 97-108.

(Page 97) "As a study in the history of geographic thought, this paper is concerned with the possible origin, or origins, of the concept of geography as a spatial or chorological science and its significance to geography during the past century and a half." The concept may have originated with Immanuel Kant, but "its importance in current geographic thought stems from the writings of Alfred Hettner."

125. Hartshorne, Richard. "Geography--The Field," International Encyclopedia of the Social Sciences, vol. 6 (New York: The Macmillan Company and The Free Press, 1968), 115-116.

A general statement on the nature of geography. Of some historical interest.

126. Hartshorne, Richard. The Nature of Geography: A Critical Survey of Current Thought in the Light of the Past. Lancaster, Pennsylvania: Association of American Geographers, 1939. vi + 482 p.

A major methodological treatise, with emphasis on regional geography and on German works. Of great value to the historian of geography. First published in the Annals of the Association of American Geographers, vol. 29, no. 3 (September 1939), 171-658.

127. Herstel, Theodor. "Die allgemeine Geographie des Menschen in den Gesamtdarstellungen der Erdkunde,

Part One--General and Topical 29

ihre Entwicklung von Büschings 'Neuer Erdbeschreibung' bis zu Ratzels 'Anthropogeographie.'" Doctoral thesis, University of Cologne (Köln), 1954. 205 p.

The development of general geography from Anton Friedrich Büsching (1724-1793) to Friedrich Ratzel (1844-1904). [HB]

128. Hettner, Alfred. "Die Entwicklung der Geographie im 19. Jahrhundert," Geographische Zeitschrift, vol. 4, no. 6 (June 1898), 305-320.

Development of geography in 19th-century Germany.

129. Hettner, Alfred. Die Geographie: Ihre Geschichte, Ihr Wesen und Ihre Methoden. Breslau: Ferdinand Hirt, 1927. viii + 463 p.

History and methodology of geography. The first part ("Buch"), "The History of Geography," pp. 1-109, treats the history of geography from ancient times to modern. The last chapter of Book 1 covers the 19th and 20th centuries and is a revised version of Hettner's 1898 paper (above).

130. Holt-Jensen, Arild. Geography: Its History and Concepts. A Student's Guide. London, etc.: Harper & Row, Publishers, 1981. xi + 171 p.

The history of geography is covered only in Chapter 2, "The Foundations of Scientific Geography," pp. 9-36, and the remainder of the book deals with the author's interpretation of recent trends in human geography (as a social science).

131. Hooson, David. "National Cultures and Academic Geography in an Urbanizing Age," pp. 157-178 in The Expanding City: Essays in Honour of Professor Jean Gottmann, ed. John Patten (London, etc.: Academic Press, 1983).

The emergence of academic geography in the last part of the 19th century and the first part of the 20th with a comparison of 4 national traditions--Britain, France, Russia, and the United States.

132. Hooson, David. "The Role of the Historical Context in the Development of National Schools of Geography," pp. 29-31 in Les écoles géographiques, ed. Józef Babicz (Warsaw: PWN--Polish Scientific Publishers, 1980).

Calls for a fuller treatment of the historical context in studies of the history of geography. Compares the development of geography in Great Britain and Russia in the formative decades before 1914.

133. Hudson, Brian. "The New Geography and the New Imperialism: 1870-1918," Antipode, vol. 9, no. 2 (September 1977), 12-19.

    Explains the emergence of modern geography by the need to serve the growing imperialism in Europe, the United States, and Japan.

134. Humboldt, Alexander von. "Geschichte der physikalischen Weltanschauung," pp. 135-520 in Kosmos: Entwurf einer physischen Weltbeschreibung, vol. 2 (Stuttgart and Tübingen: J.G. Cotta'scher Verlag, 1847).

    Evolution of geographical and cosmological views from ancient times to the age of Newton and Leibniz.

135. James, Preston E. "Geography," pp. 144-160 in Encyclopaedia Britannica, vol. 10 (Chicago, etc.: Encyclopaedia Britannica, Inc., 1972).

    General sketch of the development of geography since the ancient Greeks.

136. James, Preston E. "On the Origin and Persistence of Error in Geography," Annals of the Association of American Geographers, vol. 57, no. 1 (March 1967), 1-24.

    A review of the origin and persistence of error in geographical writings, with examples coming mostly from 19th- and 20th-century works.

137. James, Preston E., and Martin, Geoffrey J. All Possible Worlds: A History of Geographical Ideas. 2nd ed. New York: John Wiley & Sons, 1981. xv + 508 p.

    A general textbook covering the history of geography from ancient times to the present. The first 111 pages cover the story before Humboldt and Ritter. First edition (Indianapolis: The Odyssey Press, 1972) was written by James alone.

Part One--General and Topical   31

138. Joerg, W.L.G. "Recent Geographical Work in Europe," Geographical Review, vol. 12, no. 3 (July 1922), 431-484.

   Current status and recent developments in geography in Europe, excluding Russia. Mostly contemporary data, but in some cases he reached back more than a few years--e.g., in the case of Belgium (pp. 458-460) he devoted considerable space to "Elisée Reclus' Belgian Sojourn" (1892-1905).

139. Johnson, Douglas. "The Geographic Prospect," Annals of the Association of American Geographers, vol. 19, no. 4 (December 1929), 167-231.

   Development of geography and of university systems in France, Great Britain, Germany, and Belgium. This is not so much an historical paper as it is one that deals with problems and prospects.

140. Johnston, Ronald J. Geography and Geographers: Anglo-American Human Geography since 1945. London: Edward Arnold (Publishers) Ltd., 1979. 232 p.

   A survey of published works in Great Britain and North America, with overwhelming emphasis on the "spatial science" and behavioral aspects of human geography.

141. Keltie, John Scott. "A Half-Century of Geographical Progress," Scottish Geographical Magazine, vol. 31, no. 12 (December 1915), 617-636.

   Progress in geography since 1860, with emphasis on exploration and on the British.

142. Keltie, John Scott, and Howarth, O.J.R. History of Geography. ("A History of the Sciences.") New York: G.P. Putnam's Sons, 1913. vii + 208 p.

   Textbook on history of geography from ancient times to the end of the 19th century. Emphasis on exploration. The book is largely a "summary of the development of geographical knowledge among European peoples" (p. 3).

143. Kish, George, ed. A Source Book in Geography. ("Source Books in the History of the Sciences," ed. Edward H. Madden.) Cambridge: Harvard University Press, 1978. xvi + 453 p.

A selection of "geographical writings from Hesiod to Humboldt," with brief editorial commentary.

144. Krebs, Norbert. "Die Entwicklung der Geographie in den letzten fünfzehn Jahren," Frankfurter Geographische Hefte, no. 1 (1927), 5-19.

    Recounts the significant events in exploration and geography in the 15-year period from 1911 to 1926.

145. Kretschmer, Konrad. Geschichte der Geographie. ("Sammlung Göschen," 624.) Berlin and Leipzig: G.J. Goschen'sche Verlagshandlung G.m.b.H., 1912. 163 p.

    A history of geography from ancient times to the end of the 19th century, with emphasis on exploration. "Scientific geography" (19th-century German geographers, especially Humboldt and Ritter) treated in only the last five pages.

146. Kretschmer, Konrad. "Geschichte der Geographie als Wissenschaft," pp. 1-22 in Handbuch der geographischen Wissenschaft, ed. Fritz Klute: Allgemeine Geographie, Part 1, Physikalische Geographie, No. 1 (Potsdam: Akademische Verlagsgesellschaft Athenaion M.B.H., 1933).

    An essay on the history of geography, ancient times to modern, with emphasis on exploration and mapmaking. Only gets to Humboldt and Ritter, the "zwei Heroen der deutschen Wissenschaft," in the last paragraph. Followed by valuable essay by Hermann Lautensach (below).

147. Kuls, Wolfgang. "Über einige Entwicklungstendenzen in der geographischen Wissenschaft seit der zweiten Halfte des 19. Jahrhunderts," Mitteilungen der Geographischen Gesellschaft in Munchen, vol. 55 (1970), 11-30.

    Traces some of the major trends in geography since 1869, with heavy emphasis on Germany.

148. Lampe, Felix. Grosse Geographen. Bilder aus der Geschichte der Erdkunde. ("Prof. Dr. Bastian Schmids naturwissenschaftliche Bibliothek, Serie A: Für reifere Schüler, Studierende und Naturfreunde," 28.)

Leipzig and Berlin: Verlag von B.G. Teubner, 1915. 287 p.

A history of geography illustrated through the lives of some of the major explorers and geographers from ancient times onward. Some of the later chapters are devoted to James Cook, Alexander von Humboldt, Carl Ritter, and Ferdinand von Richthofen. The last chapter covers German geographers of the 19th and early 20th centuries, beginning with J.G. Kohl and C. Neumann and ending with Albrecht Penck, Hermann Wagner, and Joseph Partsch.

149. Lautensach, Hermann. "Wesen und Methoden der geographischen Wissenschaft," pp. 23-56 in Handbuch der geographischen Wissenschaft, ed. Fritz Klute: Allgemeine Geographie, Part 1, Physikalische Geographie, no. 2 (Potsdam: Akademische Verlagsgesellschaft Athenaion M.B.H., 1933).

An essay on the shaping of geography from Humboldt and Ritter onward. Deals largely with German geographers.

150. Lavrov, S.B., and Dmitrevskiy, Yu. D. "Problems in the History of Geographical Thought at the 23rd International Geographical Congress," Soviet Geography: Review and Translation, vol. 19, no. 1 (January 1978), 54-59.

Review of papers presented at the I.G.U. symposium on the history of geographical thought in Leningrad in July 1976. Many of the papers were published in the Polish journal Organon.

151. Lemosof, Paul. Le livre d'or de la géographie. Essai de biographie géographique. Paris: Librairie Ch. Delagrave, 1902. viii + 223 p.

Brief biographical entries on geographers, cartographers, and explorers from ancient times to the end of the 19th century. Emphasis on 19th century. Section A-E appeared in Revue de géographie, April 1900-March 1901.

152. Lewthwaite, Gordon. "Geography," pp. 437-443 in Encyclopedia Americana, vol. 12 (1979).

Mostly an essay on the history of geography, ancient to modern.

153. Livingstone, David N. "Some Methodological Problems in the History of Geographical Thought," Tijdschrift voor Economische en Sociale Geografie, vol. 70, no. 4 (1979), 226-231.

    Discusses problems of selection and interpretation of data by historians of geography. Argues for greater concern for the context within which geographical ideas developed.

154. Löwenberg, Julius. Geschichte der Geographie von den ältesten Zeiten bis auf die Gegenwart. 2nd ed. Berlin: Haude- und Spener'sche Buchhandlung, 1866. xii + 475 p. (1st ed. 1840.)

    Geography from ancient times to Humboldt and Ritter. Considerable emphasis on exploration.

155. Lüdde, Johann Gottfried. Die Geschichte der Erdkunde. Eine Abhandlung über ihr Wesen und ihre Literatur. Berlin: Stackebrandt'sche Buchhandlung, 1841. vii + 108 p.

    Largely a bibliographical work on the history of geography and allied fields--exploration, geographical methodology, and geographical instruction.

156. Lüdde, Johann Gottfried. Geschichte der Methodologie der Erdkunde. Leipzig: Hinrichs, 1849.

    Indispensable guide to the history of geographical methodology in Germany and to the geographical literature. [HB]

157. Mackinder, Halford J. "Modern Geography, German and English," Geographical Journal, vol. 6, no. 6 (October 1895), 367-379.

    Development of geography in Germany and England since 1735.

158. Marinelli, Giovanni. "Concetto e limiti della geografia," pp. 143-179 in Marinelli's Scritti minori, vol. 1 (Florence: Tipografia di M. Ricci, 1908).

The nature of geography as it developed in the 19th century. Paper first published in Rivista Geografica Italiana, vol. 1, no. 1 (March 1893), 6-32.

159. Marinelli, Giovanni. Scritti minori. Vol. 1: Metodi e storia della geografia. Florence: Tipografia di M. Ricci, 1908. xlviii + 640 p.

    This posthumous collection of essays by the Italian geographer Giovanni Marinelli (1846-1900) was edited by Attilio Mori, who wrote the biographical sketch of Marinelli that opens the volume. The book contains 13 essays by Marinelli on themes in the history of geography and cartography. [LG]

160. Marshall, John U. "Geography and Critical Rationalism," pp. 75-171 in Rethinking Geographical Inquiry, ed. J. David Wood ("Geographical Monographs," 11) (Toronto: York University, Atkinson College, Department of Geography, 1982).

    Philosophical currents in 20th-century human geography, with particular attention to the relevance of Karl Popper's critical rationalism.

161. Martonne, Emmanuel de. "Le développement et l'avenir de la géographie," Bulletin de la Société de géographie de Lyon et de la région lyonnaise, 2nd series, vol. 1, no. 1 (1908), 1-11.

    Article summarizes the ideas exposed in the first chapter of the author's Traité de géographie physique (1909). Retraces the stages through which geography has gone, emphasizing the dual nature of geography from its beginning; it has been both descriptive (regional geography) and explanatory (general geography). Its history has depended on political circumstances and the state of scientific knowledge. Modern geographical method is characterized by the application of three principles: the principle of spread or extent (recognized especially by Ratzel), the principle of general coordination (coordinating regional and general geography), and the principle of causality (which emphasizes the importance of the historical study of nearly all phenomena). The last 2 principles were elaborated by Ritter and Humboldt. [MCR]

162. Martonne, Emmanuel de. "L'évolution de la géographie," pp. 3-26 in de Martonne's Traité de géographie physique (Paris: Librairie Armand Colin, 1909).

   General essay on the history of geography, ancient times to modern.

163. May, Joseph A. "On Orientations and Reorientations in the History of Western Geography," pp. 31-72 in Rethinking Geographical Inquiry, ed. J. David Wood ("Geographical Monographs," 11) (Toronto: York University, Atkinson College, Department of Geography, 1982).

   Philosophical traditions in geography, from the ancient Greeks to the present.

164. Melón, Amando, and Gordejuela, Ruiz de. "Esquema sobre los modeladores de la moderna ciencia geográfica," Estudios Geográficos, vol. 6, nos. 20-21 (August-November 1945), 393-442.

   Leading figures in the development of modern geography, from Varenius to Vidal de la Blache and his disciples. Emphasis on 19th century.

165. Mill, Hugh Robert. "On Research in Geographical Science," Geographical Journal, vol. 18, no. 4 (October 1901), 407-424.

   History and present state of geography, with emphasis on Great Britain.

166. Mitchell, J.B., and Davies, W.K.D. "Geography," pp. 1035-1053 in The New Encyclopaedia Britannica, 15th ed., vol. 7 (Macropedia) (Chicago: Encyclopaedia Britannica, Inc., 1980).

   The first part (pp. 1036-1045) is "A Survey of Geographical Exploration" by Mitchell. The second part (pp. 1045-1052) is "The Modern Discipline of Geography" by Davies. The latter is a survey of current or recent trends, but for an historical sketch of modern geography one would have to go back to the articles on geography in earlier editions of the Britannica (see P.E. James).

167. Neff, Ernst. "Geographie--einmal anders gesehen," Geographische Zeitschrift, vol. 70, no. 4 (Fourth Quarter 1982), 241-260.

Describes the fragmentation of geography since the time of Humboldt and suggests a more integrated approach. Proposes to apply the principle of complementarity of the physicists Niels Bohr and Werner Heisenberg.

168. Nelson, Helge. "Geografien som vetenskap. En överblick av dess utveckling till 1900-talets början," Svensk Geografisk Arsbok, vol. 20 (1944), 208-222.

    Survey of geography from Herodotus to the beginning of the 20th century, as a background to the discussion of trends in 20th-century Swedish geography.

169. Oberlander, Hermann. Der geographische Unterricht nach den Grundsätzen der Ritterschen Schule historisch und methodologisch beleuchtet. 6th ed., ed. by Paul Weigeldt. Leipzig: Verlag von Dr. Steele & Co., 1900. viii + 332 p.

    Part 1 (pp. 3-199), "History and Method of Geographical Instruction," concerns the teaching and writings of Carl Ritter and his predecessors and successors. The last 50 pages of Part 1 concern writings in various fields of geography. Part 2 (pp. 203-332), "Detailed Examination of the Principal Elements of Comparative Geography," treats location, space, geology, water, climate, plants, animals, and Man.

170. Pereira da Silva, Clodomiro. "Ensaio de uma sintese da evolução da geografia." Instituto brasileiro de geografia e estatística, Conselho nacional de geografia, Diretorio regional, Estado de São Paulo, Boletim, no. 3 (1943). 135 p.

    History of geography from ancient times to 1942. The last 2 chapters summarize the author's ideas about the evolution of geography and fundamental criteria about the makeup of the discipline. He also discusses the limits of geography and attempts a "classification and grouping of the constituent elements of geography." [CRE]

171. Peschel, Oscar. Geschichte der Erdkunde bis auf A. v. Humboldt und Carl Ritter. ("Geschichte der Wissenschaften in Deutschland: Neuere Zeit," vol. 4.)

Munich: Literarisch-artistische Anstalt der J.G. Cotta'schen Buchhandlung, 1865. xx + 706 p.

History of geography, cartography, and exploration from ancient times through Humboldt and Ritter. Emphasis on period from about 1650 onward ("Das Zeitalter der Messungen"). 2nd ed. (1877), ed. Sophus Ruge.

172. Peschel, Oscar. "Ueber die Aufgaben einer Geschichte der Geographie," Das Ausland, vol. 37, no. 34 (20 August 1864), 793-799.

Outline and justification for planned volume in the history of geography. Part of program proposed in 1860 by Ausschuss für deutsche Geschichtsforschung under auspices of King Max of Bavaria.

173. Pinchemel, Geneviève, and Pinchemel, Philippe. "Réflexions sur l'histoire de la géographie: histoires de la géographie, histoires des géographies," France, Ministère des universités, Comité des travaux historiques et scientifiques, Bulletin de la section de géographie, no. 84 (1979-- pub. 1981), 221-231.

Some remarks on the changing conceptions of geography and of geographers through time.

174. Pinchemel, Philippe. "Géographie--L'histoire de la géographie, évolution chronologique, les tendances de la pensée géographique," pp. 621-625 in Encyclopaedia Universalis, vol. 7 (Paris, 1968).

Comprehensive encyclopedia article on the history and present state of geography.

175. Pinchemel, Philippe, ed. "Histoire et épistémologie de la géographie," France, Ministère des universités, Comité des travaux historiques et scientifiques, Bulletin de la section de géographie, no. 84 (1979--pub. 1981). 231 p.

Contains 11 articles on the history of geography and on the connections between history and geography. 6 of the chapters are cited separately in this bibliography.

176. Pinchemel, Philippe. "Réflexions sur une problématique des écoles de géographie," pp. 5-13 in Les écoles

géographiques, ed. Józef Babicz (Warsaw: PWN--Polish Scientific Publishers, 1980).

Discusses national and personal schools and the rapid postwar changes.

177. Plewe, Ernst. "Untersuchungen über den Begriff der 'vergleichenden' Erdkunde und seine Anwendung in der neueren Geographie," Zeitschrift der Gesellschaft für Erdkunde zu Berlin, Ergänzungsheft, 4 (1932). 92 p.

Follows the term "comparative geography" from Bernhard Varenius (1622-1650) to Friedrich Ratzel (1844-1904). [HB]

178. Porena, Filippo. "La geografia nel secolo decimonono," Bollettino della Società Geografica Italiana, series 4, vol. 2, no. 1 (January 1901), 10-23.

A review of geography (including exploration, cartography, and the earth sciences generally) in the 19th century.

179. Porena, Filippo. "Sistema scientifica e sistema scolastico della geografia," Bollettino della Società Geografica Italiana, series 4, vol. 1, no. 12 (December 1900), 1104-1125.

Historical essay on the evolution of the divisions of geography and the relations of geography with other sciences.

180. Prillinger, Ferdinand. "Geschichte der Geographie," Osterreich in Geschichte und Literatur, vol. 13 (1969), 38-46.

An essay describing four kinds of historical studies in geography: landscape history (historical geography), history of exploration, history of geographical science (history of geography), and history of geographical instruction.

181. Rahman, Shah M.H. Geography and Geographers: A Critical Analysis of the History of Geographers and Geographical Discoveries. Allahabad, India: The Urdu Publishing House, 1945. iv + 189 p.

A textbook for Indian students on the history of geography from ancient times to modern. Written from a European, rather than Indian, viewpoint.

182. Ravenstein, Ernst Georg. "The Field of Geography," Proceedings of the Royal Geographical Society, vol. 13, no. 10 (October 1891), 617-629.

   Brief historical sketch of geography from ancient times onward, with emphasis on cartography.

183. Ricchieri, Giuseppe. "Gli studi geografici nello sviluppo della civiltà e nell'educazione moderna," Rivista Geografica Italiana, vol. 4, no. 4 (April 1897), 177-182; nos. 5-6 (May-June 1897), 249-265.

   Historical sketch of geography from ancient times onward, with emphasis on the 19th century.

184. Richthofen, Ferdinand von. "Triebkräfte und Richtungen der Erdkunde im neunzehnten Jahrhundert," Zeitschrift der Gesellschaft für Erdkunde zu Berlin, 1903, no. 9 (November), 655-692.

   Progress in geography from ancient times onward, with emphasis on the 19th century, which was the time of Humboldt and Ritter and also of increased specialization and of the growth of geography in the universities. An English abstract, "The Impetus and Direction of Geography in the Nineteenth Century," was published in the Geographical Journal, vol. 23, no. 2 (February 1904), 229-234.

185. Ritter, Carl. Geschichte der Erdkunde und der Entdeckungen. Vorlesungen an der Universität zu Berlin gehalten von Carl Ritter. Ed. H.A. Daniel. 2nd ed. Berlin: Druck und Verlag von G. Reimer, 1880. vi + 265 p.

   From the earliest Egyptians and Hebrews to Columbus and Vasco da Gama. Conclusion (pp. 262-265) mentions some exploration and colonization in subsequent centuries. 1st ed. pub. 1861.

186. Ruge, Sophus. Abhandlungen und Vorträge zur Geschichte der Erdkunde. Dresden: G. Schonfeld's Verlagsbuchhandlung, 1888. 268 p.

   Collection of 12 diverse papers on the history of geography and exploration, including essays on the significance of the years 1781 and 1863-1888 for the development of geography, the "Sturm-und-Drang" period

Part One--General and Topical   41

    of German geography (18th century), and the African
    Association (founded in London in 1788). [HB]

187. Sanchez, Pedro C. "Evolución de la geografía,"
     Instituto panamericano de geografía e historia
     (Mexico City), Publicación, no. 12 (1935). 24 p.

     A brief essay on the history of geography (mostly
     exploration) from ancient times to about the end of the
     19th century.

188. Schmithüsen, Josef. Geschichte der geographischen
     Wissenschaft von den ersten Anfängen bis zum Ende des
     18.Jahrhunderts. ("Hochschultaschenbücher," 363/363a.)
     Mannheim: Bibliographisches Institut, 1970. 190 p.

     A history of geography from the ancient Greeks to the
     time of Humboldt, Ritter, and Goethe. Heavy emphasis
     on German geographers in the latter parts of the book.

189. Schwarz, Gabriele, ed. Die Entwicklung der
     geographischen Wissenschaft seit dem 18. Jahrhundert.
     ("Quellensammlung zur Kulturgeschichte," ed. Wilhelm
     Treue, Schrift 5.) Berlin: Wissenschaftliche
     Editionsgesellschaft M.B.H., 1948. 119 p.

     The book consists of 28 excerpts from the
     methodological writings of various geographers, all
     Germans except J.G. Granö, Carl Sauer, Paul Vidal de la
     Blache, and Albert Demangeon. The earliest was Anton
     Busching (1760). The editor contributed an 18-page
     "Historical Introduction."

190. Stoddart, D.R. "Geography--A European Science,"
     Geography, vol. 67, part 4 (October 1982), 289-296.

     (From author's abstract, p. 289) "The dominant
     themes of modern geography were ... largely established
     in Europe between 1785 and 1885 and still provide a
     guide to modern action."

191. Stoddart, David R., ed. Geography, Ideology and Social
     Concern. Oxford: Basil Blackwell, 1981. vi + 250 p.

     (Page 1) "This book has two purposes. The first is
     to demonstrate that the history of geography is more
     than simply the chronological listing of the
     achievements of a few great scholars arrayed in

national schools, and that both the ideas and the structure of the subject have developed in response to complex social, economic, ideological and intellectual stimuli. The second is to show that throughout its recent history geographers have been not only concerned with narrowly academic issues, but have also been deeply involved with matters of social concern." The chapters are listed separately in this bibliography. The book grew out of a colloquium, "The History of Geography and the History of Science," in Edinburgh in 1977.

192. Stoddart, D.R. "Growth and Structure of Geography," *Transactions of the Institute of British Geographers*, no. 41 (June 1967), 1-18.

(From author's abstract) "Tentative quantitative measures of the growth and structure are identified. Three growth parameters--the number of periodicals, of learned societies, and of higher degrees--show exponential growth rates with doubling periods of 30, 40 and 8 years respectively.... An index of the internal structure of geographical work is given by the citations in geographical publications. Citation-age structures are used to identify research frontiers in the literature."

193. Stoddart, D.R. "The Paradigm Concept and the History of Geography," pp. 70-80 in *Geography, Ideology and Social Concern*, ed. D.R. Stoddart (Oxford: Basil Blackwell, 1981).

Discusses the uses (largely polemical) to which geographers have put Kuhn's paradigm concept since 1967. Sees its value not so much as a means of understanding the complexities of change as an object of interest to the historian of science.

194. Tatham, George. "Geography in the Nineteenth Century," pp. 28-69 in *Geography in the Twentieth Century*, ed. Griffith Taylor, 3rd ed. (New York: Philosophical Library, 1957).

The first 14 pages of the essay deal with ancient, medieval, Renaissance, and 18th-century geography, and the next 17 pages treat Humboldt and Ritter. Emphasis on German geographers.

Part One--General and Topical 43

195. Taylor, Griffith, ed. Geography in the Twentieth Century: A Study of Growth, Fields, Techniques, Aims and Trends. 3rd ed. New York: Philosophical Library; London: Methuen, 1957. xi + 674 p.

   A fairly comprehensive survey of the field of geography and many of its branches. See especially Part 1, "Evolution of Geography and Its Philosophical Basis," pp. 1-162. Several of the chapters are listed separately in this bibliography.

196. Tichy, Franz. "Die vom Menschen gestaltete Erde. Auffassung und Darstellung im 19. Jahrhundert," Die Erde, vol. 91, no. 4 (December 1960), 241-257.

   Recognition of the influence of Man upon the earth, beginning with Buffon and Herder, continuing through Humboldt and Ritter, and emphasizing George Perkins Marsh.

197. Turnock, David. "The Region in Modern Geography," Geography, vol. 52, part 4 (November 1967), 374-383.

   The development of geography, and regional geography in particular, since the time of Immanuel Kant. Development shown graphically on p. 375 ("The History of Geography: A Suggested Model").

198. Vallaux, Camille. Les sciences géographiques. Paris: Librairie Félix Alcan, 1925. xxvii + 413 p.

   Not a history of geography but does contain some explanation of how geography came to its present state. Divided into two parts: Part 1, "La géographie comme science autonome," takes up the first three quarters of the book; Part 2, "Les géographies comme sciences auxiliaires," treats biological, historical, and sociological geography, as well as the geographical aspects of geology, meteorology, and oceanography.

199. Valls Taberner, Fernando, and Reparaz, Gonzalo de. "Historia de la geografía," pp. 532-562 of Geografía Universal, ed. Fernando Valls Taberner, vol. 1 (Barcelona: Instituto Gallach de Librería y Ediciones, 1928).

   Historical sketch of geography and exploration from ancient times onward. The last section, "Modern

Geographical Science" (pp. 560-562), emphasizes Humboldt and Ritter.

200. Vaugondy, Robert de. Essai sur l'histoire de la géographie. Paris: Antoine Boudet, 1755. xii + 422 p. + unpaged (4 p.) table of contents.

History of geography, cartography, and exploration from ancient times onward, with major emphasis on the cartographic sources for Vaugondy's new atlas.

201. Vivien de Saint-Martin, Louis. Histoire de la géographie et des découvertes géographiques depuis les temps les plus reculés jusqu'à nos jours. Paris: Hachette et Cie, 1873. xvi + 615 p.

From the ancient Egyptians and Moses to David Livingstone. Emphasis on exploration and mapping. Concluding chapter, "L'état actuel de la science, études et lacunes."

202. Wagner, Hermann. "Geschichte der Methodik der Geographie als Wissenschaft," pp. 17-25 in Wagner's Lehrbuch der Geographie, vol. 1, Allgemeine Erdkunde, part 1, Einleitung, 10th ed. (Hannover: Hahnsche Buchhandlung, 1920).

Geography from the ancient Greeks to the end of the 19th century.

203. Warntz, William, and Wolff, Peter, eds. Breakthroughs in Geography. New York: The New American Library, 1971. 266 p.

The 8 chapters contain original writings (with editorial commentary) illustrating conceptual breakthroughs in "theoretical-predictive" geography during the last two millennia: Claudius Ptolemy, Gerhard Kremer (Mercator), Edmund Halley, Arthur Cayley, J.H. von Thünen, Arnold Guyot, Francis Galton, William Morris Davis, Robert Horton, and John Q. Stewart.

204. Wilcock, Arthur A. "The Geographer before 1800," Area, vol. 7, no. 1 (1975), 45-46.

Uses of the terms "geographer" and "cosmographer" in the 16th to 18th centuries.

205. Wisotzki, Emil. Zeitströmungen in der Geographie. Leipzig: Verlag von Duncker & Humblot, 1897. viii + 467 p.

A history of geographical ideas from the 16th century to the end of the 19th. Mostly German sources.

206. Wolkenhauer, W. "Die Entwicklung der Geographie im XIX. Jahrhundert in einigen Merkzahlen," Deutsche Geographische Blätter, vol. 23, no. 1 (1901), 58-63.

Chronological list of significant events in geography and exploration in the 19th century.

207. Wood, J. David. "Rethinking Geographical Inquiry: Prologue, Chorus, Epilogue," pp. 1-27 in Rethinking Geographical Inquiry, ed. J. David Wood ("Geographical Monographs," 11) (Toronto: York University, Atkinson College, Department of Geography, 1982).

Introduction to the volume. Gives particular attention to the seminal works of Richard Hartshorne.

208. Wright, John K. "The History of Geography: A Point of View," Annals of the Association of American Geographers, vol. 15, no. 4 (December 1925), 192-201.

(Page 201) "The history of geography is the history of the images of the geographical environment that have been reflected in the minds of men through the ages."

209. Wright, John Kirtland. Human Nature in Geography: Fourteen Papers, 1925-1965. Cambridge: Harvard University Press, 1966. xx + 361 p.

A collection of 14 essays, of which 10 had been published previously, by John K. Wright (1891-1969), a leading historian of geography. Several of the chapters are listed separately in this bibliography.

210. Wright John K. "A Plea for the History of Geography," Isis, vol. 8, no. 3 (July 1926), 477-491.

Author calls for greater attention to the history of geography, which he thinks is a neglected field in America. Discusses previous scholarship in the field and calls for studies of not only "scientific" geography but also of "non-scientific" geographical ideas as expressed in popular works of travel and description. A revised version of this paper was published in Wright's Human Nature in Geography (above).

211. Wrigley, E.A. "Changes in the Philosophy of Geography," pp. 3-20 in Frontiers in Geographical Teaching, ed. Richard J. Chorley and Peter Haggett (London: Methuen and Co. Ltd., 1965).

    Some problems of geographical methodology, beginning with Humboldt and Ritter. Special attention is given to the French geographer Paul Vidal de la Blache (1845-1918).

212. Zeune, Johann August. Erdansichten oder Abriss einer Geschichte der Erdkunde vorzüglich der neuesten Fortschritte in dieser Wissenschaft. Berlin: Mauersche Buchhandlung, 1815.

    One of the earliest German works in the history of geography. Zeune was Carl Ritter's predecessor at the University of Berlin. [HB]

213. Zondervan, H. "De Richting in de Beoefening der Aardrijkskunde vóór A. von Humboldt en C. Ritter," Tijdschrift van het Koninklijk Nederlandsch Aardrijkskundig Genootschap, 2nd series, vol. 12 (1895), 741-756.

    Geography from ancient times through Immanuel Kant and Carl Ritter, but only barely mentions Humboldt.

III. DISCOVERY AND EXPLORATION

214. Anderson, John R.L. The Ulysses Factor: The Exploring Instinct in Man. London: Hodder and Stoughton Ltd., 1970. 352 p.

    The author has coined the term "Ulysses factor," which he does not exactly define but which he uses to cover not only exploration but "any risk-taking in pursuit of a goal of no apparent practical value." The book chronicles exploits from the fictional Ulysses himself, Abraham, and Pytheas to recent figures such as Francis Chichester. The book concludes with a lengthy

Part One--General and Topical   47

"Index to Twentieth Century Adventure," an alphabetical listing of individuals and their major achievements.

215. Baker, John Norman Leonard. A History of Geographical Discovery and Exploration. ("Harrap's New Geographical Series.") London: George G. Harrap & Co. Ltd., 1931. 543 p.

   A comprehensive history of exploration from the ancient Minoans to Richard E. Byrd. Exploration after A.D. 1800 takes up about 60% of the book.

216. Banse, Ewald. Grosse Forschungsreisende. Ein Buch von Abenteuren, Entdeckern und Gelehrten. Munich: J.F. Lehmanns Verlag, 1933. 284 p.

   The first part (pp. 9-47) is a general history of exploration from ancient times onward, and the rest of the book consists of biographical sketches of individual explorers from Marco Polo to Roald Amundsen. Some treatment of figures who could be classified as geographers but not as explorers--e.g., J.G. Kohl and Friedrich Ratzel. Emphasis on Germans in the modern era but includes some non-Germans, such as Livingstone, Stanley, Nansen, and Amundsen.

217. Barbier, J.-V. "Le rôle de la femme dans la géographie. Les voyageuses," Bulletin de la Société de géographie de l'Est (Nancy), vol. 17 (1895), 1-87.

   Women as travelers and explorers from the 13th century onward, with emphasis on the 19th century. Women have an advantage in seeing things that a male explorer would not be able to see. Concluding remarks: "Grâce à nos voyageuses, nous connaîtrons peut-être un jour le secret de l'éternel féminin de l'Orient; mais quelles Orientales viendront jamais nous révéler celui de l'éternel féminin de l'Occident?"

218. Beaglehole, J.C. "Eighteenth Century Science and the Voyages of Discovery," New Zealand Journal of History, vol. 3, no. 2 (October 1969), 107-123.

   Emphasis on James Cook and Joseph Banks. Concluding sentences: "Were we (if I may identify ourselves with our countries) in this southern hemisphere discovered by the Eighteenth Century? We were also discovered by the Middle Ages and by ancient Athens."

219. Beaglehole, J.C. The Exploration of the Pacific. ("The Pioneer Histories.") London: A. & C. Black Ltd., 1934. xv + 410 p.

The exploration of the Pacific Basin, from Magellan to Cook's third voyage.

220. Beaubois, Henry, et al. Les explorations au XXe siècle. Paris: Librairie Larousse, 1960. 391 p.

Popular history of 20th-century exploration--land, sea, and atmosphere--including mountaineering and speleology.

221. Beck, Hanno. "Geographie und Reisen im 19. Jahrhundert. Prolegomena zu einer allgemeinen Geschichte der Reisen," Petermanns Geographische Mitteilungen, vol. 101, no. 1 (February 1957), 1-14.

Considers the history of travel (journeys of discovery, scientific expeditions, and other journeys) to demonstrate its dependence on scientific geography, from about 1750 to the beginning of World War I. Emphasizes German travelers and geographers.

222. Beck, Hanno. Germania in Pacifico. Der deutsche Anteil an der Erschleissung der Pazifischen Beckens. Akademie der Wissenschaften und Literatur, Mainz, Abhandlungen der Mathematisch-Naturwissenschaftlichen Klasse, 1970, no. 3, pp. 233-327.

German geographers and travelers in the description and exploration of the Pacific Ocean and its borderlands, 16th century to the present.

223. Beck, Hanno. Grosse Reisende. Entdecker und Erforscher unserer Welt. Munich: Verlag Georg D.W. Callwey, 1971. 436 p.

21 chapters, covering as many explorers, from Pytheas (c. 330 B.C.) to Sven Hedin (1865-1952), with emphasis on Germans and Scandinavians in the modern era.

224. Beckman, Leif Olof, and Ohlmarks, Åke. Vår väg genom världen; de geografiska upptäckternas historia. 3 vols. Stockholm: Hugo Gebers Forlag, 1947-1951. Vol. 1, 1947, ix + 613 p.; vol. 2, 1948, vii + 679; vol. 3, 1951, viii + 706.

History of exploration from ancient times onward.

Part One--General and Topical   49

225. Behrmann, Walter. "Die Entschleierung der Erde," <u>Frankfurter geographische Hefte</u>, 16th year, sole number (Whole no. 23) (1948). 56 p. + 11 maps and a chart.

   History of exploration from ancient times to about A.D. 1900.

226. Berghaus, Heinrich. <u>Abriss einer Geschichte der geographischen Entdeckungen von den ältesten Zeiten bis zur Gegenwart</u>. Berlin: Hasselberg'sche Verlagshandlung, 1857. 208 p.

   History of discovery and exploration from ancient times onward (to 1856).

227. Beriot, Agnès. <u>Grands voiliers autour du monde. Les voyages scientifiques, 1760-1850</u>. Paris: Editions du Pont Royal, 1962. 295 p.

   Illustrated popular history of voyaging to the time of James Ross and Charles Wilkes.

228. Bettex, Albert. <u>The Discovery of the World: The Great Explorers and the Worlds They Found</u>. London: Thames and Hudson, 1960. 379 p.

   Illustrated history of exploration from ancient times to the attainment of the South Pole.

229. Broc, Numa. "Voyages et géographie au XVIIIe siècle," <u>Revue d'histoire des sciences et de leurs applications</u>, vol. 22, no. 2 (April-June 1969), 137-154.

   Traces the influence of travel literature on geography in the 18th century. Emphasis on French examples.

230. Buck, Peter H. (Te Rangi Hiroa). <u>Explorers of the Pacific: European and American Discoveries in Polynesia</u>. Bernice P. Bishop Museum, Honolulu, Special Publication, no. 43 (1953). viii + 125 p.

   European and American exploration in the Pacific (Polynesia, Micronesia, and Melanesia), 1519-1850.

231. Buschick, Richard. <u>Die Eroberung der Erde. Dreitausend Jahre Entdeckungsgeschichte</u>. Hannover: Fackelträger-Verlag, 1930. 400 p.

Popular history of exploration, ancient times to modern. (The copy I saw was dated 1930 but described events to 1949).

232. Butze, Herbert. Die Entdeckung der Erde. 5000 Jahre Abenteuer, Reisen und Forschen. Gütersloh: C. Bertelsmann Verlag, 1962. 367 p.

 Popular history of exploration, ancient to modern.

233. Dainelli, Giotto. La conquista della terra. Storia delle esplorazioni. Turin: Unione Tipografico-Editrice Torinese, 1950. xii + 745 p.

 Popular history of exploration, from Babylonians to present. Well illustrated, no documentation.

234. Day, Alan Edwin. Discovery and Exploration: A Reference Handbook, Vol. 1, The Old World. New York, etc.: K.G. Saur and Clive Bingley, 1980. 295 p.

 Mostly consists of alphabetical entries (by explorers' names, region, book title, etc.).

235. Debenham, Frank. Discovery and Exploration. An Atlas-History of Man's Journeys into the Unknown. London: Paul Hamlyn, 1960. 272 p.

 Popular illustrated history of exploration, from speculation about the "exploration" by early Man to the recent exploration of space. Not really an atlas, because the space devoted to maps is exceeded by that of text and other illustrations. Useful summary, "Famous Explorers and Their Routes," at end (pp. 213-236), along with maps showing spheres of exploration A.D. 150-1550 and the extent of exploration of the world's six major landmasses, 1700-1900.

236. Delpar, Helen, ed. The Discoverers: An Encyclopedia of Explorers and Exploration. New York: McGraw-Hill Book Company, 1980. 471 p.

 Articles by 28 contributors (plus editor) on exploration from ancient times to the present (space exploration).

237. Deprez, Eugène. "Les grands voyages et les grandes découvertes jusqu'à la fin du XVIIIe siècle. Origines, développement, conséquences," Bulletin of

Fig. 1. Areas Unknown to Europeans, 1700–1900. Adapted from four maps published in *Die Entschleierung der Erde* by Walter Behrmann (1948). Behrmann had eleven maps showing the progressive discovery of the world by Europeans, from 400 B.C. to A.D. 1900. The present map is a conflation of the last four of Behrmann's maps but using a different projection and eliminating Antarctica and the oceans. Like the parent maps, this one is not perfectly accurate in detail, but it is useful for giving an overall impression to the reader.

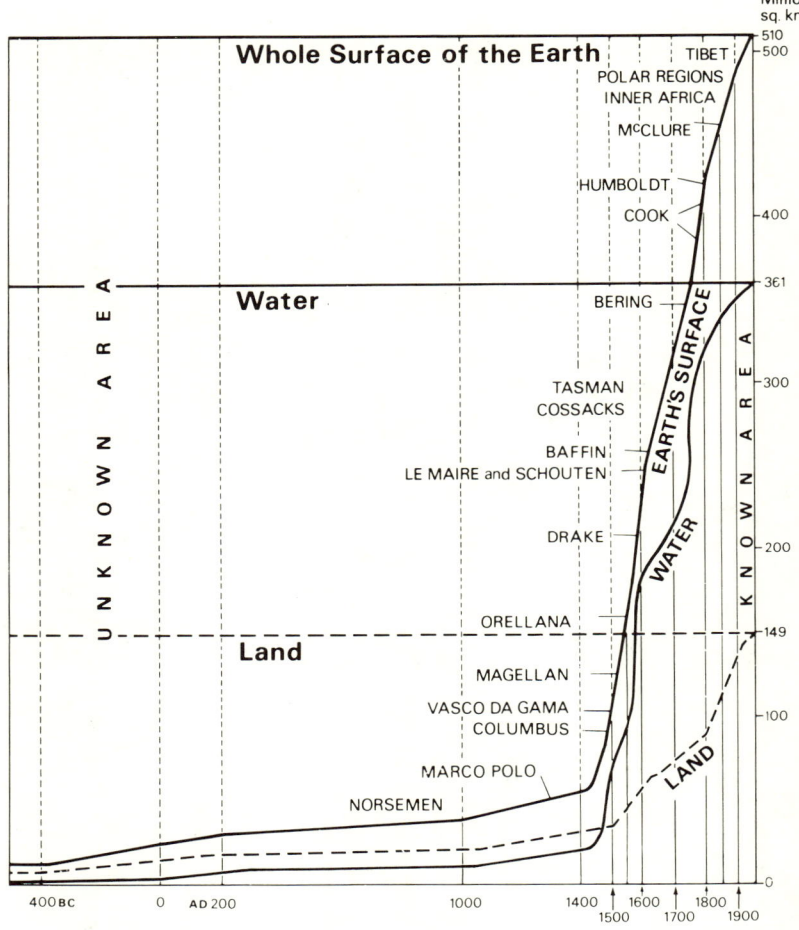

Fig. 2. The Progressive Unveiling of the Earth. Adapted from Walter Behrmann, *Die Entschleierung der Erde* (1948).

the International Committee of Historical Sciences, vol. 2, part 4 (no. 9) (June 1930), 555-614.

European discoveries from the 13th century to about 1740.

238. Dunmore, John. French Explorers in the Pacific. 2 vols. Oxford: Clarendon Press, 1965-1969. Vol. 1, 1965, vi + 356 p.; vol. 2, 1969, 428 p.

After a brief introduction that mentions some events in the 16th and 17th centuries, Vol. 1 treats several French navigators of the 18th century, from about 1768 onward (Bougainville to Marchand). Vol. 2 covers French explorers in the early part of the 19th century, ending with Dumont d'Urville's last voyage (1837-1840).

239. Fairchild, Wilma B. "The Explorers: Men and Motives," Geographical Review, vol. 38, no. 3 (July 1948), 414-425.

Concerned with the nature of biography as well as the nature of exploration. Article based on more than 40 biographies of explorers. In conclusion calls for biographies of geographers as well and suggests William Morris Davis as a subject.

240. Falkenstein, Constantin Karl. Geschichte der geographischen Entdeckungsreisen. 5 vols. in 1. Dresden: P.G. Hildchersche Buchhandlung, 1828-1829. Vol. 1, 1828, vi + 170 p.; vol. 2, 1828, 130 p.; vol. 3, 1828, viii + 152 p.; vol. 4, 1828, 135 p.; vol. 5, 1829, 231 p.

Vol. 1 goes from ancient times to the death of Columbus; Vol. 2 from the death of Columbus to the discovery of Australia; Vol. 3 from the discovery of Australia to Cook's first voyage; Vol. 4 from Cook to Humboldt (1800); Vol. 5 from Humboldt (1800) to Champollion (1828).

241. Feeken, Erwin H.J., and Feeken, Gerda E.E. The Discovery and Exploration of Australia. Melbourne: Thomas Nelson (Australia) Ltd., 1970. 318 p.

Illustrated popular history of Australian exploration, 1606-1901.

242. Forbes, Vernon S. Pioneer Travellers of South Africa: A Geographical Commentary upon Routes, Records, Observations and Opinions of Travellers at the Cape, 1750-1800. Cape Town: A.A. Balkema, 1965. 177 p.

European travelers in the Cape Province in the second half of the 18th century.

243. Friis, Herman R. "Cartographic and Geographic Activities of the Lewis and Clark Expedition," Journal of the Washington Academy of Science, vol. 44, no. 11 (November 1954), 338-351.

The geographical and cartographic contributions of the expedition of Meriwether Lewis and William Clark to the Pacific coast of North America, 1804-1806.

244. Friis, Herman R., ed. The Pacific Basin, A History of Its Geographical Exploration. American Geographical Society, Special Publication, no. 38 (1967). xi + 457 p.

A history of exploration of the Pacific Ocean and its islands and borders, beginning before 200 B.C. in the case of the Chinese, 7th century A.D. for the Japanese, and 16th century for Europeans. Concludes with geographical exploration in the 20th century. 16 contributors, most of whom were geographers.

245. Gilbert, Edmund William. The Exploration of North America, 1800-1850: An Historical Geography. New York: Macmillan; Cambridge: University Press, 1933. xiii + 233 p.

Exploration of the Trans-Mississippi West in the first half of the 19th century. (Page xi) "This book does not attempt to give a detailed historical description of every expedition; it does attempt to draw a picture of the geographical setting of western America, as it appeared to the explorers of the time."

246. Gillespie, James Edward. A History of Geographical Discovery, 1400-1800. ("Berkshire Studies in European History.") New York: Henry Holt and Company, 1933. viii + 111 p.

Mostly pre-18th century but includes a little on Bering, Cook, and other 18th-century explorers.

Part One--General and Topical   53

247. Glaser, Hugo. Die Entdecker der Welt. Von Marco Polo bis zur Gegenwart. Vienna: Schönbrunn-Verlag, 1951. 323 p.

History of exploration from Marco Polo to the present. Brief introductory chapter on travel in the ancient world. Popular work with no documentation.

248. Goetzmann, William H. Exploration and Empire: The Explorer and the Scientist in the Winning of the American West. New York: Alfred A. Knopf, 1966. xxii + 656 + xviii (index).

Exploration of the Trans-Mississippi portions of the United States from about 1805 to 1900.

249. Goodman, Edward J. The Explorers of South America. New York: The Macmillan Company, 1972. viii + 408 p.

Exploration of the continent of South America from Columbus (1498) to P.H. Fawcett (1925).

250. Grosvenor, Gilbert H. "Earth, Sea, and Sky: Twenty Years of Exploration by the National Geographic Society," Scientific Monthly, vol. 78, no. 5 (May 1954), 296-302.

Recounts exploration and scientific investigations sponsored by the NGS since 1933.

251. Grosvenor, Gilbert H. "The Geographic Conquests of the Nineteenth Century," Annual Report of the Board of Regents of the Smithsonian Institution ... for the Year Ending June 30, 1900 (Washington, D.C.: Government Printing Office, 1901), pp. 417-430.

Resume of 19th-century exploration. (Page 417) "In 1800 ... about one-fifth of the earth's land surface was known.... In 1900 approximately ten-elevenths of the earth's land surface may be described as known."

252. Hassert, Kurt. Die Polarforschung. Geschichte der Entdeckungsreisen zum Nord- und Südpol von den ältesten Zeiten bis zur Gegenwart. ("Aus Natur und Geisteswelt," 38.) 3rd ed. Leipzig and Berlin: B.G. Teubner, 1914. 134 p.

History of polar exploration from ancient times to 1912. 1st ed. pub. 1902.

253. Henze, Dietmar. Enzyklopädie der Entdecker und Erforscher der Erde. Graz: Akademische Druck- und Verlagsanstalt. 1st vol. 1975. Latest volume ("Lieferung") is no. 9 (to Herodotus) (1982).

Biographical dictionary of travelers and explorers from ancient times to modern. Entries long (Sven Hedin, pp. 484-560) and short (a Mr. Hardwicke, 3 lines).

254. Hobbs, William Herbert. "The Progress of Discovery and Exploration within the Arctic Region," Annals of the Association of American Geographers, vol. 27, no. 1 (March 1937), 1-22.

Arctic exploration from Pytheas to the present.

255. Hugues, Luigi. Cronologia delle scoperte e delle esplorazioni geografiche dall'anno 1492 a tutto il secolo XIX. ("Manuali Hoepli.") Milan: Ulrico Hoepli, 1903. viii + 487 p.

Chronicle of exploration, year by year, from 1492 to the end of the 19th century. Gets to 19th century on p. 162.

256. Hugues, Luigi. Le esplorazioni polari nel secolo XIX. Milan: Ulrico Hoepli, 1901. xx + 373 p.

19th-century polar exploration: Arctic (pp. 3-335) and Antarctic (pp. 339-361).

257. Key, Charles E. The Story of Twentieth-Century Exploration. London: George G. Harrap & Co. Ltd., 1937. 285 p.

The adventures of explorers and mountain-climbers in the first third of the 20th century, featuring such events as the attainment of the North and South poles.

258. Kirwan, L.P. The White Road: A Survey of Polar Exploration. London: Hollis & Carter, 1959. x + 374 p.

Polar exploration from Pytheas (c. 320 B.C.) to the present. Emphasis on British exploration, but also treats American, Russian, and Scandinavian. Abridged version appeared as A History of Polar Exploration (Harmondsworth, England: Penguin Books, 1962).

Part One--General and Topical   55

259. Krämer, Walter. Die Entdeckung und Erforschung der
     Erde. Mit einem ABC der Entdecker und Erforscher.
     3rd ed. Leipzig: F.A. Brockhaus Verlag, 1961. 395 p.

   Regional treatment of the history of exploration.
   Initial chapter (pp. 7-41) covers "The Evolution of Our
   World-View." About half of the book (pp. 191-375) is
   taken up by "ABC der Entdecker und Erforscher," a
   biographical dictionary of explorers and scientific
   travelers from ancient times to the present.

260. [Anonymous]. "Lady Travellers," Blackwood's Edinburgh
     Magazine, vol. 160, no. 969 (July 1896), 49-66.

   Concerns several European lady travelers of the 19th
   century. (Page 66) "It goes without saying that our
   story reflects high credit on the courage, the
   perseverance, and the benevolence of the gentler sex;
   it is a record of which women may well be proud. And
   there is this further to be said--that in no case has
   their travelling enthusiasm involved the sacrifice of
   obvious domestic duty; nor has it brought out any
   qualities inconsistent with the modesty, the grace, and
   the gentleness that must always be regarded as the
   fitting ornaments of the sex."

261. Langley, Michael. When the Pole Star Shone: A History
     of Exploration. London, etc.: George G. Harrap & Co.
     Ltd., 1972. 176 p.

   Popular history of exploration since the time of
   Homer.

262. La Roncière, Charles de. Histoire de la découverte de
     la terre. Explorateurs et conquérants. Paris:
     Librairie Larousse, 1938. viii + 304 p.

   Illustrated popular history of exploration from
   ancient times onward, with some material on
   mountaineering, oceanography, and marine biology.

263. Mason, Kenneth. "Kishen Singh and the Indian
     Explorers," Geographical Journal, vol. 62, no. 6
     (December 1923), 429-440.

   Reviews work of Indian explorers in the Himalayas and
   Tibet in the 19th century.

264. Meynen, Emil. "Ausgewählte Daten der Entdeckung und verkehrstechnischen Raumüberwindung, 1492-1952," Geographisches Taschenbuch 1953 (Stuttgart: Reise- und Verkehrsverlag, 1952), pp. 31-48.

Chronological list of some of the most important dates and events dealing with exploration and transport technology since 1492.

265. Middleton, Dorothy. Victorian Lady Travellers. London: Routledge & Kegan Paul, 1965. xiii + 182 p.

The stories of several late Victorian lady travelers, mostly English. (Pp. 3-4) "From about 1870 onwards more women than ever before or perhaps since undertook journeys to remote and savage countries; travelling as individuals, and for a variety of reasons, they were mostly middle-aged and often in poor health, their moral and intellectual standards were extremely high and they left behind them a formidable array of travel books.... Travel was an individual gesture of the house-bound, man-dominated Victorian woman."

266. Newby, Eric. The Rand McNally World Atlas of Exploration. Chicago: Rand McNally, 1975. 288 p.

Exploration from ancient times to the present, including a chapter on "The Last Frontier" (space). Text predominates over maps in this "atlas."

267. Nyström, Johan Fredrik. Geografiens och de geografiska upptäckternas historia till början af 1800-talet. Stockholm: C.E. Fritzes Kongl. Hofbokhandel, 1899. viii + 414 p.

History of geography and exploration from ancient times to the beginning of the 19th century--but, in fact, the author includes some material on personalities and events later in the century in the last part of the book.

268. Outhwaite, Leonard. Unrolling the Map: The Story of Exploration. London: Constable, 1935; New York: John Day, 1935. xiv + 351 p.

A history of exploration from ancient times to 1935. Aimed at a teenage audience but can be profitably read by adults.

Part One--General and Topical   57

269. Overton, J.D.   "A Theory of Exploration," Journal of
     Historical Geography, vol. 7, no. 1 (January 1981),
     53-70.

     (From abstract, p. 53) "Exploration, as a topic for
     study, has received much attention from geographers and
     historians but hitherto it has been examined with
     reference to a small number of outstanding journeys in
     isolation from their wider social and economic context.
     This paper aims to re-examine exploration in its
     broader realms seeing it as an interacting process
     closely linked to the economic development of a pioneer
     economy, rather than as an isolated series of events."
     As a case study, the author analyzes the inland
     exploration of the Nelson region of New Zealand from
     1841 to 1865 (pp. 60-70).

270. Parias, L.-H., ed. Histoire universelle des explora-
     tions. 4 vols. Paris: Nouvelle Librairie de France,
     1955-1956. Vol. 1, De la préhistoire à la fin du
     moyen âge, by Louis-René Nougier, Jean Beaujeu, and
     Michel Mollat, 1955, 416 p.; Vol. 2, La Renaissance
     (1415-1600), by Jean Amsler, 1955, 414 p.; Vol. 3,
     Le temps de grands voiliers, by Pierre-Jacques
     Charliat, 1955, 366 p.; Vol. 4, Epoque contemporaine,
     by J. Rouch, Paul-Emile Victor, and Haroun Tazieff,
     1956, 446 p. and separately paged 16-p. appendix,
     "Lexique alphabétique des explorateurs."

     General history of exploration from the Paleolithic
     to the present.

271. Plischke, Hans. Entdeckungsgeschichte von Altertum bis
     zur Neuzeit. ("Wissenschaft und Bildung," 290.)
     Leipzig: Verlag von Quelle & Meyer, 1933. 160 p.

     History of exploration from ancient times to modern,
     with emphasis on the period from the 15th through the
     19th centuries.

272. Plischke, Hans. "Kulturgeschichtliche Studien über die
     Grundlagen der Entdeckungserfolge der Seereisen am
     Ende des 18. Jahrhunderts," Petermanns Geographische
     Mitteilungen, vol. 84, no. 4 (April 1938), 120-127.

     European discoveries and their consequences in the
     Pacific from 1520 to 1830, with emphasis on the latter
     half of the 18th century. Includes discussion of
     introductions of domesticated plants and animals.

273. Porena, Filippo. "Le scoperte geografiche nel secolo XIX," Rivista Geografica Italiana, vol. 7, no. 5 (May 1900), 254-272; nos. 6-7 (June-July 1900), 335-352; no. 9 (November 1900), 501-516; no. 10 (December 1900), 583-598; and vol. 8, no. 1 (January 1901), 18-41.

Exploration in the 19th century. The first 2 parts concern Africa, the third includes Australia and the Americas, the fourth treats Asia, and the fifth covers the polar regions.

274. Rawat, Indra Singh. Indian Explorers of the Nineteenth Century: Account of Explorations in the Himalayas, Tibet, Mongolia and Central Asia. New Delhi: Government of India, Ministry of Information and Broadcasting, Publications Division, 1973. x + 228 p.

Indians ("Pundits") employed in the Survey of India, c. 1865-1885, with special attention to Nain Singh and Kishen Singh.

275. Rotberg, Robert I., ed. Africa and Its Explorers: Motives, Methods, and Impact. Cambridge: Harvard University Press, 1970. 351 p.

Nine authors contributed as many chapters on 19th-century African explorers: Heinrich Barth, David Livingstone, Richard Burton, John Hanning Speke, Samuel White Baker, Gerhard Rohlfs, Henry Morton Stanley, Verney Lovett Cameron, and Joseph Thomson.

276. Sievers, W. "Die geographische Erforschung Südamerikas im 19. Jahrhundert," Petermanns Geographische Mitteilungen, vol. 46, no. 6 (June 1900), 121-142.

Scientific exploration of South America in the 19th-century, beginning with Humboldt's arrival in 1799.

277. Skelton, Raleigh A. Explorers' Maps. Chapters in the Cartographic Record of Geographical Discovery. London, etc.: Spring Books, 1970. xi + 337 p.

Book has 2 themes: it is both a history of cartography and a history of exploration. Begins with Marco Polo and ends with the polar regions in the 19th century. Originally published in 1958 by Routledge & Kegan Paul Ltd.

Part One--General and Topical   59

278. Spilhaus, Margaret Whiting. <u>The Background of Geography</u>. Philadelphia: J.B. Lippincott Company, 1935. 286 p.

   Largely a history of exploration from ancient times to the end of the 17th century, with a brief concluding chapter on the Northwest and Northeast passages and Antarctica.

279. Surdich, Francesco. <u>Le grandi scoperte geografiche e la nascita del colonialismo</u>. Florence: La Nuova Italia Editrice, 1975. 134 p.

   After a lengthy introduction, the book consists of excerpts from various writings on exploration and colonialism, with emphasis on the 15th and 16th centuries. Brief mentions of slavery in the 18th and 19th centuries.

280. Sykes, Percy. <u>A History of Exploration from the Earliest Times to the Present Day</u>. London: Routledge; New York: Macmillan, 1934. xiv + 374 p.

   History of exploration from the Sumerians onward. Third edition (1950) includes appendix covering developments after 1934.

281. Ulrich, Johannes. "Grosse Meereskundliche Forschungsfahrten, 1920-1974," <u>Geographisches Taschenbuch 1975/976</u> (Wiesbaden: Franz Steiner Verlag GMBH, 1975), pp. 1-8.

   A list of the important oceanographic expeditions from 1920 onward.

282. Wood, Herbert J. <u>Exploration and Discovery</u>. London: Hutchinson's University Library, 1951. 192 p.

   A history of exploration, from ancient times to the present, with emphasis on the Renaissance period. Appendix (pp. 175-186), "The History of Navigation," by Eila M.J. Campbell.

283. Wright, John K. "Some Broader Aspects of the History of Exploration. A Review," <u>Geographical Review</u>, vol. 25, no. 2 (April 1935), 317-320.

   Review article based on several recent works on the history of exploration (by J.N.L. Baker, Percy Sykes,

Leonard Outhwaite, Hans Plischke, J.C. Beaglehole, J.B. Brebner, and H.R. Wagner).

284. Zavatti, Silvio. Dizionario degli esploratori e delle scoperte geografiche. Milan: Feltrinelli Editore, 1967. vi + 360 p.

An historical dictionary of explorers and exploration. Explorers are treated alphabetically on pp. 5-302; exploration, by continent and chronologically, on pp. 305-328; and route maps on pp. 331-356.

IV. CARTOGRAPHY, SURVEY, AND NAVIGATION

285. Alinhac, G. Historique de la cartographie. 2 vols. Paris: Institut géographique national, Ecole nationale des sciences géographiques, 1965. (New ed. 1973.) Vol. 1 (Text), v + 109 + 43 p. "Livre second" on the 1:80,000 map of France; Vol. 2 (Plates), 77 p.

Vol. 1 is a history of cartography from ancient times to modern. After Mercator, the text is concerned solely with French topographic maps.

286. Andrews, John H. A Paper Landscape: The Ordnance Survey in Nineteenth-Century Ireland. Oxford: Clarendon Press, 1975. xxiv + 350 p.

Official mapping of Ireland from 1824 onward, with some consideration of earlier developments, such as the establishment of the Ordnance Survey of Great Britain in 1791.

287. Arden-Close, Charles. The Early Years of the Ordnance Survey. A reprint with a new introduction by J.B. Harley. Newton Abbot, England: David & Charles; New York: Augustus M. Kelley, Publishers, 1969. xxxv + 164 p.

History of official mapping in Great Britain, 1746-1846. Originally appeared as a series of articles in the Royal Engineers Journal and was subsequently published in book form by the Institution of Royal Engineers in 1926. Charles Close (1865-1952) took the name Charles Arden-Close in 1938.

Part One--General and Topical 61

288. Arnberger, Erik. "Beiträge zur Geschichte der angewandten Kartographie und ihrer Methoden in Österreich," pp. 1-43 in *Festschrift zur Hundertjahrfeier der Geographischen Gesellschaft in Wien 1856-1956*, ed. Konrad Wiche (Vienna: Geographische Gesellschaft, 1957).

Cartography in Austria from 16th century to the present.

289. Bagrow, Leo. *Die Geschichte der Kartographie*. Berlin: Safari-Verlag, 1951. 383 p.

History of cartography from ancient times to just beyond the middle of the 18th century. Also some consideration of the maps of non-Western peoples. Bagrow completed the writing in 1943. The work was first printed in 1944 but was destroyed by fire before release. D.L. Paisey translated the work into English in 1960, and it was published in 1964 as *History of Cartography*, revised and edited by R.A. Skelton (London: Watts; Cambridge: Harvard University Press) (312 p.). The revised English translation was translated back into German and published in 1963 as *Meister der Kartographie* by Bagrow and Skelton (Berlin: Safari-Verlag) (579 p.).

290. Bagrow, Leo. *A History of the Cartography of Russia up to 1600* and *A History of Russian Cartography up to 1800*. 2 vols. Ed. Henry W. Castner. Wolfe Island, Ontario, Canada: The Walker Press, 1975. Vol. 1, xv + 139 p.; vol. 2, xiv + 311 p.

History of Russian cartography from the time of Herodotus. Approximately the last 40 pages (pp. 196-238) of the text of Vol. 2 have to do with the latter half of the 18th century.

291. Bay, Helmuth. "The History and Technique of Map Making," *Bulletin of The New York Public Library*, vol. 47, no. 11 (November 1943), 795-809.

Map-making and printing from ancient times to the present, with emphasis on the United States.

292. Bernleithner, Ernst. "The Development of Cartography in Austria," pp. 296-298 in Vol. 4 of *Actes du XIe Congrès International d'Histoire des Sciences* (Warsaw, etc., 1965) (Warsaw, etc.: Ossolineum, 1968).

Cartography in Austria, particularly since the 15th century.

293. Bertacchi, Cosimo. "Della storia della geografia con speciale riferimento alla geografia matematica," Rivista Geografica Italiana, vol. 10, nos. 6-7 (June-July 1903), 297-313.

    Mathematical geography (i.e., survey and measurement) as basis for mapping. Brief sketch of mathematical geography from classical times onward.

294. Biddle, C.A. "The Bicentenary of the Survey of India," The Chartered Surveyor, vol. 100, no. 10 (April 1968), 500-505.

    Official mapping of India from 1767 onward. James Rennell's appointment by Clive "in January 1767 as Surveyor-General of the provinces of Bengal, Bihar and Orissa ... is the event which may be taken as the real beginning of the Survey of India" (p. 500).

295. Blakemore, M.J., and Harley, J.B. "Concepts in the History of Cartography: A Review and Perspective," ed. Edward H. Dahl, Cartographica, Monograph no. 26 (vol. 17, no. 4 Winter 1980). 120 p.

    The authors define the history of cartography, describe trends in the literature, and make suggestions for future work.

296. Bonacker, Wilhelm. "Globen, einst und jetzt," Kartographische Nachrichten, vol. 10, no. 1 (1960), 19-22.

    A brief history of globes.

297. Bonacker, Wilhelm. "Globenmacher aller Zeiten," Globusfreund, no. 5 (December 1956), 17-28.

    Biographical dictionary of globemakers.

298. Bonacker, Wilhelm. Kartenmacher aller Länder und Zeiten. Stuttgart: Anton Hiersemann, 1966. 243 p.

    A biographical dictionary of 6350 people who were concerned with maps (cartographers, geographers, surveyors, publishers, map-sellers, collectors, etc.) from ancient times onward. Introduction (pp. 7-21) in both German and English.

Part One--General and Topical 63

299. Bonacker, Wilhelm. "Das Schrifttum zur Globenkunde," Janus, vol. 48 (1959), 81-132.

    List of 660 works dealing with globes--9 manuscript items (c. 1272 to the 18th century) and 651 printed works (1509-1958). Also issued separately by E.J. Brill of Leiden in 1960 (58 p.).

300. Brown, Lloyd A. The Story of Maps. Boston: Little, Brown and Company, 1949. 393 p.

    A history of cartography from Strabo to World War II.

301. Burky, Charles. "Kümmerly & Frey et la cartographie suisse," Geographica Helvetica, vol. 7, no. 3 (September 1952), 169-172.

    Brief history of Swiss cartography, beginning with the Karte der Eidgenossenschaft of Türst (1495-1499). Gottfried Kümmerly began in 1852 with a small lithographic establishment in Bern.

302. Cortesão, Armando. History of Portuguese Cartography. 2 vols. Lisbon: Junta de Investigações do Ultramar, 1969. Vol. 1, xxiii + 323 p.; vol. 2, xv + 469 p.

    History of Portuguese cartography through the 15th century. Vol. 1, Chapter 1, pp. 1-70, "Cartography and Its Historians," concerns the Viscount de Santarém (1791-1856) and other historians of cartography. This is the English edition of the work, which was also published in Portuguese.

303. Crone, Gerald R. Maps and Their Makers: An Introduction to the History of Cartography. London: Hutchinson's University Library, 1953. 181 p.

    A textbook on the history of cartography from ancient times onward. A 5th edition (152 p.) was published in 1978.

304. Curnow, Irene J. The World Mapped: Being a Short History of Attempts to Map the World from Antiquity to the Twentieth Century. London: Sifton Praed & Co. Ltd., 1930. 104 p.

    A history of cartography for the general reader. Only the last 7 pages of the text concern the period after 1800.

305. Dainville, François de. "De la profondeur à l'altitude," International Yearbook of Cartography, vol. 2 (1962), 150-160.

Submarine contours (isobaths) were in use before the idea was applied to land areas, so that "one must seek the origins of the modern representation of terrestrial relief in the realm of the hydrographer." English translation by Arthur H. Robinson published in Surveying and Mapping, vol. 30 (1970), 389-403.

306. Dunbar, Gary S. "Elisée Reclus and the Great Globe," Scottish Geographical Magazine, vol. 90, no. 1 (April 1974), 57-66.

Describes the abortive plan of the French geographer Elisée Reclus to construct globes at scales from 1:50,000 to 1:500,000 in the period 1895-1900. Also mentions other large globes, actual and projected, from 1823 to the 1960s.

307. Eckert, Max. Die Kartenwissenschaft: Forschungen und Grundlagen zu einer Kartographie als Wissenschaft. 2 vols. Berlin and Leipzig: Walter de Gruyter & Co., 1921-1925. Vol. 1, 1921, xvi + 640 p.; vol. 2, 1925, xiv + 880 p.

Comprehensive treatment of the field of cartography, with much attention to historical aspects.

308. Engelmann, Gerhard. "Alexander von Humboldt und die Geographische Kunstschule von Heinrich Berghaus in Potsdam," pp. 101-107 in Alexander von Humboldt: Vorträge und Aufsätze Anlässlich der 100. Wiederkehr Seines Todestages am 6. Mai 1959, ed. Johannes F. Gellert (Geographische Gesellschaft der Deutschen Demokratischen Republik, Wissenschaftliche Abhandlungen, vol. 2) (Berlin: VEB Deutscher Verlag der Wissenschaften, 1960).

Berghaus' Geographische Kunstschule lasted from 1839 to 1848. Its principal product was Berghaus' Physikalische Atlas (1837-1848), the great thematic atlas instigated by Humboldt.

309. Engelmann, Gerhard. "Carl Ritters Produktenkarten 1800-1836. Ein Beitrag zur Geschichte der thematischen Karten," International Yearbook of Cartography, vol. 6 (1966), 41-46.

Part One--General and Topical

Carl Ritter's maps of economic (especially agricultural) production, 1800-1836.

310. Engelmann, Gerhard. "Carl Ritters 'Sechs Karten von Europa,'" Erdkunde, vol. 20, no. 2 (May 1966), 104-110.

(From English abstract, p. 104) "Carl Ritter's 'Sechs Karten von Europa' (six maps of Europe), Schnepfenthal, 1806, is the earliest physical atlas devoted to a continent." The atlas consists of two physical maps and four economic maps.

311. Engelmann, Gerhard. "Deutscher National-Atlas. Entwurf von Heinrich Berghaus, Ablehnung durch Carl Ritter," Forschungen und Fortschritte, vol. 34, no. 6 (June 1960), 161-164.

Berghaus' plan for a national atlas in 1847-1848.

312. Engelmann, Gerhard. "Der Physikalische Atlas des Heinrich Berghaus. Die kartographische Technik der ältesten thematischen Kartensammlung," International Yearbook of Cartography, vol. 4 (1964), 154-161.

Heinrich Berghaus' Physikalische Atlas (1845) was the first thematic atlas in the world.

313. Engelmann, Gerhard. "Das Seekartenwerk des Heinrich Berghaus," Petermanns Geographische Mitteilungen, vol. 110, no. 4 (December 1966), 310-320.

Heinrich Berghaus' abortive plans for nautical atlases.

314. Engelmann, Gerhard. "Traugott Bromme und der 'Atlas zu Alex. v. Humboldt's Kosmos,'" Forschungen und Fortschritte, vol. 36, no. 11 (November 1962), 334-337.

Traugott Bromme (1802-1866) and his role as editor of the atlas accompanying Humboldt's Kosmos.

315. Engelmann, Gerhard. "Zeittafel der Kartographie 1700-1850," Geographisches Taschenbuch 1966/69 (Wiesbaden: Franz Steiner Verlag GMBH, 1968), pp. 1-20.

A list of facts about European cartography, 1700-1850: topographic maps, pp. 1-15; thematic maps, pp. 15-20.

316. Fauser, Alois. *Die Welt in Händen. Kurze Kulturgeschichte des Globus.* Stuttgart: Schuler Verlagsgesellschaft, 1967. 184 p.

History and description of terrestrial and celestial globes, almost all of which were made before 1750.

317. Fead, Margaret Irene. "Notes on the Development of the Cartographic Representation of Cities," *Geographical Review*, vol. 23, no. 3 (July 1933), 441-456.

Essay on the mapping of cities from the Babylonians to the present.

318. Forbes, Eric G. "Die Entwicklung der Navigationswissenschaft im 18. Jahrhundert," *Rete*, vol. 2, no. 4 (November 1975), 307-321.

Development of navigational science in the 18th century.

319. Fordham, Herbert G. *Some Notable Surveyors and Map-Makers of the Sixteenth, Seventeenth, and Eighteenth Centuries and Their Work: A Study in the History of Cartography.* Cambridge: Cambridge University Press, 1929. xii + 99 p.

Survey and cartography in Great Britain and France from Elizabethan times to the end of the 18th century.

320. Freitag, Ulrich. "Zeittafel zur Geschichte der Kartennetzlehre," *Kartographische Nachrichten*, vol. 18, no. 3 (July 1968), 92.

Chronological table of landmark events in the history of map projections, 550 B.C. to A.D. 1957.

321. Friis, Herman R. "A Brief Review of the Development and Status of Geographical and Cartographical Activities of the United States Government: 1776-1818," *Imago Mundi*, vol. 19 (1965), 68-80.

(Page 68) "The forty-two years 1776-1818 were the formative period in the history of geographical and cartographical activities of the Federal government. Much surveying and mapping and some geographical exploration was accomplished, but ... it was for the most part sporadic and special purpose, with surprisingly little attempt to coordinate the operations and results with a national plan until 1818."

Part One--General and Topical   67

322. Friis, Herman R. "Statistical Cartography in the United States Prior to 1870 and the Role of Joseph C.G. Kennedy and the U.S. Census Office," The American Cartographer, vol. 1, no. 2 (October 1974), 131-157.

(Abstract, p. 131) "This paper (1) briefly describes ... the development of statistical cartography in Europe and the United States during the first six decades of the 19th century [and] (2) discusses Joseph Camp Griffith Kennedy's role as Superintendent of the Census Office (1845-1865) in ... compiling the first forms of statistical cartography produced in that office."

323. Fryer, D.H. "Cartography and Aids to Navigation," pp. 438-465 in A History of Technology, vol. 5, The Late Nineteenth Century, c1850 to c1900, ed. Charles Singer, E.J. Holmyard, A.R. Hall, and Trevor I. Williams (New York and London: Oxford University Press, 1958).

Cartography, geodesy, surveying (land and sea), and navigation in the latter half of the 19th century.

324. Gardiner, Leslie. Bartholomew 150 Years. Edinburgh: John Bartholomew & Son Ltd., 1976. 111 p.

An illustrated popular history of the Edinburgh cartographic establishment, John Bartholomew & Son Ltd., founded in 1826.

325. Gardiner, R.A. "William Roy, Surveyor and Antiquary," Geographical Journal, vol. 143, no. 3 (November 1977), 439-450.

William Roy (1726-1790) was a British surveyor and mapmaker who paved the way for the establishment of the Ordnance Survey in 1791, a year after his death.

326. Goode, J. Paul. "The Map as a Record of Progress in Geography," Annals of the Association of American Geographers, vol. 17, no. 1 (March 1927), 1-14.

History of mapmaking from ancient Babylon to the present.

327. Grob, Richard. "Geschichte der schweizerischen Kartographie," Jahresbericht der Geographischen

Gesellschaft von Bern, vol. 33 (1937-1939), 1-98; vol. 34 (1940-1941), 97 [sic]-183.

History of cartography in Switzerland from Ptolemy to the present. Period before A.D. 1500 covered in 6 pages. Also published separately by Verlag Kümmerly und Frey, Bern, in 1941 (194 p. + 28 plates).

328. Harms, Hans. Künstler des Kartenbilds. Biographien und Porträts. Oldenburg: Ernst Völker, Kartographie und Verlag, 1962. 245 p.

Mostly consists of portraits and brief biographical sketches of European cartographers, from Conrad Celtis (Pickel) (1459-1508) to Jean Dominique Cassini (1748-1845).

329. Harvey, P.D.A. The History of Topographical Maps: Symbols, Pictures and Surveys. London: Thames and Hudson Ltd., 1980. 199 p.

(Page 9) "This book is about the early development of topographical mapping. By a topographical map we mean a large-scale map." Chronicles the changes from symbols to pictures and finally to precise surveys. Although some recent maps are mentioned, the author emphasized the periods before the early 17th century, when "the map and the bird's-eye view parted company"-- i.e., the "change from the picture-map to the scale-map based on measured survey."

330. Heaney, G.F. "Rennell and the Survey of India," Geographical Journal, vol. 134, part 3 (September 1968), 318-327.

History of the Survey of India from 1767 to 1947. Marked 200th anniversary of the appointment of James Rennell as first Surveyor General of Bengal.

331. Hill, Gillian. Cartographical Curiosities. London: British Museum Publications for The British Library, 1978. 63 p.

Booklet written to accompany an exhibition at the British Library in 1978. Describes maps that were created for some purpose other than practical or scientific. These maps are "literary, satirical, frivolous, or simply decorative." Describes maps from late medieval times to the present.

Part One--General and Topical 69

332. Hodgkiss, Alan G. Understanding Maps: A Systematic History of Their Use and Development. Folkestone, England: Wm. Dawson & Son Ltd., 1981. 209 p.

 A history of cartography from ancient times to modern: world maps, regional maps, nautical charts, route maps, town plans and views, and various other kinds of thematic maps.

333. Horn, Werner. "Die Geschichte der Isarithmenkarten," Petermanns Geographische Mitteilungen, vol. 103, no. 3 (1959), 225-232.

 A history of isarithmic maps. Isarithms are lines drawn on maps to connect points of equal value. The first isarithms were contour lines, isobaths (submarine contours), and isogonic (magnetic) lines. The greatest use of isarithms has been in the field of climatology, beginning with Humboldt's isotherms (1817).

334. Horn, Werner. "Das kartographische Gesamtwerk Adolf Stielers," Petermanns Geographische Mitteilungen, vol. 111, no. 4 (December 1967), 312-326.

 The cartographic work of Adolf Stieler (1775-1836), who was especially known for Stielers Handatlas (maps produced 1816-1836).

335. Irwin, B. St. G. "The Ordnance Survey--Roy's Legacy," Geographical Journal, vol. 143, part 1 (March 1977), 14-26.

 Brief history of Great Britain's Ordnance Survey from the 18th century to the present.

336. Irwin, Daniel. "The Historical Development of Terrain Representation in American Cartography," International Yearbook of Cartography, vol. 16 (1976), 70-83.

 Terrain representation in American cartography, 1770-1920.

337. Jarcho, Saul. "The Contributions of Heinrich and Hermann Berghaus to Medical Cartography," Journal of the History of Medicine and Allied Sciences, vol. 24, no. 4 (October 1969), 412-415.

 Concerns the maps showing distributions of diseases in the works of the German cartographer Heinrich

Berghaus (1797-1884) and his nephew Hermann (1828-1890).

338. Jervis, W.W. The World in Maps: A Study in Map Evolution. 2nd ed. New York: Oxford University Press, 1938. 208 p.

    A book on cartography for the general reader. Historical emphasis throughout. 1st ed. pub. 1936.

339. Kish, George. La carte: image des civilisations. Paris: Seuil, 1980. 287 p.

    A history of cartography from ancient times to the present, with emphasis on developments before 1800. Text (55 p.) followed by 116 pages of plates and 103 pages of notes on the plates.

340. Kish, George. "Early Thematic Mapping: The Work of Philippe Buache," Imago Mundi, vol. 28 (1976), 129-136.

    The French geographer and cartographer Philippe Buache (1700-1773) was a pioneer of thematic maps, best known of which were a bathymetric chart of the English Channel and a map showing the world's mountain systems as a single, interconnected whole.

341. Lauf, G.B. The Origin and Development of Cartography. Johannesburg: Witwatersrand University Press, 1955. 34 p.

    Cartography and survey from ancient times to modern, with special treatment of South Africa.

342. Laussedat, A. "Histoire de la cartographie," Revue scientifique, vol. 49, no. 23 (4 June 1892), 705-714; no. 24 (11 June 1892), 742-751.

    Describes 4 chronological phases in the development of cartography: origins, Latin and Arab cartography, era of the great discoveries, and modern cartography. Places the history of cartography in the general context of the history of cosmography and geography and even of scientific progress generally. Author gives detailed treatment of some topics, such as hydrography and relief representation. Also issued separately as a 56-page monograph. [GP]

Part One--General and Topical   71

343. Lehmann, Edgar. "Carl Ritters kartographische Leistung," Die Erde, vol. 90, no. 2 (April 1959), 184-222.

    The cartographic works of the German geographer Carl Ritter (1779-1859).

344. Lejeaux, A. "Notice historique succincte sur la cartographie," Revue de géographie, vol. 46 (1900), 277-283, 349-355, 416-423; vol. 47 (1900), 5-11.

    The history of cartography, with emphasis on France. Treats topographic cartography (scales larger than 1:200,000), geographic cartography (smaller scales), relief models, globes, etc. In the modern period, special attention is given to German cartographic productions in Gotha and to the revival of cartography in France after 1870. [MCR]

345. Libault, André. Histoire de la cartographie. Paris: Chaix, 1959. 86 p.

    A history of cartography from ancient Babylon to the present. For the general reader.

346. Lynam, Edward. British Maps and Map-Makers. London: Collins, 1947. 47 p.

    A popular history of mapmaking in Britain, c. 1250-1935. First published 1944.

347. Lynam, Edward. The Mapmaker's Art: Essays on the History of Maps. London: The Batchworth Press, 1953. ix + 140 p.

    Essays on the history of English cartography, mostly 16th and 17th centuries. Reprint of articles previously published (1939-1950), with one new essay, "William Hack and the South Sea Buccaneers."

348. MacEachren, Alan M. "The Evolution of Thematic Cartography: A Research Methodology and Historical Review," The Canadian Cartographer, vol. 16, no. 1 (June 1979), 17-33.

    Special purpose or thematic mapping from the 18th century onward. Evolution of the basic symbolization (point, line, and area symbols) of thematic cartography.

349. Margot-Duclos, Jean-Luc. La France à l'échelle: histoire de la cartographie. Paris: Solar Editeur, 1978. 199 p.

    History of cartography, with emphasis on France's Institut géographique national and its predecessors.

350. Markham, Clements R. "The History of the Gradual Development of the Groundwork of Geographical Science," Geographical Journal, vol. 46, no. 3 (September 1915), 173-187.

    History of navigation and survey. (Page 173) "The gradual development, from the earliest times, of a knowledge of methods for fixing positions, of map making, of the construction and use of instruments, of all that constitutes the groundwork of our science ought surely not only to interest geographers, but to form an essential part of geographical education."

351. Mathieson, John. "Geodesy: A Brief Historical Sketch," Scottish Geographical Magazine, vol. 42, no. 6 (November 1926), 328-347.

    From the earliest notions of the shape of the earth to 19th-century triangulation by the Ordnance Survey of Great Britain.

352. Melón, Amando. "'Curriculum vitae' de la cartografía moderna," Estudios geográficos (Madrid), vol. 26, no. 99 (May 1965), 169-206.

    History of cartography, from Guillaume Delisle (1675-1726) to the present.

353. Muris, Oswald, and Saarmann, Gert. Der Globus im Wandel der Zeiten. Eine Geschichte der Globen. Berlin and Beutelsbach bei Stuttgart: Columbus Verlag Paul Oestergaard KG, 1961. 287 p.

    History of globes and globemaking, from earliest times to the present, with emphasis on the Renaissance period.

354. Nischer, Ernst. Osterreichische Kartographen. Ihr Leben, Lehren und Wirken. Vienna: Osterreichischer Bundesverlag für Unterricht, Wissenschaft und Kunst, 1925. 192 p.

Austrian cartographers from the beginning of the 16th century through the 19th century.

355. [Anonymous]. One Hundred Years of Map Making. The Story of W. & A.K. Johnston. Edinburgh: W. & A.K. Johnston, Ltd., n.d. (1925). 20 p.

   A brief history of the Johnston cartographic establishment in Edinburgh, founded 1825.

356. Pearson, Karen S. "The Nineteenth-Century Colour Revolution: Maps in Geographical Journals," Imago Mundi, vol. 32 (1980), 9-20.

   (Page 9) "The essence of this colour revolution in cartography has been the transformation of colour from a non-essential decorative supplement into an integral and functional element of design, indispensable to the cartographic objective." Color lithography was employed in Vienna by 1826, but its adoption by journals was delayed. Article uses examples from three geographical journals, one each from Germany, France, and Great Britain, from the 1850s onward.

357. Raisz, Erwin. General Cartography. 2nd ed. New York: McGraw-Hill Book Company, Inc., 1948. xv + 354 p.

   Textbook on cartography, with 50-page introductory section, "The History of Maps," covering the history of cartography from ancient times to modern. 1st ed. pub. 1938.

358. Raisz, Erwin. "Outline of the History of American Cartography," Isis, vol. 26, no. 2 (March 1937), 373-391.

   Concerns maps made in America from 1677 onward. Includes "Time Charts of Historical Cartography--United States" (c. 1750-1933).

359. Raisz, Erwin. "Time Charts of Historical Cartography," Imago Mundi, vol. 2 (1937), 9-16.

   Includes 4 charts illustrating the history of cartography, to be published in author's textbook on cartography (above). A fifth chart, showing the development of cartography in the United States, was published in Isis (above).

360. Raudzens, G. "The British Ordnance Department and the Advancement of Geographic Science," Cartography, vol. 7, no. 3 (June 1971), 106-109.

Brief history of the British Ordnance Department and its various activities, including the mapping of the United Kingdom, surveying the boundary between the U.S. and Canada, aiding land surveys in Australia, and conducting meteorological observations.

361. Reeves, Edward A. "The Mapping of the Earth--Past, Present, and Future," pp. 421-433 in the Report of the Eighty-Sixth Meeting of the British Association for the Advancement of Science, Newcastle-on-Tyne, 1916 (London: John Murray, 1917).

A brief history of surveying and mapping, with considerable attention to instruments, from the time of Thales onward.

362. Ristow, Walter W., comp. A la Carte: Selected Papers on Maps and Atlases. Washington, D.C.: Library of Congress, 1972. x + 232 p.

Reprint of 20 papers, of which 18 were first published in the Quarterly Journal of the Library of Congress. Seven papers concern maps and atlases of the 16th and 17th centuries, and the other 13, American maps of the 17th to 19th centuries.

363. Ristow, Walter W., comp. Guide to the History of Cartography: An Annotated List of References on the History of Maps and Mapmaking. 3rd ed. Washington, D.C.: Library of Congress, Reference Department, Geography and Map Division, 1973. 96 p.

Annotated bibliography of 398 books or monographs (no articles) on the history of cartography. Approximately 6 times as many references as in 2nd edition. 1st ed. (with Clara LeGear) pub. 1954; 2nd ed. (with LeGear) pub. 1960.

364. Ritchie, G.S. The Admiralty Chart: British Naval Hydrography in the Nineteenth Century. London: Hollis & Carter, 1967. xi + 388 p.

English hydrographic mapping, from Edmund Halley in the 1690s to the beginning of the 20th century.

Part One--General and Topical 75

365. Ritchie, G.S. "500 Years of Graphical and Symbolical Representation on Marine Charts," International Hydrographic Review, vol. 53, no. 1 (January 1976), 141-153.

Article concerns "the development of the various forms of graphic representations of ... the components of the marine chart over the 500 years prior to 1900, together with the ever-growing acceptance of recognised chart symbols" (p. 141).

366. Robinson, A.H.W. Marine Cartography in Britain: A History of the Sea Chart to 1855. Leicester: Leicester University Press, 1962. 222 p.

From the manuscript charts of the 16th century to 1855, the date of Francis Beaufort's retirement as Hydrographer. Hydrographic Office founded 1795.

367. Robinson, Arthur H. Early Thematic Mapping in the History of Cartography. Chicago: University of Chicago Press, 1982. xiv + 266 p.

History of thematic cartography from the late 17th century to about 1860, with emphasis on northwestern Europe.

368. Robinson, Arthur H. "The 1837 Maps of Henry Drury Harness," Geographical Journal, vol. 121, part 4 (December 1955), 440-450.

Harness (1804-1883) was a pioneer in statistical cartography. The maps that he produced in 1837 to accompany the Second Report of the Irish Railway Commissioners were the first to employ a number of techniques: (1) graduated circles for city populations, (2) urban and rural populations on the same map, (3) density of population, (4) flow lines to show movement, and (5) the dasymetric technique.

369. Robinson, Arthur H. "The Genealogy of the Isopleth," Cartographic Journal, vol. 8, no. 1 (June 1971), 49-53.

(Page 49) "Defining isometric lines as those portraying basic numerical distributions [absolute values, such as elevation, depth, temperature, etc.] and isopleths as portraying values [relative values, such as density, correlations, etc.], this paper traces the growth of the 'iso' family from the use of the isobath in 1584, through other members of the isometric family to the conception of the isopleth in 1845 [Lalanne], its birth in 1857 [Ravn], and its christening in 1859 [von Sydow]."

370. Robinson, Arthur H. "The Thematic Maps of Charles Joseph Minard," *Imago Mundi*, vol. 21 (1967), 95-110.

   Minard (1781-1870) was a French engineer and cartographer whose 51 maps "show a combination of cartographic ingenuity and concern with the graphic portrayal of statistical data that was almost unique during the central portion of the [19th] century."

371. Robinson, Arthur H., and Wallis, Helen M. "Humboldt's Map of Isothermal Lines: A Milestone in Thematic Cartography," *The Cartographic Journal*, vol. 4, no. 2 (December 1967), 119-123.

   The German geographer Alexander von Humboldt invented the isotherm (line drawn on a map to connect points of equal temperature) in 1816, and this proved to be a landmark event in geography and cartography, because it stimulated the invention of numerous other isolines.

372. Seymour, W.A., ed. *A History of the Ordnance Survey*. Folkestone, England: Wm. Dawson & Sons Ltd., 1980. xvi + 394 p.

   Chapters by 10 authors on the history of the Ordnance Survey, the official mapping agency in Great Britain, founded in 1791, with considerable background material on 18th-century military mapping.

373. Shalowitz, Aaron L. "Nautical Charting (1807-1957)," *Scientific Monthly*, vol. 84, no. 6 (June 1957), 290-301.

   The history of nautical chartmaking in the United States, beginning with the establishment of the Coast Survey in 1807.

374. Shibanov, Fyodor A. *Studies in the History of Russian Cartography*. Trans. by L.H. Morgan and ed. by James R. Gibson. *Cartographica*, nos. 14-15 (1975) (Supplements nos. 2-3 to *The Canadian Cartographer*, vol. 12 [1975]).

Part One--General and Topical 77

>Some aspects of survey and cartography in Russia from the 16th to the early 20th centuries. Part 1 deals with Russian cartography in the 16th and 17th centuries; Part 2 with the 18th, 19th, and early 20th centuries. Includes studies, bibliographies, and archival materials written and collected between 1947 and 1970.

375. Skelton, Raleigh A. "Cartography," pp. 596-628 in A History of Technology, vol. 4, The Industrial Revolution, c1750 to c1850, ed. Charles Singer, E.J. Holmyard, A.R. Hall, and Trevor I. Williams (New York and London: Oxford University Press, 1958).

>Cartography, surveying, hydrography, oceanography, and map-printing from the 1660s to the middle of the 19th century.

376. Skelton, Raleigh A. "Cook's Contribution to Marine Survey," Endeavour, vol. 27, no. 100 (January 1968), 28-32.

>James Cook laid the foundation of scientific hydrography and oceanography in his survey of the Newfoundland coasts (1763-1767) and his three Pacific voyages (1768-1779). Article also discusses earlier navigation guides and charts, from the time of Scylax of Caryanda (4th century B.C.).

377. Skelton, R.A. "Landmarks in British Cartography, III: The Origins of the Ordnance Survey of Great Britain," Geographical Journal, vol. 128, part 4 (December 1962), 415-426.

>After a discussion of English county surveying in the 18th century, paper treats the origins of the national survey of Great Britain (Ordnance Survey), emphasizing the work of the Duke of Richmond and General William Roy.

378. Skelton, R.A. Maps: A Historical Survey of Their Study and Collecting. Chicago: University of Chicago Press, 1972. xv + 138 p.

>Lectures given by Skelton (1906-1970) at the Newberry Library in Chicago in 1966. As the title indicates, the lectures dealt not so much with the history of maps as with the history of the study and collecting of

maps. Contains bibliography of Skelton's publications, 1946-1974 (pp. 111-131).

379. Stams, Werner. "Die Kartographie in den ersten 30 Jahrgängen von 'Petermanns Geographische Mitteilungen': Zum 100. Todestag von August Petermann," Petermanns Geographische Mitteilungen, vol. 122, no. 3 (May 1978), 185-202; no. 4 (August 1978), 271-284.

Analyzes the cartographic work in the first 30 volumes of PGM, emphasizing the contribution of August Petermann (1822-1878). See especially pp. 195-196, which contain a time chart showing 3 generations of German cartographers in the 19th century (actually c. 1749-c. 1929) and a table listing German geographical periodicals in the 19th century (actually from late 1790s to about World War I).

380. Stavenhagen, W. "Frankreichs Kartenwesen in geschichtlicher Entwickelung," Mittheilungen der K.K. Geographischen Gesellschaft in Wien, vol. 45, nos. 7-8 (1902), 173-212.

French cartography from 1528 onward.

381. Stavenhagen, W. "Die geschichtliche Entwickelung des preussischen Militärkartenwesens," Geographische Zeitschrift, vol. 6, no. 8 (1900), 435-449; no. 9, 504-512; no. 10, 549-565.

History of Prussian mapping from 1566 onward, with emphasis on military maps.

382. Stavenhagen, W. "Russlands Kartenwesen in Vergangenheit und Gegenwart," Petermanns Geographische Mitteilungen, vol. 48, no. 10 (1902), pp. 224-229; no. 11, 254-260; no. 12, 274-278.

Russian cartography from 1549 onward.

383. Stevenson, Edward L. Terrestrial and Celestial Globes. 2 vols. New Haven: Yale University Press, 1921. Subtitle: "Their History and Construction, including a Consideration of Their Value as Aids in the Study of Geography and Astronomy." Vol. 1, xxvi + 218 p.; vol. 2, xi + 291 p.

Part One--General and Topical 79

Vol. 1 treats globes from ancient times to the end of the 16th century; vol. 2 covers globes and globemakers of the 17th and 18th centuries. The terminal chapter of Vol. 2 (pp. 196-219) concerns "The Technic of Globe Construction--Materials and Methods." At end (pp. 249-273) is "Index of Globes and Globe Makers" (including present locations of the globes).

384. Stine, Gordon E. "Milestones in the History of Cartography," Surveying and Mapping, vol. 30, no. 2 (June 1970), 285-288.

Brief survey of cartography from ancient times to the present.

385. Taylor, Eva G.R. The Mathematical Practitioners of Hanoverian England, 1714-1840. Cambridge: Cambridge University Press, 1966. xv + 502 p.

The cartographers, surveyors, navigators, and instrument-makers of England in the time of the Hanoverian monarchs, 1714-1840. Brief biographical notices of 2282 people make up the largest part of the book (pp. 107-483). Sequel to the author's Mathematical Practitioners of Tudor and Stuart England (1954).

386. Thrower, Norman J.W. Maps and Man: An Examination of Cartography in Relation to Culture and Civilization. Englewood Cliffs, New Jersey: Prentice-Hall, Inc., 1972. vii + 184 p.

A textbook on the history of cartography from ancient times to the present.

387. Tooley, Ronald V. Maps and Mapmakers. 6th ed. London: B.T. Batsford Ltd., 1978. xv + 140 p.

History of cartography from early times to mid-19th century. Very little on 19th-century maps. 1st ed. 1949.

388. Tooley, R.V. Tooley's Dictionary of Mapmakers. New York: Alan R. Liss, Inc.; Amsterdam: Meridian Publishing Company, 1979. xii + 684 p.

(Foreword, p. v) "The aim of this Dictionary is to give information ... on persons associated with the production of maps from the earliest times to the year

1900.... Originally, the first half of the work appeared in parts in Map Collectors' Circle [1965-1975], which has now been discontinued. Those portions have been revised and considerably amplified, in some cases more than doubling the number of entries. For the first half of the alphabet the present volume is thus a new edition; the second half is here published for the first time."

389. Wagner, Hermann. "Zur Geschichte der Gothaer Kartographie," Petermanns Geographische Mitteilungen, vol. 58, no. 1 (January 1912), 12-15; no. 2 (February 1912), 76-79.

Cartographic publications of the Justus Perthes Anstalt in Gotha, Germany, in the 19th century, with emphasis on the work of Adolf Stieler, Heinrich Berghaus and his nephew Hermann Berghaus, and August Petermann. English adaptation by Martha Krug Genthe, "Notes on the History of Gotha Cartography," Bulletin of the American Geographical Society, vol. 45, no. 1 (1913), 33-38.

390. Wallis, Helen, ed. Map-Making to 1900: An Historical Glossary of Cartographical Innovations and Their Diffusion. London: The Royal Society, 1976. xviii + 52 p.

Historical glossary of cartographic techniques or innovations: Section 1, Innovations in purpose (e.g., globe, road map); Section 2, Innovations in detail and design (cartouche, grid); and Section 3, Production techniques and materials (cerography, woodcut). "The study contains twenty-six sample entries which have been selected as the most suitable to demonstrate the form and scope of the project. Those not selected are being held over for inclusion in the comprehensive edition, 1980."

391. Wilford, John Noble. The Mapmakers. New York: Alfred A. Knopf, 1981. xi + 414 p.

History of cartography from ancient times to the present, with emphasis on the modern period (Prince Henry onward). Includes chapters on extraterrestrial mapping (Moon and Mars).

Part One--General and Topical 81

392. Wolkenhauer, Wilhelm. "Aus der Geschichte der Kartographie," <u>Deutsche Geographische Blätter</u>, vol. 27, no. 2 (1904), 95-116; vol. 33, no. 4 (1909), 239-264; vol. 35, nos. 1-2 (1912), 29-47; vol. 36, nos. 3-4 (1913), 136-158; vol. 38, no. 1 (1916-1917), 101-128; and vol. 38, no. 2 (1917), 157-201.

History of cartography from Ptolemy to the present. Last two installments concern the period since 1750. Vol. 38, no. 1, "Die Periode der Triangulation und topographischen Aufnahmen (1750-1840)," is arranged chronologically and gives names of cartographers and significant achievements. The last installment, "Die moderne Kartographie" (1840-1917), is a chronological bibliography of significant maps and atlases and also works in the history of cartography.

393. Wolkenhauer, W. "Zeittafel zur Geschichte der Kartographie," <u>Deutsche Geographische Blätter</u>, vol. 16, no. 4 (1893), 319-348.

A chronological list of important events in the history of cartography from 600 B.C. to A.D. 1893.

394. Wolter, John Amadeus. "The Emerging Discipline of Cartography." Unpublished Ph.D. thesis in Geography, University of Minnesota, 1975. xii + 345 p.

History of cartography, historians of cartography, and map collections (pp. 15-94); bibliographies (pp. 95-149); journals and societies (pp. 150-207); textbooks and manuals (pp. 208-251); and education and training (252-280).

395. Wolter, John A. "Geographical Libraries and Map Collections," pp. 236-266 in Vol. 9 of <u>Encyclopedia of Library and Information Science</u> (New York: Marcel Dekker, Inc., 1973).

History of geographical libraries and map collections, with emphasis on the 19th and 20th centuries.

396. Woodward, David. "The Study of the History of Cartography: A Suggested Framework," <u>The American Cartographer</u>, vol. 1, no. 2 (October 1974), 101-115.

History of cartography is defined as (p. 114) "the study of maps, mapmakers, and mapmaking techniques in their human context through time. The field includes the study of creation of the maps by survey, compilation, engraving, printing and coloring; the history of their distribution by publishing and selling; and the history of their acquisition, care, cataloging, and use."

397. Wroth, Lawrence C. Some American Contributions to the Art of Navigation, 1519-1802. Providence, Rhode Island: The Associates of the John Carter Brown Library, 1947. 41 p.

    From the Spanish-American contributions, beginning with Enciso's Suma de geografía in 1519, to Nathaniel Bowditch's The New American Practical Navigator in 1802. Preprint of article in Proceedings of the Massachusetts Historical Society, vol. 68 (1947).

398. Zögner, Lothar. "Die kartographische Darstellung der Polargebiete bis in das 19. Jahrhundert," Die Erde, vol. 109, no. 2 (1978), 136-152.

    Cartographic representation of the polar regions from the 16th through the 19th centuries.

V. PHYSICAL GEOGRAPHY AND BIOGEOGRAPHY

This section consists of a selection of materials on the history of geomorphology, climatology, and biogeography, with brief mention of the related fields of geology, meteorology, and ecology.

399. Baker, Victor R., and Pyne, Stephen. "G.K. Gilbert and Modern Geomorphology," American Journal of Science, vol. 278, no. 2 (February 1978), 97-123.

    The work of the American geologist and geomorphologist Grove Karl Gilbert (1843-1918) shows "continued relevance for the future conduct of geomorphic research."

400. Baumgärtel, Hans. "Alexander von Humboldt: Remarks on the Meaning of Hypothesis in His Geological Researches," pp. 19-35 in Toward a History of Geology, ed. Cecil J. Schneer (Cambridge: The M.I.T. Press, 1969).

Essay describes the geological research of the German geographer and naturalist Alexander von Humboldt and gives examples of his scientific hypotheses. Although "his manner of working was predominantly empirical and inductive" (p. 21), he tried to generalize and to formulate hypotheses.

401. Beck, Hanno. "Alexander von Humboldt und die Eiszeit," Gesnerus, vol. 30, no. 3/4 (1973), 105-121.

    Theories of glaciation in the 19th century, with special reference to the ideas of the German geographer Alexander von Humboldt.

402. Beguinot, A. "Pensieri intorno all'origine, alla storia dello sviluppo ed allo stato attuale della geografia botanica," Bollettino della Società Geografica Italiana, series 4, vol. 7, no. 11 (November 1906), 1048-1065; no. 12 (December 1906), 1170-1191.

    Plant geography from Aristotle and Theophrastus to Haeckel, de Candolle, Warming, and Schimper.

403. Brown, Ralph H. "The First Century of Meteorological Data in America," Monthly Weather Review, vol. 68, no. 5 (May 1940), 130-133.

    The collecting of meteorological data in the area that is now the United States from 1738 (when recordkeeping began in Charleston, South Carolina) to 1838.

404. Browne, Janet. The Secular Ark: Studies in the History of Biogeography. New Haven: Yale University Press, 1983. x + 273 p.

    The study of the distribution of plants and animals from Athanasius Kircher in the 17th century to Charles Darwin in the 19th.

405. Cannon, Walter F. "The Uniformitarian-Catastrophist Debate," Isis, vol. 51, part 1 (March 1960), 38-55.

    Controversies in British geology in the 1830s and 1840s.

406. Chorley, Richard J. "Diastrophic Background to Twentieth-Century Geomorphological Thought," Bulletin of the Geological Society of America, vol. 74, no. 8 (August 1963), 953-970.

Traces the rise and fall of the eustatic theory of the Austrian geologist Eduard Suess (1832-1914).

407. Chorley, Richard J.; Dunn, Antony J.; and Beckinsale, Robert P. The History of the Study of Landforms; or, The Development of Geomorphology, Vol. 1, Geomorphology before Davis. London: Methuen; New York: John Wiley & Sons Inc., 1964. xvi + 678 p.

History of geomorphology to 1889 (publication of W.M. Davis' first statements on the cycle of erosion). The time before 1800 is covered in about 50 pages, and the rest of the book deals with the 19th century. Vol. 2 is a biography of Davis and is cited in the American biographical section (below).

408. Claval, Paul. "Problèmes d'histoire de la géographie naturelle," pp. 37-58 in Studia z dziejów geografii i kartografii / Etudes d'histoire de la géographie et de la cartographie, ed. Józef Babicz (Warsaw: Polish Academy of Science, 1973).

Physical geography in the 19th and 20th centuries.

409. Court, Arnold. "Climatic Classification and Plant Geography in 1842," Weather, vol. 22, no. 7 (July 1967), 276-288.

An English naval surgeon, Richard Brinsley Hinds, proposed systems of climatic and vegetation classification in 1842.

410. Daly, Charles P. "On the History of Physical Geography," Journal of the American Geographical Society, vol. 22 (1890), 1-55.

Observations and speculation on cosmography and physical geography from ancient times to Alexander von Humboldt, Carl Ritter, and Mary Somerville.

411. Davies, Gordon L. "From Flood and Fire to Rivers and Ice--Three Hundred Years of Irish Geomorphology," Irish Geography, vol. 5, no. 1 (1964), 1-16.

Fig. 3. Alexander von Humboldt lecturing on physical geography in Berlin in 1827–1828. Among the listeners was Carl Ritter (bottom right). Reproduced from Hanno Beck, *Grosse Geographen* (1982), through the courtesy of Professor Beck.

Fig. 4. Carl Ritter lecturing at the University of Berlin, 1834–1835. The listener in the foreground was Alexander von Humboldt. Reproduced from Hanno Beck, *Grosse Geographen* (1982), with the permission of Professor Beck.

Part One--General and Topical 85

Study of Irish landforms by naturalists between 1600 and 1900.

412. Davis, William Morris. The Coral Reef Problem. American Geographical Society, Special Publication, no. 9. New York: American Geographical Society, 1928. 596 p.

First part (pp. 1-141) is devoted to criticism of all previous theories of coral reef formation, beginning with Charles Darwin (1837).

413. Davis, William Morris. The Physical Geography (Geomorphology) of William Morris Davis. Ed. Philip B. King and Stanley A. Schumm. Norwich, England: Geo Abstracts Limited, 1980. xxii + 217 p.

Davis' lectures at the University of Texas, 1926-1927, as transcribed by Philip King (pp. 1-151). Also contains reprint of Reginald Daly's biographical memoir of Davis, published in 1944 (pp. 175-217).

414. Dittrich, Mauritz. "Alexander von Humboldt und die Pflanzengeographie," pp. 25-42 in Alexander von Humboldt: Vorträge und Aufsätze Anlässlich der 100. Wiederkehr seines Todestages am 6. Mai 1859, ed. Johannes F. Gellert (Geographische Gesellschaft der Deutschen Demokratischen Republik, Wissenschaftliche Abhandlungen, vol. 2) (Berlin: VEB Deutscher Verlag der Wissenschaften, 1960).

The contributions of Alexander von Humboldt to plant geography, with some consideration of his predecessors and successors.

415. Dresch, Jean. "Un demi-siècle de recherches géomorphologiques," Hérodote, no. 12 (4th quarter 1978), 11-51.

Geomorphology in the half-century from 1928 to 1978, with emphasis on French work.

416. Egerton, Frank N. "Ecological Studies and Observations before 1900," pp. 311-351 in Issues and Ideas in America, ed. Benjamin Taylor and Thurman White (Norman: University of Oklahoma Press, 1976).

Ecological observations from Thomas Hariot and Captain John Smith around the beginning of the 17th century to C. Hart Merriam. (Page 311) "Ecology became a formal science at the turn of the twentieth century." Followed by McIntosh's article (below) on ecology since 1900.

417. Engler, A. "Die Entwickelung der Pflanzengeographie in den letzten hundert Jahren," Wissenschaftliche Beiträge zum Gedächtniss der hundertjährigen Wiederkehr des Antritts von Alexander von Humboldt's Reise nach Amerika am 5. Juni 1799 (Issued by the Gesellschaft für Erdkunde au Berlin on the occasion of the 7th International Geographical Congress) (Berlin: W.H. Kühl, 1899). 247 p. (3 separately paged monographs make up this volume.)

Plant geography from Tournefort (1717) onward, with special mention of Humboldt's Essai sur la géographie des plantes (1807) and other publications. A major part of the monograph consists of 19th-century references on the vegetation of various regions of the world.

418. Fenneman, Nevin M. "The Rise of Physiography," Bulletin of the Geological Society of America, vol. 50, no. 3 (1 March 1939), 349-359.

An essay on the history of physiography (geomorphology) from ancient times to the present.

419. Frisinger, H. Howard. The History of Meteorology: to 1800. (American Meteorological Society, Historical Monograph Series.) New York: Science History Publications, 1977. 148 p.

History of meteorology from ancient times to A.D. 1800. Author sees 3 major periods in the history of meteorology: (1) 600 B.C. to A.D. 1600, "The Period of Speculation"; (2) A.D. 1600 to 1800, "The Dawn of Scientific Meteorology"; (3) after 1800.

420. Gross, Walter E. "The American Philosophical Society and the Growth of Meteorology in the United States: 1835-1850," Annals of Science, vol. 29, no. 4 (December 1972), 321-338.

(Page 338) "Interest in meteorology [in the APS] ... reached a high point during the 1835-1850 era," after which the federal government assumed the dominant role. "The institutionalization of science in this country that was occurring in the nineteenth century was rendering poorly-financed, localized organizations obsolete in many fields of science."

421. Günther, Siegmund. "Die atmosphärische Physik bei Leopold v. Buch," Beiträge zur Geophysik: Zeitschrift für physikalische Erdkunde, vol. 5, no. 2 (1901), 171-205.

Leopold von Buch (1774-1853) and his work in meteorology and climatology, including observations on the effects of climate on vegetation.

422. Günther, Siegmund. "Die Bedeutung De la Métherie's für die Entwicklung der physikalischen Erdkunde," Mittheilungen der Kais. Konigl. Geographischen Gesellschaft in Wien, vol. 43 (1900), 3-14.

Discusses significance of Jean-Claude de La Métherie's Théorie de la terre (1795). Used German translation published 1797-1798.

423. Hall, D.H. History of the Earth Sciences during the Scientific and Industrial Revolutions with Special Emphasis on the Physical Geosciences. Amsterdam, etc.: Elsevier Scientific Publishing Company, 1976. xii + 297 p.

Considers the Scientific Revolution to be the period from 1450 to 1700, and the Industrial Revolution to be from 1760 to 1830. Also chapter on the mid-19th century, 1830-1870. Emphasis on Europe. Alexander von Humboldt treated on pp. 222-231.

424. Hellmann, G. "Hundert Jahre meteorologische Gesellschaften," Meteorologische Zeitschrift, vol. 40, no. 11 (November 1923), 321-329.

Meteorological societies of the past one hundred years, beginning with the Meteorological Society of London, founded in 1823.

425. Higgins, Charles G. "Theories of Landscape Development: A Perspective," pp. 1-28 in Theories of Landform Development, eds. Wilton N. Melhorn and Ronald C. Flemal (Publications in Geomorphology) (Binghamton: State University of New York, n.d. [1976?]).

   A history of geomorphology from 1877 onward, with emphasis on the United States and on William Morris Davis.

426. Hoffmeister, J. Edward. "James Dwight Dana's Studies of Volcanoes and of Coral Islands," Proceedings of the American Philosophical Society, vol. 82, no. 5 (29 June 1940), 721-732.

   Dana's observations on the Wilkes Exploring Expedition, 1838-1842, and later. Dana's ideas about coral reef formation compared with those of Charles Darwin and William Morris Davis.

427. Humboldt, Alexander von. Alexander von Humboldts Vorlesungen über physikalische Geographie nebst Prolegomenen über die Stellung der Gestirne. Berlin im Winter von 1827 bis 1828. Berlin: Miron Goldstein, 1934. 190 p.

   Humboldt's lectures on physical geography in 1827-1828. Not essentially an historical work, except for pp. 17-25, where Humboldt covers 6 major periods in the acquisition of knowledge of the natural world, but there are frequent allusions to earlier writers (travelers, geographers, naturalists, etc.), especially from the period from the 16th to the early 19th centuries but also going back to the ancient Greeks.

428. Hume, Edgar Erskine. "The Foundation of American Meteorology by the United States Army Medical Department," Bulletin of the History of Medicine, vol. 8, no. 2 (February 1940), 202-238.

   The first official weather records in America began in 1814, when the Surgeon General of the Army, James Tilton, ordered hospital surgeons to record the weather. This work continued for 60 years.

Part One--General and Topical 89

429. Janssens, Emile. "Esquisse d'une histoire critique des théories zoogéographiques," Bulletin de la Société Royale Belge de Géographie, vol. 76, nos. 3-4 (March 1953), 33-50.

Zoogeography interests both the geographer and the zoologist. In the author's opinion, the former can touch only the exterior, and his work will be exclusively descriptive. Valuable results can be obtained only by the zoologist. The first true zoogeographers were the English zoologists of the mid-19th century: Darwin, Wallace, Sclater, and Bates. Last work treated is Hesse's Tiergeographie (1924).

430. Jaudel, L., and Tricart, J. "Les précurseurs Anglo-Saxons de la notion Davisienne de cycle d'érosion," Revue générale des sciences pures et appliquées, vol. 65 (1958), 237-247.

Some of the geomorphological concepts of Charles Lyell, John Peter Lesley, Andrew Ramsay, and Grove Karl Gilbert, as background to William Morris Davis' concept of the erosional cycle.

431. King, Cuchlaine A.M., ed. Landforms and Geomorphology: Concepts and History. ("Benchmark Papers in Geology," vol. 28.) Stroudsburg, Pennsylvania: Dowden, Hutchinson & Ross Inc., 1976. xv + 404 p.

Part 1, "History of Geomorphology," pp. 11-117, consists of the editor's comments on several papers, her historical sketch of the history of geomorphology from Werner to W. Penck (pp. 13-35), and several previously published papers (or excerpts) dealing with geomorphology from John Playfair to Walther Penck (pp. 36-117).

432. Kington, J.A. "The Societas Meteorologica Palatina: An Eighteenth-Century Meteorological Society," Weather, vol. 29, no. 11 (November 1974), 416-426.

The SMP, the world's first meteorological society, was founded in Mannheim, Germany, in 1780. Published Ephemerides (1781-1785).

433. Koelsch, William A. "The New England Meteorological Society, 1884-96: A Study in Professionalism," pp. 89-104 in The Origin of Academic Geography in the

United States, ed. Brian Blouet (Hamden, Connecticut: Archon Books, 1981).

The NEMS, from its founding in 1884 until its demise in 1896, illustrates the transformation of laymen's science into professional science. "The key to the rise and decline of the society ... lay in the changing career strategies of American academic geography's greatest entrepreneur, William Morris Davis." After 1895, Davis concentrated on geomorphology and turned "away from meteorological research and institution-building," allowing the NEMS to become extinct.

434. Landsberg, Helmut E. "Early Stages of Climatology in the United States," Bulletin of the American Meteorological Society, vol. 45, no. 5 (May 1964), 268-275.

Studies of weather before 1870 in the area that is now the United States. From Thomas Jefferson to C.A. Schott.

435. Landberg, H.E. "Roots of Modern Climatology," Journal of the Washington Academy of Sciences, vol. 54, no. 4 (April 1964), 130-141.

Development of climatology, with special emphasis on the 19th century.

436. Leighly, John. "Climatology since the Year 1800," American Geophysical Union, Transactions, vol. 30, no. 5 (October 1949), 658-672.

Discusses the main classes of investigation or presentation in climatology during the previous century and a half: empirical formulation of climatic data, descriptive climatology, climatologic cartography, organization of observational data by synoptic categories, investigation of the physical bases of climate, definition of climatic types and the delineation of climatic regions, and reconstruction of past climates.

437. Marcinek, Joachim, and Nitz, Bernhard. "Hundert Jahre Eiszeitforschung und ihre Vorgeschichte: Zur Kenntnisentwicklung im nördlichen Mitteleuropa," Geographische Berichte, vol. 20, no. 3 (1975), 179-191.

Pleistocene researches of the last 100 years, from the time of Otto Torell (1875) onward, with emphasis on German work. Mention of earlier studies, back to the time of Horace-Bénédict de Saussure (1779+).

438. Marinelli, Olinto. "Del moderno sviluppo della geografia fisica e della morfologia terrestre," Bollettino della Società Geografica Italiana, series 4, vol. 9, no. 3 (March 1908), 226-248.

 Development of physical geography in the 19th century from Humboldt and Maury, "the two founders of modern physical geography," onward.

439. Martin, Geoffrey J. "A Fragment on the Penck(s)-Davis Conflict," Special Libraries Association, Geography and Map Division, Bulletin, no. 98 (December 1974), 11-27.

 Reception of William Morris Davis' geomorphological ideas in Germany, particularly by Albrecht Penck and his son Walther. Further light is shed on the Davis-Penck controversy by the papers of the American geographer Isaiah Bowman. Article consists mostly of excerpts from Bowman's correspondence.

440. McIntosh, Robert P. "Ecology since 1900," pp. 353-372 in Issues and Ideas in America, ed. Benjamin Taylor and Thurman White (Norman: University of Oklahoma Press, 1976).

 Growth of ecology as a formal discipline in the 20th century. (Page 354) "The early development of ecology as a normal and recognized science is commonly attributed to botanists." Studies in animal ecology followed. Companion-piece to Egerton's paper (above).

441. Meinardus, Wilhelm. "Die Entwickelung der Karten der Jahres-Isothermen von Alexander von Humboldt bis auf Heinrich Wilhelm Dove," Wissenschaftliche Beiträge zum Gedächtniss der hundertjährigen Wiederkehr des Antritts von Alexander von Humboldt's Reise nach Amerika am 5. Juni 1799 (Issued by the Gesellschaft für Erdkunde zu Berlin on the occasion of the 7th International Geographical Congress)(Berlin: W.H. Kühl,

1899). 32 p. (3 separately paged monographs make up this volume.)

The mapping of isotherms from Humboldt (1817) to Dove (1852).

442. Morris, A.S. "Mikhail Lomonosov and the Study of Landforms," Institute of British Geographers, Transactions, no. 41 (June 1967), 59-64.

The Russian scientist M. Lomonosov (1711-1765) supported uniformitarianism but had little appreciation for erosional agents. A full appreciation of his geomorphological work has only come in the 20th century.

443. Munzar, Jan. "Alexander von Humboldt and His Isotherms," Weather, vol. 22, no. 9 (September 1967), 360-363.

(Page 363) "Humboldt's map of isotherms 1817 represented the beginning of cartographical representation of climate."

444. Peltier, Louis C. "Events in the Development of Geomorphology," pp. 25-42 in Thresholds in Geomorphology, eds. Donald R. Coates and John D. Vitek (London: George Allen & Unwin Ltd., 1980).

Shows how social, as well as natural, events have influenced the study of landforms, with examples from Europe and the United States during the last 2 centuries. (Page 34) "... the objectives and scope of geomorphology have been expanded by social demand. It is no longer sufficient to postulate a natural history alone. The emphasis is shifting to a concern for the meaning of landforms in a socioeconomic context."

445. Ramakers, Günter. "Die 'Géographie des Plantes' des Jean-Louis Giraud-Soulavie (1752-1813): Ein Beitrag zur Problem- und Ideengeschichte der Pflanzengeographie," Die Erde, vol. 107, no. 1 (1976), 8-30.

Examines the work of the French naturalist Giraud-Soulavie in plant geography (1780-1784) and speculates about direct influences on Humboldt.

Part One--General and Topical 93

446. Raup, Hallock F. "Continental Glaciation Hypotheses before Louis Agassiz," Scientific Monthly, vol. 55, no. 1 (July 1942), 66-70.

(Page 66) "Without detracting from the accomplishment of Agassiz, it should be noted that his glacial theories [Etudes sur les glaciers, 1840] had been postulated to a considerable extent by foremost American scientists of the early nineteenth century" (beginning with Benjamin DeWitt in 1793).

447. Raup, Hugh M. "Trends in the Development of Geographic Botany," Annals of the Association of American Geographers, vol. 32, no. 4 (December 1942), 319-354.

Plant geography since the time of Theophrastus, with emphasis on the 19th and 20th centuries. 4 pages on "Development of Geography as a Whole."

448. Schmithüsen, Josef. "Streifzug durch die Vor- und Frühgeschichte der Biogeographie," Geographisches Taschenbuch 1979/1980 (Wiesbaden: Franz Steiner Verlag GMBH, 1979), pp. 72-82.

Biogeography from ancient times (Hippocrates, Aristotle, and Theophrastus) to the end of the 18th century.

449. Schmitthenner, Heinrich. "Die Entstehung der Geomorphologie als geographische Disziplin (1869-1905)," Petermanns Geographische Mitteilungen, vol. 100, 4th quarter (15 November 1956), 257-268.

Review of German geomorphology from Peschel's Neue Probleme (1869) to the death of Richthofen (1905), but also mentions many subsequent personalities and events down to 1937.

450. Schneider-Carius, Karl. "Alexander von Humboldt in seinen Beziehungen zur Meteorologie und Klimatologie," pp. 17-24 in Alexander von Humboldt: Vorträge und Aufsätze Anlässlich der 100. Wiederkehr seines Todestages am 6. Mai 1959, ed. Johannes F. Gellert (Geographische Gesellschaft der Deutschen Demokratischen Republik, Wissenschaftliche Abhandlungen vo. 2) (Berlin: VEB Deutscher Verlag der Wissenschaften, 1960).

Alexander von Humboldt's contributions to climatology and meteorology. Mentions previous work, beginning with Boyle and Townley (1662).

451. Schneider-Carius, Karl. Wetterkunde-Wetterforschung. Geschichte ihrer Probleme und Erkenntnisse in Dokumenten aus drei Jahrtausenden. ("Orbis Academicus, Problemgeschichten der Wissenschaft in Dokumenten und Darstellungen," II/9.) Freiburg and Munich: Verlag Karl Alber, 1955. xvi + 423 p.

The history of meteorology from ancient times to the present (from Homer and Hesiod to V. Bjerknes and Napier Shaw), with excerpts from the writings of the time. 15-p. biographical register at end.

452. Schuepp, Max. "Klimatologie gestern, heute und morgen," Geoforum, no. 18 (1974), 72-76.

Brief review of climatology from the invention of the thermometer and barometer in the 17th century to the present. Definitions of climate, beginning with Humboldt (1831). Climatic classification from Köppen onward. Followed by brief discussion of present (computers, weather satellites, etc.) and future (weather modification) climatology.

453. Sestini, Aldo. "La geografia fisica negli ultimi cinquant'anni," Scientia (Asso, Italy), vol. 92, no. 546 (October 1956), 261-264.

A brief overview of developments in physical geography during the past half-century. Published in series "Mezzo secolo di progresso scientifico." French translation, pp. 170-173, in Supplément: Traductions Françaises.

454. Staszewski, Josef. "Alexander von Humboldts Gedanke der isothermen Linien," Wissenschaftliche Zeitschrift der Humboldt-Universität zu Berlin, Mathematisch-Naturwissenschaftliche Reihe, vol. 8, nos. 4-5 (1958-1959), 509-518.

Alexander von Humboldt originated the isotherm in 1817 while pursuing studies of plant distributions.

455. Stearn, William T. "Humboldt's Essai sur la géographie des plantes," Journal of the Society for the Bibliography of Natural History, vol. 3, part 7 (October 1960), 351-357.

Alexander von Humboldt's work in plant geography, with emphasis on his Essai sur la géographie des plantes (Paris, 1807).

456. Thornthwaite, C. Warren. "Problems in the Classification of Climates," Geographical Review, vol. 33, no. 2 (April 1943), 233-255.

Historical survey of climatic classifications in the 19th and 20th centuries, with particular attention to the 72-year career of Wladimir Köppen (1846-1940).

457. Tilley, Philip. "Early Challenges to Davis' Concept of the Cycle of Erosion," Professional Geographer, vol. 20, no. 4 (July 1968), 265-269.

Responses to William Morris Davis' work by German geographers--Alfred Hettner, in particular.

458. Turrill, William B. Pioneer Plant Geography: The Phytogeographical Researches of Sir Joseph Dalton Hooker. The Hague: Martinus Nijhoff, 1953. x + 267 p.

The work of the English botanist J.D. Hooker (1817-1911) in plant geography. Turrill published a biography of Hooker in 1963.

459. Weber, Heinrich. "Die Entwickelung der physikalischen Geographie der Nordpolarländer bis auf Cooks Zeiten," Münchener Geographischer Studien, no. 4 (1898). 250 p.

Progressive unfolding of knowledge of the physical features of the Arctic from ancient times to about 1770.

460. Wilcock, Arthur A. "Köppen after Fifty Years," Annals of the Association of American Geographers, vol. 58, no. 1 (March 1968), 12-28.

Chronicles the vicissitudes of the climatic classification published by Wladimir Köppen in 1918. Includes a biographical sketch of Köppen (1846-1940).

461. Willis, Bailey. "American Geology, 1850-1900," Proceedings of the American Philosophical Society, vol. 86, no. 1 (25 September 1942), 34-44.

Mentions geological work of Benjamin Franklin and William Maclure but concentrates on latter half of the 19th century. The geomorphological work of Powell, Gilbert, and Davis is treated on pp. 38-39.

462. Zittel, Karl Alfred von. Geschichte der Geologie und Paläontologie bis Ende des 19. Jahrhunderts. ("Geschichte der Wissenschaften in Deutschland," vol. 23.) Munich and Leipzig: Druck und Verlag von R. Oldenbourg, 1899. xi + 868 p.

History of geology from ancient times to the end of the 19th century. Time before 1790 is covered in the first 75 pages. See especially chapters on physiographic geology, dynamic geology, and topographic geology (regional geology of various parts of the world) (pp. 260-567). English translation appeared in 1901.

VI. HUMAN GEOGRAPHY

463. Almagià, Roberto. "La geografia umana nell'ultimo mezzo secolo," Scientia (Asso, Italy), vol. 92, no. 541 (May 1957), 160-166.

Human (or anthropo-) geography, beginning with Friedrich Ratzel (1844-1904). Almost exclusively concerned with European geographers. Published in series "Mezzo secolo di progresso scientifico." French translation, pp. 98-104, in Supplément: Traductions Françaises.

464. Brunhes, Jean. "Human Geography," pp. 55-105 in The History and Prospects of the Social Sciences, ed. Harry Elmer Barnes (New York: Alfred A. Knopf, 1925).

Scope and nature of human geography as practiced in Europe and the United States, with particular emphasis on Friedrich Ratzel and the French geographers.

465. Buttimer, Anne. "Social Geography," pp. 134-145 in International Encyclopedia of the Social Sciences,

Part One--General and Topical 97

vol. 6 (New York: The Macmillan Company and The Free Press, 1968).

Includes a lengthy section, pp. 135-139, on "The Development of Social Geography," beginning with Herodotus and Thucydides but getting very quickly to the 19th century. Emphasizes the work of French geographers and sociologists.

466. Claval, Paul. Essai sur l'évolution de la géographie humaine. ("Cahiers de géographie de Besançon," no. 12; "Annales littéraires de l'Université de Besançon," vol. 67.) Paris: Les Belles Lettres, 1964. 162 p.

Evolution of human geography from Humboldt and Ritter onward. Makes basic division between traditional or "classical" geography and prospective geography. Revised edition published 1976.

467. Claval, Paul. "La naissance de la géographie humaine," pp. 355-376 in La pensée géographique française contemporaine: mélanges offerts à André Meynier (Saint-Brieuc, France: Presses Universitaires de Bretagne, 1972).

The early development of human geography, from about the end of the 18th century to the end of the 19th, with emphasis on Volney, Humboldt, Ritter, Levasseur, Ratzel, and Vidal de la Blache.

468. Dawson, John A. "Some Early Theories of Settlement Location and Size," Journal of the Town Planning Institute, vol. 55, no. 10 (December 1969), 444-448.

Some early statements of central-place theory, from Machiavelli and Botero in the 16th century to von Thünen and Lalanne in the 19th.

469. Dockès, Pierre. L'espace dans la pensée économique du XVIe au XVIIIe siècle. ("Nouvelle Bibliothèque Scientifique," ed. Fernand Braudel.) Paris: Flammarion, Editeur, 1969. 461 p.

A study of mercantilism and spatial variations in economic matters, as presented in the writings of French and British economists from the 16th to the 18th centuries (Jean Bodin to Adam Smith). Von Thünen supplies a narrow bridge linking the spatial theories of the 18th century with those of today.

470. Dunbar, Gary S. "Some Early Occurrences of the Term 'Social Geography,'" Scottish Geographical Magazine, vol. 93, no. 1 (April 1977), 15-20.

An historical sketch of the uses of the term "social geography' by English- and French-speaking geographers and sociologists, from 1884 onward.

471. Duncan, James S. "The Superorganic in American Cultural Geography," Annals of the Association of American Geographers, vol. 70, no. 2 (June 1980), 181-198.

The theory of culture as a superorganic entity was outlined by anthropologists Alfred Kroeber and Robert Lowie and was then passed on to Carl Sauer and his students of cultural geography. Paper reviews the uses of the culture concept by American geographers and casts considerable doubt on its explanatory power.

472. Fairbairn, Kenneth J., and Barr, Brenton M. "Acknowledging the Past: Richard Cantillon's Pattern of Urban Settlement Location," Area, vol. 6, no. 3 (1974), 208-210.

Richard Cantillon made observations on the origin, location, growth, and size of urban settlements in his Essai sur la nature de commerce en général (1755). In many ways, Cantillon foreshadowed work of later theoreticians, such as Christaller and Lösch.

473. Feldman, Douglas A. "The History of the Relationship between Environment and Culture in Ethnological Thought: An Overview," Journal of the History of the Behavioral Sciences, vol. 11, no. 1 (January 1975), 67-81.

Statements about the relationship between culture and environment from Polybius onward, with emphasis on the views of modern anthropologists, but with some consideration of geographers such as Friedrich Ratzel, Ellen Churchill Semple, and Carl Sauer.

474. Johnston, Ronald J. "Paradigms, Revolutions, Schools of Thought, and Anarchy: Reflections on the Recent

Part One--General and Topical   99

History of Anglo-American Human Geography," pp. 303-317 in *The Origins of Academic Geography in the United States*, ed. Brian Blouet (Hamden, Connecticut: Archon Books, 1981).

(Page 303) "This paper attempts an evaluation of anglo-American human geography since 1945 using [Thomas] Kuhn's framework as the context." Concludes that Kuhn's paradigm concept has little explanatory power in human geography.

475. Koelsch, William A. "The Historical Geography of Harlan H. Barrows," *Annals of the Association of American Geographers*, vol. 59, no. 4 (December 1969), 632-651.

Concerns Harlan Barrows (1877-1960), professor of geography in the University of Chicago, and his teaching of historical geography.

476. Kramer, Fritz L. "Eduard Hahn and the End of the 'Three Stages of Man,'" *Geographical Review*, vol. 57, no. 1 (January 1967), 73-89.

A review of the "Three-Stage Theory" of culture history, which sees Mankind as progressing through three stages (usually something like L.H. Morgan's "savagery to barbarism to civilization"), from the classical period until its refutation by the German geographer Eduard Hahn (1856-1928).

477. Kraus, Alois. *Versuch einer Geschichte der Handels- und Wirtschaftsgeographie*. Frankfurt: J.S. Sauerlandersverlag, 1905. viii + 103 p.

History of economic and commercial geography from ancient times onward, with emphasis on German sources and on the 18th and 19th centuries.

478. Kristof, Ladis K.D. "The Origins and Evolution of Geopolitics," *Journal of Conflict Resolution*, vol. 4, no. 1 (March 1960), 15-51.

An historical sketch of geopolitics and political geography from Aristotle to the present.

479. McKinney, William M. "Carey, Spencer, and Modern Geography," *Professional Geographer*, vol. 20, no. 2 (March 1968), 103-106.

The American Henry Carey (1793-1879) and the Englishman Herbert Spencer (1820-1903) and their pioneering attempts to explore "the relationships between space, distance, population and movement."

480. McManis, Douglas R. "A Prism to the Past: The Historical Geography of Ralph Hall Brown," Social Science History, vol. 3, no. 1 (Fall 1978), 72-86.

    Ralph Brown (1898-1948) was one of the foremost American historical geographers. His work, notably Mirror for Americans (1943) and Historical Geography of the United States (1948), is analyzed in this essay. Mention is also made of some of Brown's precursors in the study of historical geography.

481. Mikesell, Marvin W. "Landscape," pp. 575-580 in Vol. 8 of International Encyclopedia of the Social Sciences (New York: The Macmillan Company and The Free Press, 1968).

    Concept of "landscape" in geography. Emphasis is on the use of the term in American and German (landschaft) geography, especially since the publication of Carl Sauer's "The Morphology of Landscape" in 1925.

482. Mikesell, Marvin W. "The Rise and Decline of 'Sequent Occupance': A Chapter in the History of American Geography," pp. 149-169 in Geographies of the Mind: Essays in Historical Geosophy in Honor of John Kirtland Wright, ed. David Lowenthal and Martyn J. Bowden (New York: Oxford University Press, 1975).

    Traces the rise and decline of the concept of "sequent occupance," as used by American historical geographers, from the time of its invention by Derwent Whittlesey in 1929 to its virtual disappearance in recent years. Reprinted in The Nature of Change in Geographical Ideas ("Perspectives in Geography," vol. 3), ed. Brian Berry (DeKalb, Illinois: Northern Illinois University Press, 1978), pp. 2-15.

483. Mikesell, Marvin W. "Tradition and Innovation in Cultural Geography," Annals of the Association of American Geographers, vo. 68, no. 1 (March 1978), 1-16.

History and current status of cultural geography, which derives essentially from the work of German geographers of the late 19th and early 20th centuries, such as August Meitzen, Eduard Hahn, Friedrich Ratzel, and Robert Gradmann. Other pioneers included George Perkins Marsh, Maximilien Sorre, and Carl Sauer.

484. Müller-Wille, Christopher F. "The Forgotten Heritage: Christaller's Antecedents," pp. 38-64 in The Nature of Change in Geographical Ideas ("Perspectives in Geography," vol. 3), ed. Brian J.L. Berry (DeKalb, Illinois: Northern Illinois University Press, 1978).

    German work in urban geography before the publication of Walter Christaller's Die zentralen Orte in Süddeutschland (1933).

485. Olwig, Kenneth Robert. "Historical Geography and the Society/Nature 'Problematic': The Perspective of J.F. Schouw, G.P. Marsh and E. Reclus," Journal of Historical Geography, vol. 6, no. 1 (January 1980), 29-45.

    Contrary to a prevailing conception of geography that "dichotomizes human society and nature into fixed exclusive categories," the author explores an alternative "problematic" or disciplinary framework, with examples from the works of "three prominent socially-engaged nineteenth-century geographers"-- Joachim Frederik Schouw (1789-1852) of Denmark, George Perkins Marsh (1801-1882) of the United States, and Elisée Reclus (1830-1905) of France.

486. Quaini, Massimo. La costruzione della geografia umana. ("Strumenti," 41.) Florence: La Nuova Italia Editrice, 1975. 185 p.

    Essays on the history of human geography. Companion volume to, and in some ways a reaction against, the works of Paul Claval and Lucio Gambi (q.v.). Goes back to Bodin, Varenius, Vauban, and Rousseau and forward to present-day French philosophers such as Foucault and Althusser.

487. Robic, Marie-Claire. "Cent ans avant Christaller: une théorie des lieux centraux," L'espace géographique, vol. 11, no. 1 (January-March 1982), 5-12.

Features the French author Jean Reynaud (1806-1863) as a forerunner of Walter Christaller (1893-1969) in central-place theory. Contains long quotations from Reynaud's article, "Villes," in the Encyclopédie Nouvelle (Vol. 8, 1841).

488. Sapper, Karl. "Geography--Economic," pp. 626-629 in Vol. 6 of Encyclopaedia of the Social Sciences (London: Macmillan and Co. Ltd., 1932).

    Economic geography since the time of Herodotus, with emphasis on German- and English-language works of the 19th and 20th centuries.

489. Sauer, Carl. "Geography--Cultural," pp. 621-624 in Vol. 6 of Encyclopaedia of the Social Sciences (London: Macmillan and Co. Ltd., 1932).

    The development of cultural geography in the 19th and 20th centuries, with emphasis on Germany.

490. Sauer, Carl. "Recent Developments in Cultural Geography," pp. 154-212 in Recent Developments in the Social Sciences, ed. Edward Cary Hayes (Philadelphia: J.B. Lippincott, 1927).

    Development of geography, especially human geography, in Europe and the United States, from Humboldt and Ritter onward.

491. Schmidt, Peter Heinrich. Wirtschaftsforschung und Geographie. Jena: Verlag von Gustav Fischer, 1925. ix + 239 p.

    A sort of history of geography and economics, beginning with Renaissance voyaging and mercantilism.

492. Spate, O.H.K. "Environmentalism," pp. 93-97 in Vol. 5 of International Encyclopedia of the Social Sciences (New York: The Macmillan Company and The Free Press, 1968).

    An essay on the nature of geographical environmentalism, including environmental determinism and possibilism. Uses mostly 19th- and 20th-century examples but also mentions Hippocrates and Montesquieu.

Fig. 6. The French geographer Elisée Reclus making a short flight in a balloon in Brussels in 1896. This was apparently Reclus' first and only balloon ascension, although he was a longtime friend of Nadar, the famous balloonist and photographer, from 1870 onward. Reproduced through the courtesy of the Bibliothèque Nationale and Mme. Louise Rapacka.

Fig. 5. The American geomorphologist Grove Karl Gilbert on horseback in the Colorado piedmont in 1894. Reproduced through the courtesy of the National Academy of Sciences, Washington, D.C. The nature of research in geography has depended greatly on the state of technology, transport technology in particular, and several of the photographs were chosen to illustrate various forms of transportation used by geographers in the late 19th and early 20th centuries. It is also interesting to note the styles of field dress.

Fig. 7. The American geographer Isaiah Bowman in the Atacama Desert, probably in 1907. Reproduced through the courtesy of Professor Geoffrey Martin.

Part One--General and Topical

493. Spate, O.H.K. "Toynbee and Huntington: A Study in Determinism," Geographical Journal, vol. 118, part 4 (December 1952), 406-428.

Similarities and differences in the ideas of the English historian Arnold Toynbee and the American geographer Ellsworth Huntington.

494. Tatham, George. "Environmentalism and Possibilism," pp. 128-162 in Geography in the Twentieth Century, 3rd ed., ed. Griffith Taylor (New York: Philosophical Library, 1957).

An historical essay on the concept of environmentalism in geography from Hippocrates and Aristotle to the modern era (Humboldt, Ritter, Haeckel, Buckle, Demolins, Semple, and finally Griffith Taylor).

495. Thomale, Eckhard. Sozialgeographie: Eine disziplingeschichtliche Untersuchung zur Entwicklung der Anthropogeographie. Marburger geographische Schriften, no. 53 (1972). 95 p.

Historical treatment of the terminology of anthropogeography and social geography from the time of Friedrich Ratzel onward. Also treats related concepts in sociology, beginning with Frédéric Le Play.

496. Zelinsky, Wilbur, ed. "Human Geography: Coming of Age," American Behavioral Scientist, vol. 22, no. 1 (September-October 1978), 3-167.

A special issue devoted to articles by 11 geographers on developments in human geography during the last quarter-century, with almost exclusive emphasis on English-language publications.

VII. MISCELLANEOUS SUBFIELDS OR ANCILLARY

DISCIPLINES

497. Berry, Brian J.L. "Statistical Geography," pp. 145-151 in Vol. 6 of International Encyclopedia of the Social Sciences (New York: The Macmillan Company and The Free Press, 1968).

Statistical geography "attempts to identify and measure regularities observable in spatial distributions." Emphasis on English-language works of the period 1950-1965.

498. Deacon, Margaret. Scientists and the Sea, 1650-1900: A Study of Marine Science. London: Academic Press Inc. (London) Ltd., 1971. xvi + 445 p.

    Marine science (physical oceanography and marine biology) from ancient times to the beginning of the 20th century. Pp. 3-65 cover the periods before 1650 (ancient world, middle ages, Renaissance).

499. Dean, James R. Down to the Sea: A Century of Oceanography. Glasgow: Brown, Son & Ferguson, Ltd., 1966. xvii + 128 p.

    Oceanography from Prince Albert I of Monaco (1848-1922) onward, with an introductory chapter covering "The Beginnings of Marine Science" (starting with the ancient Phoenicians).

500. Dunbar, Gary S. "What Was Applied Geography?," Professional Geographer, vol. 30, no. 3 (August 1978), 238-239.

    Reply to an earlier article, "What Is Applied Geography?," by James Harrison (PG, August 1977). Instead of being a new term, as Harrison implies, "applied geography" was in use by 1890. Cites references to the term in publications, 1890-1904.

501. Linklater, Eric. The Voyage of the Challenger. London: John Murray (Publishers) Ltd., 1972. 288 p.

    Voyage of the British corvette Challenger (1872-1876), a signal event in the establishment of the science of oceanography. Report published in 50 vols., 1885-1895.

502. Meynier, André. "Trois dictionnaires géographiques: 1842, 1907, 1970," Bulletin de l'Association de géographes français, vol. 56, nos. 462-463 (May-October 1973), 233-236.

    Comparison of 3 French geographical dictionaries with observations about trends in scientific vocabulary.

Part One--General and Topical   105

503. Schlee, Susan.   The Edge of an Unfamiliar World: A
     History of Oceanography.   New York: E.P. Dutton &
     Co., Inc., 1973. 398 p.

     Oceanography in the 19th and 20th centuries.

504. Sears, Mary, and Merriman, Daniel, eds. Oceanography:
     The Past.   (Proceedings of the Third International
     Congress on the History of Oceanography, 1980.) New
     York, etc.: Springer-Verlag, 1980. xx + 812 p.

     69 essays on the history of oceanography, with
     emphasis on the 19th and 20th centuries.

505. Wright, John K.   "The Heights of Mountains: An
     Historical Notice," Special Libraries Association,
     Geography and Map Division, Bulletin, no. 31
     (February 1958), 4-15.

     Treats 3 topics: (page 5) "1) the development of
     methods of measuring mountain heights; 2) the question
     of the world's highest known mountain; and 3) certain
     early mountain measurements and estimates in the
     Americas." Reprinted in Wright's Human Nature in
     Geography (above in Section II), pp. 140-153.

                VIII. GEOGRAPHICAL SOCIETIES

506. Agostini, Enrico de.   "L'attivita della Società
     Geografica Italiana nel primo decennio postbellico,"
     Bollettino della Società Geografica Italiana, series
     8, vol. 8, nos. 9-10 (September-October 1955), 393-414.

     Activities of the SGI in the period 1945-1955.

507. Agostini, Enrico de.   La Reale Società Geografica
     Italiana e la sua opera dalla fondazione ad oggi
     (1867-1936). Rome: Reale Società Geografica Italiana,
     1937. 149 p.

     Brief history of the RSGI, which was founded in
     Florence in 1867 and moved to Rome in 1872. Lists of
     publications, medalists, honorary and corresponding
     members, bylaws, and regulations.

508. Arnold, Adolf. "Hundert Jahre Geographische Gesellschaft zu Hannover 1878-1978," pp. 1-17 in Hannover und sein Umland: Festschrift zur Feier des 100jährigen Bestehens der Geographischen Gesellschaft zu Hannover 1878-1978 (Jahrbuch der Geographischen Gesellschaft zu Hannover, 1978), ed. Wolfgang Eriksen and Adolf Arnold.

Centennial history of Hannover Geographical Society.

509. Asúa, Miguel de. "Reseña de las tareas de la Corporación en sus primeros cincuenta años de vida," Boletin de la Real Sociedad Geográfica (Madrid), vol. 66 (1926), 220-262.

Activities of the Royal Geographical Society (Madrid), founded 1876.

510. Bader, Frido J. Walter. "Die Gesellschaft für Erdkunde zu Berlin und die koloniale Erschliessung Afrikas in der zweiten Hälfte des 19. Jahrhunderts bis zur Gründung der ersten deutschen Kolonien," Die Erde, vol. 109, no. 1 (1978), 36-48.

Role played by the Berlin Geographical Society and its members in African colonization up to 1885.

511. Baetens, R. Het Koninklijk Aardrijkskundig Genootschap van Antwerpen 1876-1976. Antwerp: Koninklijk Aardrijkskundig Genootschap, 1976. 80 p.

A history of the Royal Geographical Society of Antwerp, Belgium, during the first century of its existence. [JAvG]

512. Barton, Thomas Frank. "Leadership in the Early Years of the National Council of Geography Teachers, 1916-1935," Journal of Geography, vol. 63, no. 8 (November 1964), 345-355.

The first 2 decades of the development of the National Council of Geography Teachers, renamed the National Council for Geographic Education in 1957.

513. Baschin, Otto. "Die Gesellschaft für Erdkunde zu Berlin 1828 bis 1928: Vorgeschichte, Begründung und Entwicklung," Die Naturwissenschaften, vol. 16, no. 21 (25 May 1928), 369-374.

Part One--General and Topical   107

   An historical sketch of the first 100 years of the
   Berlin Geographical Society, founded in 1828. Also
   treats earlier societies: Kosmographische Gesellschaft
   in Nürnberg (1740s), Société de géographie de Paris
   (1821), and the African Association (1788).

514. Baschin, Otto.   "Royal Geographical Society, 1830-
     1930," Die Naturwissenschaften, vol. 18, no. 29 (18
     July 1930), 659-662.

   Historical sketch of the Royal Geographical Society,
   founded in London in 1830. Brief treatment of earlier
   British interest in geography, beginning with Richard
   Eden, "The Father of English Geography," in the mid-
   16th century.

515. Bassin, Mark.   "The Russian Geographical Society, the
     'Amur Epoch,' and the Great Siberian Expedition 1855-
     1863," Annals of the Association of American
     Geographers, vol. 73, no. 2 (June 1983), 240-256.

   (From Abstract, p. 240) "This paper attempts to
   demonstrate that the genesis and the very purpose of
   the Great Siberian Expedition cannot be understood
   apart from the political and ideological climate of the
   1840s and 1850s in Russia. It shows that the rapid and
   powerful growth of nationalist sentiment at this time
   was the main inspiration for the work of the young
   Russian Geographical Society."

516. Behrmann, Walter.   "Geschichte des Vereins für
     Geographie und Statistik zu Frankfurt am Main in den
     ersten Jahren seines Bestehens," pp. 1-34 in
     Festschrift zur Hundertjahrfeier des Vereins für
     Geographie und Statistik zu Frankfurt am Main, ed.
     Wolfgang Hartke (Frankfurt: Verlag der Geographischen
     Verlagsanstalt Ludwig Ravenstein, 1936).

   Centennial history of the Frankfurt Geographical
   Society, founded in 1836 as the Geographical
   Association ("Verein") of Frankfurt. Some discussion
   of interest in geographical matters in Frankfurt from
   16th century onward.

517. Bernleithner, Ernst.   "Die Geographische Gesellschaft
     in Wien und ihr Anteil an der Entwicklung der
     Landeskunde von Deutschland und Osterreich," Berichte

zur deutschen Landeskunde, vol. 21, no. 2 (September 1958), 294-324.

The Vienna Geographical Society and its role in the development of regional geography in Germany and Austria. Society founded 1856. Early geographical instruction in Vienna, leading up to the appointment of Friedrich Simony to the first chair of geography in the University of Vienna in 1851. Geographical societies and related organizations in Austria and Czechoslovakia, 1753-1856.

518. Bösiger, Kurt. "Die Entwicklung der Geographisch-Ethnologischen Gesellschaft Basel von ihrer Gründung bis ins Jahr 1960," Regio Basiliensis, vol. 2, no. 1 (October 1960), 12-18.

A history of the Basel (Switzerland) Geographical and Ethnological Society since its founding in 1923.

519. Bridges, R.C. Sir John Speke and the Royal Geographical Society," The Uganda Journal, vol. 26, no. 1 (March 1962), 23-43.

Role of the RGS in supporting Speke and other African explorers.

520. Bridges, R.C. "W.D. Cooley, the RGS and African Geography in the Nineteenth Century," Geographical Journal, vol. 142, part 1 (March 1976), 27-47; part 2 (July 1976), 274-286.

The geographical work of William Desborough Cooley (1795-1883), especially his association with the Royal Geographical Society, the founding of the Hakluyt Society, and promotion of African exploration.

521. Brigham, Albert Perry. "The Association of American Geographers, 1903-1923," Annals of the Association of American Geographers, vol. 14, no. 3 (September 1924), 109-116.

Brigham's reminiscences of the last 20 years, beginning with W.M. Davis' proposal for an association in 1903 and the founding of the AAG in 1904.

522. Broc, Numa. "Le rôle de la Société de géographie de Bordeaux (1874) dans les premiers Congrès nationaux

Part One--General and Topical				109

   de géographie (1878-1896)," Revue géographique des
   Pyrénées et du Sud-Ouest, vol. 49, no. 1 (January
   1978), 150-155.

   The role of the Bordeaux Geographical Society
   (founded 1874) in the creation of an annual congress of
   French geographical societies. Also concerns the
   decline of the old geographical societies and the rise
   of professional geography ("géographie des professeurs").

523. Brown, T. Nigel N. The History of the Manchester
   Geographical Society, 1884-1950. Manchester: Manchester University Press, 1971. viii + 102 p.

   A chronological history of the Society, from its
   founding in 1884 until 1950, when Nigel Brown (d. 1969)
   became editor of the Society's publications. The
   Society was involved in the long struggle to establish
   a chair in geography at Manchester University,
   culminating in the appointment of H.J. Fleure in 1930.

524. Cameron, Ian. To the Farthest Ends of the Earth: The
   History of the Royal Geographical Society 1830-1980.
   London: MacDonald, 1980. 288 p.

   Mostly a history of RGS-sponsored exploration. Also
   chapters on "The Founding of the Society" and "The
   Educational Role of the Society." Appendixes list
   officers, medalists, and expeditions.

525. Camu, Pierre. "Le quatre-vingtième anniversaire de la
   Société de géographie de Québec," Cahiers de
   géographie de Quebec, vol. 2, no. 3 (October 1957),
   135-140.

   Brief history of the Society, which was founded in
   1877, incorporated in 1879, and began publishing a
   bulletin in 1880.

526. Cantor, L.M. "The Royal Geographical Society and the
   Projected London Institute of Geography 1892-1899,"
   Geographical Journal, vol. 128, part 1 (March 1962),
   30-35.

   Halford Mackinder's plan to establish an Institute of
   Geography in London with support from the Royal
   Geographical Society, which resulted instead in the

creation of the School of Geography in the University of Oxford in 1899.

527. Carazzi, Maria. <u>La Società Geografica Italiana e l'esplorazione coloniale in Africa (1867-1900)</u>. (Pubblicazioni della Facoltà di Lettere e Filosofia dell'Università di Milano, 60, Sezione a cura dell'Istituto di Geografia Umana, 2.) Florence: La Nuova Italia Editrice, 1972. ix + 199 p.

The Italian Geographical Society's interest in North and East Africa, from the founding of the Society in 1867 until the end of the 19th century. Also concerned with the more general activities of the Society, its participation in international conferences, and its contributions to Italian life.

528. Caswell, John E. "The RGS and the British Arctic Expedition, 1875-76," <u>Geographical Journal</u>, vol. 143, part 2 (July 1977), 200-210.

The role of the Royal Geographical Society in the British Arctic Expedition under the command of George Nares. More a study of exploration than of the RGS.

529. [Anonymous]. "La celebrazione del primo centenario della Società Geografica Italiana," <u>Bollettino della Società Geografica Italiana</u>, series 9, vol. 8, nos. 1-3 (January-March 1967), 1-16.

Celebrations of the 100th anniversary of the Italian Geographical Society, founded in Florence in 1867. Also mentions earlier geographical societies, beginning with the Accademia degli Argonauti in Venice (1680).

530. [Anonymous]. "Cent-cinquantenaire de la Société de géographie, 1821-1971," <u>Acta Geographica</u>, no. 20 (March 1975), 1-52.

Brief history of the Paris Geographical Society and its activities. Lists of medalists and prizewinners. The library and archives of the Society were transferred to the Département des cartes et plans of the Bibliothèque Nationale in 1941.

Part One--General and Topical   111

531. Claparède, Arthur de. *Coup d'oeil sur la Société de géographie de Genève depuis sa fondation en 1858.* 2nd ed. Geneva: Imprimerie "Atar," 1908. 76 p.

   A history of the first half-century of the Geneva Geographical Society. 1st ed. pub. 1896. See Goegg (below) for the next 25 years (to 1933).

532. Close, Charles, et al. "The Centenary Meeting: Addresses on the History of the Society," *Geographical Journal,* vol. 76, no. 6 (December 1930), 455-476.

   Some historical notes and reminiscences of the earlier years of the Royal Geographical Society on the occasion of its 100th anniversary. The longest and most informative remarks were those of Hugh Robert Mill (pp. 458-462).

533. Crone, Gerald R. *Royal Geographical Society, A Record, 1931-1955.* London: Royal Geographical Society and John Murray (Publishers) Ltd., 1955. 36 p.

   A review of the activities of the Royal Geographical Society in the 25 years since the centenary celebrations in 1930.

534. Cumming, Duncan. "Royalty and the Royal Geographical Society," *Geographical Journal,* vol. 143, part 2 (July 1977), 171-178.

   Relationships, formal and informal, between the RGS and British royalty from the time of the founding of the Society in 1830 to the present.

535. Dalla Vedova, Giuseppe. *La Società Geografica Italiana e l'opera sua nel Secolo XIX.* Rome: La Società Geografica Italiana, 1904. 90 p.

   The work of the Italian Geographical Society (founded 1867) in the 19th century.

536. Davies, Gordon L. "Dr. Anthony Farrington and the Geographical Society of Ireland," *Irish Geography,* vol. 4, no. 5 (1963), 311-320.

   Article is mostly about Farrington (b. 1893) but it also describes the founding (1934) and early years of the Geographical Society of Ireland and its predecessor, the Irish Geographical Association (1919-1924).

537. Dunbar, Gary S. "The Rival Geographical Societies of Fin-de-Siècle San Francisco," Yearbook of the Association of Pacific Coast Geographers, vol. 40 (1978), 57-63.

The Geographical Society of the Pacific (founded in 1881) and the Geographical Society of California (1891).

538. Fick, Karl E. "Frankfurts Geographische Gesellschaft 1961-1976. Ein Zwischenbericht aus Anlass ihres 140-jährigen Jubilaums," pp. 3-43 in Festschrift zur 140-Jahrfeier der Frankfurter Geographischen Gesellschaft (Frankfurter Geographische Hefte, no. 53, 1980).

The Frankfurt Geographical Society, 1961-1976.

539. Fleiuss, Max. "The Brazilian Historical and Geographical Institute: A Well-Spent Century," Bulletin of the Pan-American Union, vol. 72, no. 10 (October 1938), 557-567.

Centennial historical sketch of the Instituto Histórico e Geográfico Brasileiro, its officers and activities. Founded 21 October 1838 in Rio de Janeiro.

540. Fleure, Herbert John. "Sixty Years of Geography and Education: A Retrospect of the Geographical Association," Geography, vol. 38, part 4 (November 1953), 231-266.

Some historical notes on the Geographical Association, a British teachers' association that emphasizes pre-university teaching, founded in 1893. Also treats geography before 1893.

541. Freeman, T.W. "The Manchester and Royal Scottish Geographical Societies," Geographical Journal, vol. 150, part 1 (March 1984), 55-62.

An historical sketch of the Manchester Geographical Society and Royal Scottish Geographical Society, both founded in 1884, with emphasis on their publications.

Part One--General and Topical 113

542. Freeman, Thomas Walter. "The Royal Geographical Society and the Development of Geography," pp. 1-99 in Geography Yesterday and Tomorrow, ed. Eric Brown (Oxford: Oxford University Press, 1980).

A chronological history of the RGS, with a fair sampling of its activities and publications.

543. Friederichsen, L. "Rückblick auf die Gründung und Entwickelung der Geographischen Gesellschaft in Hamburg während der ersten fünfundzwanzig Jahre ihre Bestehens," Mittheilungen der Geographischen Gesellschaft in Hamburg, vol. 14 (1898), 1-43.

The first 25 years of the Hamburg Geographical Society, 1873-1898. Lists of officers, medalists, honorary and corresponding members, publications, etc. For the next 25 years (1898-1923) of the Society, see Schlee (below).

544. Garnett, Alice. "I.B.G.: The Formative Years--Some Reflections," Transactions of the Institute of British Geographers, n.s. vol. 8, no. 1 (1983), 27-35.

(From Abstract, p. 27) "The author's reminiscences of geography and geographers from the early twenties to the early thirties are considered with respect to events that led to the foundation of the Institute of British Geographers" (1933). Followed by two shorter papers of reminiscences: "Recollections of a Founder Member" by S.H. Beaver (pp. 36-37) and "Some Early Impressions" by E.G. Bowen (pp. 38-40).

545. Gilbert, Edmund W. "The RGS and Geographical Education in 1871," Geographical Journal, vol. 137, part 2 (June 1971), 200-202.

The Royal Geographical Society's role in the movement to secure the recognition of geography in the universities began in 1871, when Henry Rawlinson, president of the Society, wrote to the vice-chancellors of Oxford and Cambridge to urge the importance of geography as a subject in the Universities' Schools Examinations. Letters reproduced here with commentary.

546. Goegg, Edmond. Coup d'oeil sur la Société de géographie de Genève de 1908 à 1933. Geneva: Société générale d'imprimerie, 1933. 69 p.

Events of the last quarter-century. Issued on the 75th anniversary of the founding of the Society. See Claparède (above) for the period 1858-1908.

547. Goode, J. Paul. "Geographical Societies of America," Journal of Geography, vol. 2, no. 7 (September 1903), 343-350.

Brief descriptions of 12 North American geographical societies: American Geographical Society, Appalachian Mountain Club, Quebec Geographical Society, American Climatological Association, National Geographic Society, Geographical Society of California, Sierra Club, Philadelphia Geographic Society (sic), Geographic Society of Chicago, Alaska Geographical Society, Geographical Society of Baltimore, and the Harvard Travelers' Club.

548. Graf, J.H. "Die geographische Gesellschaft in Bern 1873-1898: Ein Rückblick gelegentlich der Feier des 25jährigen Bestehens der Gesellschaft," Jahresbericht der Geographischen Gesellschaft von Bern, vol. 16 (1897--pub. 1898), 3-57.

The activities of the Bern Geographical Society in the first 25 years of its existence.

549. Grosvenor, Gilbert H. "The National Geographic Society and Its Magazine," Foreword, pp. 1-115, in National Geographic Magazine: Cumulative Index, 1899-1946 (Washington: National Geographic Society, 1948).

History of the Society and Magazine by the long-time president and editor. Lists of presidents, trustees, medalists, life members, and expeditions and distribution of members (1947).

550. Grosvenor, Gilbert H. The National Geographic Society and Its Magazine. Washington: National Geographic Society, 1957. 196 p.

Reprint of Foreword and Illustrated Prefaces to the Cumulative Index to the National Geographic Magazine, vol. 2, 2nd ed., 1947-1956. Contains articles by

Part One--General and Topical 115

> Gilbert Grosvenor ("The National Geographic Society and Its Magazine," pp. 1-115) and John Oliver La Gorce ("Geography in Human Terms," pp. 117-146) and lists of presidents, trustees, and medalists (1888-1957), etc.

551. Günther, Siegmund. "Münchens Geographische Gesellschaft im Lichte der Zeitgeschichte," pp. 1-22 in <u>Festschrift der Geographischen Gesellschaft in Munchen zur Feier ihres fünfundzwanzigjährigen Bestehens</u> ..., ed. Eugen Oberhummer (Jahresbericht der Geographischen Gesellschaft in München, vol. 15) (Munich: Theodor Ackermann, 1894).

    First 25 years of the Munich Geographical Society, founded in 1869.

552. Hallett, Robin, ed. <u>Records of the African Association 1788-1831</u>. London, etc.: Thomas Nelson and Sons Ltd., 1964. viii + 318 p.

    A collection of documents illustrating the history of the Association for Promoting the Discovery of the Interior Parts of Africa from its founding in 1788 to its merger with the Royal Geographical Society in 1831. Based on the Minute Books of the Association in the Cambridge University Library.

553. Hamelin, Louis-Edmond, and Beauregard, Ludger, eds. <u>Retrospective 1951-1976</u>. Montreal: Canadian Association of Geographers, 1979. 129 p.

    Papers commemorating the 25th anniversary of the Canadian Association of Geographers, founded in 1951. Of special interest are the papers by J. Keith Fraser ("The Development of the Canadian Association of Geographers, 1951-1976," pp. 5-17) and J. Lewis Robinson ("Geography in Canada 25 Years Ago," pp. 19-30).

554. Hellmann, Gustav. "Aus der Geschichte der Gesellschaft für Erdkunde zu Berlin im zweiten halben Jahrhundert ihres Bestehens (1879-1928)," pp. 1-14 in <u>Sonderband zur Hundertjahrfeier der Gesellschaft, Zeitschrift der Gesellschaft für Erdkunde zu Berlin, 1828-1928</u> (Berlin, 1928).

The Berlin Geographical Society during the second 50 years of its existence (1879-1928). For the first 50 years, see Koner (below).

555. [Anonymous]. "The Humboldt Society for the Study of the History of Geography and Cartography," Imago Mundi, vol. 2 (1937), 22.

Abortive effort of Hermann Wagner and others to establish a "society bearing the name of Humboldt which would make the history of geography and cartography the object of its work and study" in 1911-1912.

556. Instituto Histórico e Geográfico de São Paulo. Jubileu Social, 1894-1944. São Paulo: Imprensa Oficial do Estado São Paulo, 1944. 183 p.

First 50 years of the São Paulo Historical and Geographical Institute, founded in 1894. Lists of founders, officers, members, etc. Contents of Revista (1895-1944).

557. Jackson, S.P. "The South African Geographical Society, 1917-1977," South African Geographical Journal, vol. 60, no. 1 (April 1978), 3-12.

A brief review of the activities of the first 60 years of the Society, founded in Johannesburg on 8 June 1917.

558. James, Preston E. "The Association of American Geographers," American Council of Learned Societies, ACLS Newsletter, vol. 26, nos. 2-3 (Spring-Summer 1975), 9-18.

An historical sketch of developments in geography since 1870, with emphasis on the Association of American Geographers, founded in 1904.

559. James, Preston E., and Ehrenberg, Ralph E. "The Original Members of the Association of American Geographers," Professional Geographer, vol. 27, no. 3 (August 1975), 327-335.

Brief biographical notes on the 48 charter members of the Association of American Geographers (founded in 1904), as well as 10 nominees who declined charter

Part One--General and Topical 117

membership and 12 others who were considered for charter membership but were not elected.

560. James, Preston E., and Martin, Geoffrey J. The Association of American Geographers: The First Seventy Five Years, 1904-1979. Washington: Association of American Geographers, 1978. xii + 279 p.

A history of American geography and especially of the Association of American Geographers, founded in 1904. Recent history of the AAG written by Clyde Kohn (1949-1963, pp. 113-147) and Harm de Blij (1963-1978, pp. 149-184).

561. James, Preston E., and Martin, Geoffrey J. "On AAG History," The Professional Geographer, vol. 31, no. 4 (November 1979), 353-357.

An historical sketch of the first 75 years of the Association of American Geographers. See especially the book that had just been published by these authors in 1978 (above).

562. Jantzen, Günther. "1873-1973: 100 Jahre Geographische Gesellschaft in Hamburg," pp. 1-6 of Mitteilungen der Geographischen Gesellschaft in Hamburg, Registerband (1973).

Brief history of the Hamburg Geographical Society and its antecedents. Followed by Ilse Möller's index to the first 60 volumes of the Mitteilungen (pp. 7-51).

563. Kádár, László. "A Hundred Years of Hungarian Geographical Society," Geoforum, no. 6 (1971), 75-83.

Geography in Hungary since the 12th century, with emphasis on the last 100 years, since the founding of the Hungarian Geographical Society in 1872.

564. Kelly, Christine, comp. "The RGS Archives," Geographical Journal, vol. 141 (1975), 99-107; vol. 142 (1976), 117-130, 287-301; vol. 143 (1977), 73-85, 266-278.

Handlist of the materials in the archives of the Royal Geographical Society (London). Not in itself an historical work but an invaluable guide for historians of geography. Also available in an 80-page reprint by the RGS.

565. Kirwan, L.P. "The R.G.S. and British Exploration: A Review of Recent Trends," Geographical Journal, vol. 130, part 2 (June 1964), 221-225.

Brief but perceptive essay on the relations between the Royal Geographical Society and exploration--and, indeed, between the RGS and the field of geography.

566. Klingenberg, K.S. "Det Norske Geografiske Selskab og norsk geografisk forskning gjennem 50 år 1889-1939," Norsk Geografisk Tidsskrift, vol. 7, nos. 5-8 (1939), 3-33.

Role of the Norwegian Geographical Society, founded in 1889, in promoting geographical research, mostly polar exploration. English summary (pp. 25-33).

567. Koner, W. "Zur Erinnerung an das fünfzigjährige Bestehen der Gesellschaft für Erdkunde zu Berlin," Zeitschrift der Gesellschaft für Erdkunde zu Berlin, vol. 13 (1878), 169-250.

First 50 years of the Berlin Geographical Society, 1828-1878. For the next 50 years, see Hellmann (above).

568. [Anonymous]. "Koninklijk Nederlandsch Aardrijkskundig Genootschap 1873-1958," Tijdschrift van het Koninklijk Nederlandsch Aardrijkskundig Genootschap, series 2, vol. 75, no. 4 (October 1958), 299-318.

Describes activities of the 85th anniversary celebrations of the Royal Dutch Geographical Society in 1958.

569. "P.K." [Kropotkin, Peter]. "The Fifty Years' History of the Russian Geographical Society," Geographical Journal, vol. 10, no. 1 (July 1897), 53-56.

The first 50 years of the Russian Geographical Society, founded in 1845 on the model of the Royal Geographical Society (London). Note based on 3-vol. history of the Russian Geographical Society, published (in Russian) in St. Petersburg in 1896.

Part One--General and Topical 119

570. Leigh, Myee D. "The Manchester Geographical Society, 1884-1979: An Historical Summary," The Manchester Geographer, vol. 1, no. 1 (Autumn 1980), 7-14.

The Manchester Geographical Society from its founding in 1884. Complements T.N.L. Brown's history of the Society (above), which ends in 1950.

571. Lenz, Karl. "The Berlin Geographical Society (1828-1978)," Geographical Journal, vol. 144, part 2 (July 1978), 218-223.

Brief description of the history and current activities of the Gesellschaft für Erdkunde zu Berlin, founded in 1828.

572. Lenz, Karl. "150 Jahre Gesellschaft für Erdkunde zu Berlin," Die Erde, vol. 109, no. 1 (1978), 15-35.

The activities of the Berlin Geographical Society since its founding in 1828.

573. Léotard, Jacques. "Histoire de la Société (1876-1906)," Bulletin de la Société de géographie et d'études coloniales de Marseille, vol. 30, no. 1 (First Quarter 1906), 5-51.

The first 30 years of the Marseille Society of Geography and Colonial Studies. Lists of officers, medals, prizes, conferences, publications, corresponding societies, etc.

574. Léotard, Jacques. "Notice historique sur la Société de géographie de 1906 à 1926," Bulletin de la Société de géographie et d'études coloniales de Marseille, vol. 47 (1926), 43-75.

The fourth and fifth decades of existence of the Marseille Society of Geography and Colonial Studies. Lists of officers, medalists, publications, numbers of members, etc. See also pp. 5-9 of the same volume for an anonymous sketch of the first 50 years of the Society.

575. Lewis, G. Malcolm. "Association with a Mission," Geographical Magazine, vol. 52, no. 7 (April 1980), 455 and 457.

An essay describing the vicissitudes of the British geography teachers' society, the Geographical Association, established in 1893, with particular emphasis on recent years.

576. "M.L." [Lindeman, M.]. "25 Lebensjahre der geographischen Gesellschaft in Bremen," <u>Deutsche Geographische Blätter</u>, vol. 18, nos. 1-2 (1895), 5-11.

The Bremen Geographical Society was founded in 1870 as the Verein für die deutsche Nordpolfahrt (renamed Geographical Society in 1876).

577. Louis, Herbert. "Die Geographische Gesellschaft München, Rückblick im hundertsten Jahre ihres Bestehens" (Festschrift zur 100-Jahrfeier der Geographischen Gesellschaft München 1869-1969, part 1), <u>Mitteilungen der Geographischen Gesellschaft in München</u>, vol. 54 (1969), 5-20.

The first 100 years of the Munich Geographical Society, founded in 1869.

578. Markham, Clements R. "The Fifty Years' Work of the Royal Geographical Society," <u>Journal of the Royal Geographical Society</u>, vol. 50 (1880), 1-255 (Text, 1-126; appendixes, 127-255).

The first 50 years of the Royal Geographical Society, 1830-1880, with several introductory pages on British geography before 1830, going back to Richard Eden, "The Father of English Geography," in the 16th century.

579. Marshall-Cornwall, James. <u>The History of the Geographical Club (1826-1975)</u>. London: Royal Geographical Society, 1976. 49 p.

A history of the Raleigh Club (1826-1854), which contributed to the establishment of the Royal Geographical Society in 1830, and its successor, the Geographical Club (1854 onward). The Club functions as "the Dining Club of the Royal Geographical Society."

580. [McGee, W. J.]. "The Work of the National Geographic Society," <u>National Geographic Magazine</u>, vol. 7, no. 8 (August 1896), 253-259.

Part One--General and Topical   121

   The author's views of the development of geography
   (especially what he calls "the new geography") and the
   purposes and methods of the National Geographic
   Society.

581. Mead, William R., and Wadel, C. "Scandinavia and the
   Scandinavians in the Annals of the Royal Geographical
   Society, 1830-1914," Norsk Geografisk Tidsskrift,
   vol. 18, nos. 3-4 (1961-1962), 99-143.

   Chronicle of contacts between the Royal Geographical
   Society (London) and Scandinavia, particularly in
   exploration. See also Mead's companion-piece on
   Finland in Terra, vol. 71 (1959), 111-119 (in Finnish,
   with English summary on p. 119).

582. Middleton, Dorothy. "Guide to the Publications of the
   Royal Geographical Society, 1830-1892," Geographical
   Journal, vol. 144, part 1 (March 1978), 99-116.

   Commentary on some of the articles in the
   publications of the Royal Geographical Society, the
   Journal and Proceedings, from 1831 to 1892.

583. Mill, Hugh Robert. "Recollections of the Society's
   Early Years," Scottish Geographical Magazine, vol.
   50, no. 5 (September 1934), 269-280.

   Mill's recollections of scientific life in Edinburgh
   in the 1880s and of the early years of the Royal
   Scottish Geographical Society down to 1892.

584. Mill, Hugh Robert. The Record of the Royal
   Geographical Society, 1830-1930. London: Royal
   Geographical Society, 1930. xvi + 288 p.

   A history of the Royal Geographical Society, founded
   in 1830. First chapter, "Forerunners" (pp. 1-10),
   traces geographical interests in Great Britain from
   Julius Caesar through King Alfred and Richard Hakluyt
   to the African Association and Raleigh Club, the parent
   organizations of the RGS.

585. Miller, E. Willard. "A Short History of the American
   Society for Professional Geographers," Professional
   Geographer, n.s. vol. 2, no. 1 (January 1950), 19-40.

   The American Society for Professional Geographers
   began in 1943 as the American Society for Geographical

Research (name changed in 1944) and merged with the Association of American Geographers in 1948.

586. Mitra, Sevati. "A Short History of the Geographical Society of India," Geographical Review of India, vol. 30, no. 4 (December 1968), 106-112.

   The Geographical Society of India began as the Calcutta Geographical Society in 1933 (name changed 1951).

587. Moir, Donald G. "The Royal Scottish Geographical Society, 1884-1959: Early Days of the Society," Scottish Geographical Magazine, vol. 75, no. 3 (December 1959), 131-142.

   Recounts the early years of the Society, 1884-1891. Lists of officers and medalists, 1885-1959.

588. [Newbigin, Marion I.]. "The Royal Scottish Geographical Society: The First Fifty Years," Scottish Geographical Magazine, vol. 50, no. 5 (September 1934), 257-269.

   A chronicle of the RSGS, founded in 1884. Lists of officers and medalists.

589. Panetta, Rinaldo. "La Società Geografica Italiana," L' Universo, vol. 53, no. 6 (November-December 1973), 1179-1200.

   Essay describing the activities of the Italian Geographical Society from its founding in 1867.

590. Pauly, Philip J. "The World and All That Is in It: The National Geographic Society, 1888-1918," American Quarterly, vol. 31, no. 4 (Fall 1979), 517-532.

   The first 30 years of the National Geographic Society. (Pp. 517-518) "The National Geographic Society (NGS), founded by professionals in 1888, was captured and transformed by a group of amateurs in the first decade of this century.... The NGS was a scientific society run by and for nonspecialists.... Although the National Geographic has been primarily a magazine of natural history, its genesis must be understood in the context of developments in late nineteenth-century geography."

591. Payne, Melvin M. "75 Years Exploring Earth, Sea and Sky: National Geographic Society Observes Its Diamond Anniversary," National Geographic Magazine, vol. 123, no. 1 (January 1963), 1-43.

   The first 75 years of the National Geographic Society, founded in 1888. Emphasis on expeditions sponsored by the Society.

592. Penck, Albrecht. "Hundert Jahre Gesellschaft für Erdkunde," Zeitschrift der Gesellschaft für Erdkunde zu Berlin, 1928, nos. 5-6 (July 1928), 162-169.

   History of German geography since the 18th century (Büsching), with emphasis on the Berlin Geographical Society (founded 1828).

593. [Pergameni, Charles]. "Cinquantième anniversaire de la fondation de la Société royale belge de géographie," Bulletin de la Société royale belge de géographie, vol. 150, nos. 3-4 (1926), 174-188.

   Chronicles the first 50 years of the Royal Belgian Geographical Society, founded in 1876, and mentions earlier manifestations of interest in forming geographical societies in Belgium, back to 1856.

594. Pinchemel, Philippe. "Les sociétés savantes et la géographie," pp. 69-78 in Actes du 100e Congrès national des sociétés savantes, Paris, 1975 (Paris: Bibliothèque Nationale, 1976).

   In 1890 there were 30 French geographical societies (including 3 in Algeria), but in 1975 there were only 8. Most of the societies were founded in the period 1873-1890. These societies were of great importance to the field of geography in the 19th century.

595. Primer Centenario de la Sociedad Mexicana de Geografía y Estadística, 1833-1933. 2 vols. Mexico City: Sociedad Mexicana de Geografía y Estadística, 1933. Volumes continuously paged: Vol. 1, xv + 456 p.; Vol. 2, pp. 457-771.

   Contains various articles on geography, especially in Mexico. See especially pp. 1-6, Agustín Aragon, "Influjo de la Sociedad Mexicana de Geografía y Estadística en la cultura del país"; pp. 7-29, Francisco L. de la Barra, "La geografía, el derecho de gentes y la política internacional"; and pp. 31-50, José Ugalde, "La geografía como base del conocimiento humano." These articles contain material on the history of geography, although the Society itself is not given extensive treatment.

596. Richthofen, Ferdinand von. "Die Geographie im ersten Halbjahrhundert der Gesellschaft für Erdkunde," pp. 15-30 in Sonderband zur Hundertjahrfeier der Gesellschaft, Zeitschrift der Gesellschaft für Erdkunde zu Berlin, 1828-1928 (Berlin, 1928).

Geography during the first 50 years of the Berlin Geographical Society, 1828-1878. Previously unpublished address given by Richthofen at the 50th anniversary celebrations of the Society, on 30 April 1878.

597. Ruiz, Ernesto A. "Geography and Diplomacy: The American Geographical Society and the 'Geopolitical' Background of American Foreign Policy, 1848-1861," Unpublished Ph.D. thesis in History, Northern Illinois University, 1975. ix + 275 p.

(From Introduction, p. vi) This dissertation "attempts to trace the origin of American geopolitical thought and to explain how geography became an ideological tool rationalizing a policy of national expansion." The American Geographical Society of New York, founded in 1851, "acted as a pressure group upon the national government and its members ... played influential roles in the formulation of United States foreign policy."

598. Salmon, Pierre. "Histoire de la Société royale belge de géographie (1876-1976)," Revue belge de géographie, vol. 101, nos. 1-3 (October 1977), 7-19.

History of the Société belge de géographie, founded in 1876 ("royale" added in 1882). Considered a merger with the Société d'études coloniales in 1903 but then gave up the plan. Article followed by Christian Vandermotten's paper, "Cent ans de publication" (pp. 21-61), an index to vols. 26-100 of the Bulletin de la Société royale belge de géographie (changed to Revue belge de géographie in 1962). Vols. 1-25 (1876-1901) had been indexed by Maurice Rahir in 1902.

Part One--General and Topical   125

599. Sandru, Ion. "Le centenaire de la Société roumaine de géographie (1875-1975)," Geoforum, vol. 6, no. 1 (September 1975), 9-14.

   A history of the Romanian Geographical Society, founded in 1875.

600. Schamp, Heinz. "Einhundert Jahre 'Société de géographie d'Egypte' in Kairo," Die Erde, vol. 109, nos. 3-4 (1978), 517-519.

   A brief sketch of the activities of the Egyptian Geographical Society (founded 1875) during the first century of its existence.

601. Schlee, Paul. "Rückblick auf die Entwicklung der Geographischen Gesellschaft in Hamburg während der zweiten 25 Jahre ihre Bestehens, und zwar von 1898 bis 1923," Mitteilungen der Geographischen Gesellschaft in Hamburg, vol. 36 (1924), 1-27.

   Activities of the Hamburg Geographical Society, 1898-1923. Lists of officers, medalists, honorary and corresponding members, articles, and maps. For the first 25 years of the Society, 1873-1898, see Friederichsen (above).

602. Schrader, R. "Honderd jaar Koninklijk Nederlands Aardrijkskundig Genootschap 1873-1973," Geografisch Tijdschrift, n.s. vol. 8, no. 4 (1974), 234-402 (also paged separately, vii + 164).

   Activities of the first 100 years of the Royal Dutch Geographical Society, 1873-1973.

603. [Anonymous]. "La Société de géographie, 1821-1921," La Géographie, vol. 36, no. 2 (July-August 1921), 137-208.

   A centenary history of the Paris Geographical Society.

604. Sparn, Enrique. "Cronología, diferenciación, número de socios y distribución de las sociedades de geografía," Boletín de la Academia Nacional de Ciencias (Córdoba, Argentina), vol. 32 (1935), 323-336 (+ 5 pages of plates).

   Growth of geographical societies, from 1821 onward. In 1935 there were 137 societies in the world, with

102,700 members. Brief discussion of libraries and journals. 3 of the 5 plates are maps showing the distribution of societies.

605. Spreitzer, Hans. "Zum hundertjährigen Bestand der Geographischen Gesellschaft in Wien: Rückschau und Ausblick," pp. xv-xxxv in Festschrift zur Hundertjahrfeier der Geographischen Gesellschaft in Wien 1856-1956, ed. Konrad Wiche (Vienna: Geographische Gesellschaft, 1957).

Geography in Vienna during the last 100 years, with emphasis on the role of the Geographical Society.

606. Spreng, A. "Der Geographische Gesellschaft von Bern 1873-1923," Jahresbericht der Geographische Gesellschaft von Bern, vol. 25 (1919-1922, pub. 1923), 1-13.

Activities of the Bern Geographical Society during its first 50 years.

607. Stafford, Mary Peary, and Klimm, Lester E. Geographical Society of Philadelphia: History 1891-1960. Philadelphia: Geographical Society of Philadelphia, n.d. (1960). 92 p.

"History of the Society" by Mary Peary Stafford (pp. 5-53) and "History of the Society's Bulletin" by Lester Klimm (pp. 54-74). Founded 1891 as Geographical Club of Philadelphia ("Society" after 1896). Bulletin published 1893-1939.

608. Stoddart, David R. "Progress in Geography: The Record of the I.B.G.," Transactions of the Institute of British Geographers, n.s. vol. 8, no. 1 (1983), 1-13.

A history of the Institute of British Geographers (founded 1933), the leading professional geographical society in Great Britain. Lead article in "A Special Issue of Transactions to Mark the Fiftieth Anniversary of the Institute," compiled by D.R. Stoddart.

609. Stoddart, D.R. "The RGS and the 'New Geography': Changing Aims and Changing Roles in Nineteenth Century Science," Geographical Journal, vol. 146, part 2 (July 1980), 190-202.

Part One--General and Topical

(From Abstract, p. 190) "At its foundation in 1830 the RGS [Royal Geographical Society] was mainly a society for travellers, soldiers and sailors, though with a substantial number of natural scientists, especially geologists.... Towards the end of the century the Society's character changed markedly, partly as military men became less numerous, and partly as the natural scientists withdrew to their own increasingly professional organizations. The emergence of the 'new geography' after 1885, in both schools and universities, reflected a growing professionalization within the RGS itself, and the terms of the debate about the content of the subject were largely governed by the economic and social pressures which affected it as well as neighbouring disciplines."

610. Swedberg, Swen. "Geografiska Föreningen i Göteborg 1908-1958," Gothia, series 2, no. 8 (1958), 1-37.

An historical sketch of the Gothenburg (Göteborg, Sweden) Geographical Society from its founding in 1908.

611. Taberini, Annalena. La Società Geografica Italiana. Rome: Società Geografica Italiana, 1980. 31 p.

A pamphlet describing the history and present organization of the Italian Geographical Society, founded in 1867. Text in both Italian and English (English translation by Pietro Bindelli).

612. Taylor, Griffith, and "O.H.K.S." [Spate, O.H.K.]. "The Society's Silver Jubilee," The Australian Geographer, vol. 6, no. 1 (June 1952), 3-5.

The first 25 years of the New South Wales Geographical Society. "The Founding of the Society, 1927" by Griffith Taylor (pp. 3-4) and "The Work of the Society As Reflected in Its Journal" by O.H.K. Spate (pp. 4-5).

613. Thomas, Paul. "Le cinquantenaire de la Société de géographie de Lille," Bulletin de la Société de géographie de Lille, vol. 72 (January-March 1930), 68-106.

The first 50 years of the Lille Geographical Society: its organization, its activities, and its publications. The Society withdrew from the Union Géographique du

Nord de la France in 1882 and created sections at Roubaix and Tourcoing. [PP]

614. Torroja y Miret, José M. "La Real Sociedad Geográfica de Madrid en el LXXV anniversario de su fundación, octubre de 1952," <u>Boletín de la Real Sociedad Geográfica</u>, vol. 89, nos. 1-3 (January-March 1953), 21-32.

    Historical sketch of the Royal Geographical Society of Madrid, founded in 1876.

615. Vilá Valentí, J. "Origen y significado de la Sociedad Geográfica de Madrid," <u>Revista de geografía</u> (Barcelona), vol. 11, nos. 1-2 (January-December 1977), 5-21.

    The founding of the Madrid Geographical Society in 1876 and 4 decades of previous work leading to the establishment of the Society.

616. Wagner, Hermann. "Ueber den Plan zur Begründung einer 'Humboldt-Gesellschaft' für Geschichte der Geographie und Kartographie," <u>Verhandlungen des achtzehnten Deutsches Geographentages zu Innsbruck 1912</u> (published in Berlin, 1912), pp. 231-235.

    Concerns the abortive attempt to establish a "Humboldt Society for the History of Geography and Cartography" in Germany in 1912. [HB]

617. Wagner, Julius. "Den Verein für Geographie und Statistik zu Frankfurt a.M. im ersten Viertel des zweiten Jahrhunderts seines Bestehens," pp. 7-30 in <u>Festschrift zur 125-Jahrfeier der Frankfurter Geographischen Gesellschaft</u> (<u>Verein für Geographie und Statistik</u>) (Frankfurter Geographische Hefte, no. 37, 1961).

    The Frankfurt Geographical Society, 1936-1961.

618. Warrington, T.C. "The Beginnings of the Geographical Association," <u>Geography</u>, vol. 38, part 4 (November 1953), 221-230.

    The Geographical Association, a British organization for teachers of geography, was founded in 1893. Paper read at 50th anniversary in 1943. See also companion paper by Fleure (above).

Part One--General and Topical 129

619. Wright, John K. "The American Geographical Society: 1852-1952," Scientific Monthly, vol. 74, no. 3 (March 1952), 121-131.

Centennial history of the American Geographical Society, organized 9 October 1851.

620. Wright, John K. "The Field of the Geographical Society," pp. 543-565 in Geography in the Twentieth Century, ed. Griffith Taylor, 3rd ed. (New York: The Philosophical Library, 1957).

Geographical societies, actual and planned, from 1693 onward.

621. Wright, John K. Geography in the Making: The American Geographical Society, 1851-1951. New York: American Geographical Society, 1952. xxi + 437 p.

Centennial history of the American Geographical Society of New York, known as the American Geographical and Statistical Society until 1871.

IX. MISCELLANEOUS ORGANIZATIONS

622. Arden-Close, Charles. "International Geographical Congresses," pp. 144-179 in Geographical By-Ways and Some Other Geographical Essays by Charles Arden-Close (London: Edward Arnold & Co., 1947).

International Geographical Congresses, from the first (Antwerp, 1871) to the 15th (Amsterdam, 1938).

623. Beaver, Stanley H. "Geography in the British Association for the Advancement of Science," Geographical Journal, vol. 148, no. 2 (July 1982), 173-181.

(Page 173) "The British Association was founded in 1831; geography was at first of little importance, being included in the Geology Section, but in 1851 it was accorded a Section of its own. The paper traces the story of Section E over 130 years."

624. Beaver, S.H. "The Le Play Society and Field Work," Geography, vol. 47, part 3 (July 1962), 225-240.

The Le Play Society (1930-1960) was a British organization dedicated to geographical and sociological field work. It was named for the French sociologist Frédéric Le Play (1806-1882). Article emphasizes the lives of Le Play and his Scottish interpreter, Patrick Geddes (1854-1932).

625. Blakeney, T.S. "The Alpine Club," Geographisches Taschenbuch 1964/65 (Wiesbaden: Franz Steiner Verlag GMBH, 1964), pp. 256-269.

    The oldest of the mountaineering clubs, the Alpine Club was founded in London in 1857. It is more a learned society than a social club. Article mentions British travelers in the Alps from William Windham (1741) onward.

626. Broc, Numa. "Un musée de géographie en 1795," Revue d' histoire des sciences et de leurs applications, vol. 27, no. 1 (January 1974), 37-43.

    Describes an abortive plan for a Musée de géographie, de topographie militaire et d'hydrographie in Paris in 1795. Was to have a geographical and cartographic research establishment similar to the present Institut géographique national.

627. Club Alpin Français, Commission des travaux scientifiques. L'oeuvre scientifique du Club Alpin Français (1874-1922). Paris: Club Alpin Français, 1936. vi + 518 p.

    The activities of the French Alpine Club since its founding in 1874.

628. Crone, Gerald R. "'Jewells of Antiquitie': The Work of the Hakluyt Society," Geographical Journal, vol. 128, part 3 (September 1962), 321-324.

    Brief sketch of the Hakluyt Society, founded in 1846 for the publication of travel narratives.

629. Dunbar, Gary S. "Societies for the History of Discoveries," Terrae Incognitae, vol. 6 (1974), 65-71.

    Survey of societies concerned with the history of geographical discovery and exploration. (Page 65) "Of the twelve societies described in this paper, only the

Part One—General and Topical   131

   Society for the History of Discoveries [founded 1960] publishes a journal, holds meetings at which members read papers, and, in general, behaves like a professional or scholarly organization. The others, like the Hakluyt Society, were formed for the sole purpose of supporting the publication of travel narratives and other historical documents."

630. Georgi, Johannes. "Die 'Allgemeine Deutsche Versammlung von Freunden der Erdkunde' in Frankfurt a.M. 1865 und ihre Bedeutung für die Geographie," Petermanns Geographische Mitteilungen, vol. 112, no. 2 (15 May 1968), 104-111.

   Describes the General German Conference of Friends of Geography in Frankfurt, 23-24 July 1865, its antecedents and consequences, particularly in the area of polar research.

631. Grosjean, Georges. "Der Schweizer Alpen-Club 1863-1963 und die Erforschung der Schweizer Alpen," Geographisches Taschenbuch 1964/65 (Wiesbaden: Franz Steiner Verlag GMBH, 1964), pp. 269-284.

   Mountaineering and Alpine exploration from the 16th century onward, with emphasis on the Swiss Alpine Club, founded in 1863.

632. Horn, Werner. "Die Geschichte der Gothaer Geographischen Anstalt im Rahmen der Entwicklung von Geographie und Kartographie," Vermessungstechnik, vol. 2 (1954), 29-32.

   History of the Justus Perthes publishing house (now VEB Hermann Haack) in Gotha (German Democratic Republic) and its relations to the development of geography and cartography. [HB]

633. Horn, Werner. "Die Geschichte der Gothaer Geographischen Anstalt im Spiegel des Schrifttums," Petermanns Geographische Mitteilungen, vol. 104, no. 4 (December 1960), 271-287.

   Bibliographical essay on the Geographical Institute founded in Gotha, Germany, in 1785 by Justus Perthes (1749-1816). Brief sketches of the major personalities connected with the Institute. Pp. 281-287 consist of 402 bibliographical citations of works dealing with the Institute, its publications, and the geographers and cartographers connected with it.

634. Howarth, O.J.R. "The Centenary of Section E (Geography)," The Advancement of Science, vol. 8, no. 30 (September 1951), 151-165.

An historical sketch of the first 100 years of Section E (Geography) of the British Association for the Advancement of Science. Appendix ("Orationes Personae," pp. 162-165) contains biographical notes on the geographers mentioned in the paper.

635. Leconte, P. "Histoire de l'Union Géographique Internationale et des congrès internationaux de géographie," International Geographical Union, The IGU Newsletter, vol. 10, no. 1 (1959), 3-20; no. 2, 43-69.

Concerns the International Geographical Congresses, beginning with the first congress in Antwerp in 1871, and the International Geographical Union, which was founded in 1922.

636. Lejeune, Dominique. "Contribution à l'histoire de la Société de géographie: alpinisme et géographie au siècle dernier," Acta Geographica, 3rd series, no. 46 (2nd Quarter 1981), 1-10.

Alpinism and geography in the 19th century. The Club Alpin Français was founded in 1874, with the help of the Société de géographie de Paris.

637. Lunn, Arnold. A Century of Mountaineering, 1857-1957. London: George Allen & Unwin Ltd., 1957. 263 p.

A history of mountaineering and mountain research since the 15th century, with emphasis on the Alpine Club, founded in 1857. Describes attitudes toward mountains from classical times onward. The book is "A Centenary Tribute to the Alpine Club" by the Swiss Foundation for Alpine Research.

638. Martonne, Emmanuel de. "Brief History of the International Geographical Union," International Geographical Union, The IGU Newsletter, vol. 1, no. 2 (June 1950), 3-5.

Concerns the International Geographical Union (founded 1922) and the subsequent international congresses (Cairo 1925 to Lisbon 1949). Also printed in French.

639. [Perthes, Bernhard]. <u>Justus</u> <u>Perthes</u> <u>in</u> <u>Gotha</u>, <u>1785-1885</u>. Munich: Knorr & Hirth, n.d.[1885]. 108 p.

A centennial history of the Justus Perthes publishing firm and geographical institute in Gotha, Germany. Emphasis on maps and atlases and the individuals who produced them.

640. Pinchemel, Philippe, et al. <u>La géographie à travers un siècle de congrès internationaux / Geography through a Century of International Congresses</u>. Paris: International Geographical Union, Commission on the History of Geographical Thought, 1972. 252 p.

Chapters in French or English by various authors concerning the International Geographical Congresses from the first (Antwerp, 1871) through the 21st (New Delhi, 1968).

641. Traversi, Carlo. "I cento anni dell'Istituto Geografico Militare nella vita d'Italia," <u>L'Universo</u>, vol. 52, no. 5 (September-October 1972), 873-906.

Describes the first century of existence of the Istituto Geografico Militare (Florence), the agency that produces the topographic maps of Italy. In the same issue is an article by Tomaso Urso on the first century of the Institute's library (pp. 935-942).

X. GEOGRAPHICAL PERIODICALS

642. Bader, Frido J. Walter. "Rückblick nach 100 Bänden der Zeitschrift der Gesellschaft für Erdkunde zu Berlin," <u>Die Erde</u>, vol. 100, nos. 2-4 (1969), 93-117.

A survey of the first 100 volumes of the Journal of the Berlin Geographical Society (<u>Zeitschrift</u> first published 1853; called <u>Die Erde</u> since 1949). Followed by a list of the Society's publications, 1834-1969, compiled by Hans Leonhardy (pp. 118-123).

643. Bernhardt, Peter. "'Petermanns Geographische Mitteilungen' und die deutschsprachigen geographischen Zeitschriften bis zum Ende des 19. Jahrhunderts," <u>Petermanns Geographische Mitteilungen</u>, vol. 125, no. 3 (1981), 167-183.

19th-century German geographical periodicals, with emphasis on Petermanns Geographische Mitteilungen (founded 1855).

644. Bird, James. "Transactions of Ideas: A Subjective Survey of the Transactions during the First Fifty Years of the Institute," Transactions of the Institute of British Geographers, n.s. vol. 8, no. 1 (1983), 55-69.

(From Abstract, p. 55) "A subjective, and necessarily highly selective, survey of the Transactions since 1935 is presented, and ideas in the journal's papers are discussed in relation to eleven themes."

645. Eder, Herbert M. "The Biography of a Periodical: Geographische Zeitschrift, 1895-1963," The Professional Geographer, vol. 16, no. 3 (May 1964), 1-5.

The German periodical Geographische Zeitschrift was founded by Alfred Hettner in 1895, lapsed after 1944, and was revived in 1963. Table shows summary of contents, 1895-1944.

646. Engelmann, Gerhard. "Alexander von Humboldts Plan einer geographischen Zeitschrift," Geographische Zeitschrift, vol. 52, no. 4 (1964), 317-324.

Alexander von Humboldt's abortive plan to establish a "Magazin für Erd- und Völkerkunde" in 1824.

647. Fairchild, Wilma B. "Adventures in Longevity: Fifty Years of the 'Geographical Review,'" Geographical Review, vol. 56, no. 1 (January 1966), 1-11.

Review of the contents, authors, and readers of the Geographical Review since its inception in 1916 (as successor to the Bulletin of the American Geographical Society). Compare article by Gladys Wrigley (below).

648. Fox, Harold S.A., and Stoddart, David R. "The Original Geographical Magazines 1790 and 1874," Geographical Magazine, vol. 47, no. 8 (May 1975), 482, 485-487.

Description of earlier British journals that bore the name Geographical Magazine--a short-lived one appearing in 1790 and another that lasted from 1874 to 1879, when it was taken over by the Royal Geographical Society and merged with the Society's Proceedings.

649. Freeman, T.W. "The Scottish Geographical Magazine, Its First Thirty Years," Scottish Geographical Magazine, vol. 92, no. 2 (September 1976), 92-100.

Survey of the contents of the SGM, 1885-1914.

650. Freeman, T.W. "The Scottish Geographical Magazine in War and Peace from 1914," Scottish Geographical Magazine, vol. 92, no. 3 (December 1976), 138-147.

The journal since 1914.

651. Hohmann, Joseph. "Geographische Zeitschriften des 18. Jahrhunderts: Ein Beitrag zur Geschichte deutscher geographischer Periodika," Erdkunde, vol. 13, no. 4 (December 1959), 455-463.

25 German geographical periodicals founded between 1750 and 1798. Perhaps the first important periodical was Anton Friedrich Büsching's Magazin für die neue Historie und Geographie (1767-1788) and the most important was the Allgemeinen Geographischen Ephemeriden (1798-1831), founded by Friedrich Justin Bertuch and Franz von Zach.

652. Köhler, Franz. "125 Jahrgänge von Petermanns Geographischen Mitteilungen: Wandlungen im Profil der Zeitschrift," Petermanns Geographische Mitteilungen, vol. 125, no. 1 (1981), 1-10.

Analysis of changes in types of articles published in the PGM from its founding in 1855 to 1980. Obvious increase in interest in the Soviet Union and in Marxism-Leninism after World War II.

653. Martonne, Emmanuel de. "Le cinquantenaire des Annales de Géographie," Annales de géographie, vol. 51, no. 285 (January-March 1942), 1-6.

An historical sketch of the leading French geographical journal, the Annales de géographie, "the mirror of progress accomplished by geography": changes, contents, titles, and authors. At first dominated by naturalists, especially geologists, the journal soon began to show the growing presence of university geographers with articles on France and human geography. [PP]

654. McDonald, James R. "Publication Trends in a Major French Geographical Journal," Annals of the

Association of American Geographers, vol. 55, no. 1 (March 1965), 125-139.

Trends in French geographical research, as shown by analysis of articles in the Annales de géographie from 1912 to 1961.

655. Neff, Ernst. "Zum 100. Jahrgang von Petermanns Geographischen Mitteilungen," Petermanns Geographische Mitteilungen, vol. 100, no. 1 (15 February 1956), 1-3.

The first 100 volumes of PGM, founded in 1855.

656. Sandner, Gerhard. "Die 'Geographische Zeitschrift' 1933-1944: Eine Dokumentation über Zensur, Selbstzensur und Anpassungsdruck bei wissenschaftlichen Zeitschriften im Dritten Reich," Geographische Zeitschrift, vol. 71, no. 2 (Second Quarter 1983), 65-87; no. 3 (Third Quarter 1983), 127-149.

The Geographische Zeitschrift in Nazi Germany, 1933-1944, with emphasis on the pressures on the editors, Alfred Hettner and Heinrich Schmitthenner.

657. Toni, Y. "The 'Bulletin de la Société de Géographie d'Egypte': A Review of Its Volumes, 1875-1965," Bulletin de la Société de géographie d'Egypte, vol. 39 (1966), 83-114.

The vicissitudes of the Egyptian Geographical Society and its Bulletin, which both began in 1876.

658. Turri, Eugenio. "La moderna idea geografica nella divulgazione di una rivista americana ('The National Geographic Magazine')," L'Universo, vol. 39, no. 2 (March-April 1959), 295-306.

An essay on the changing character of the National Geographic Magazine and the images of life in different parts of the world that the magazine portrays.

659. Watson, J. Wreford. "The Development of Canadian Geography: The First Twenty-Five Volumes of the Canadian Geographer," The Canadian Geographer, vol. 25, no. 4 (Winter 1981), 391-398.

A review of the first 25 volumes of The Canadian Geographer (1950-1981) and thus an overview of the development of geography in Canada in that period.

660. Whittemore, Katheryne T. "Celebrating Seventy-Five Years of The Journal of Geography, 1897-1972," Journal of Geography, vol. 71, no. 1 (January 1972), 7-18.

The Journal of Geography, the organ of the National Council for Geographic Education, an organization devoted to promoting geographical instruction in the United States, particularly in the schools, originated as The Journal of School Geography in 1897. After merger with the Bulletin of the American Bureau of Geography, it assumed its present name in 1902.

661. Witthauer, Karl. "100 Jahre 'Ergänzungshefte zu Petermanns Geographischen Mitteilungen,'" Petermanns Geographische Mitteilungen, vol. 104, no. 4 (December 1960), 288-291.

Monograph series began 1860. 269 published by 1960. Pp. 289-291 consist of titles of monographs.

662. Wrigley, Gladys M. "Adventures in Serendipity: Thirty Years of the 'Geographical Review,'" Geographical Review, vol. 42, no. 4 (October 1952), 511-542.

An anecdotal review of the Geographical Review (which began 1916 as successor to the Bulletin of the AGS) by the long-time editor. Compare article by Fairchild (above).

XI. GEOGRAPHY IN THE UNIVERSITIES

663. Annaheim, Hans, and Bühler, Alfred. "Zum Fünfhundertjahr-Jubiläum der Alma mater basiliensis: Die Geographie und die Ethnologie an der Basler Universität," Regio Basiliensis, vol. 1, no. 2 (March 1960), 61-67.

Geographical work in the University of Basel, Switzerland, with emphasis on the period since 1912. Geography is described by Annaheim on pp. 62-65, ethnology by Bühler on pp. 65-67.

664. Arnhold, Helmut. "Die Pflege der Geographie in Leipzig, 1860-1870," Leipziger Geographische Beiträge, 1965, pp. 7-15.

History of geography in Leipzig, 1860-1870. [HB]

665. Atwood, Wallace W. The Graduate School of Geography in Clark University, 1920-1945. ("Publications of the Clark University Library.") Worcester, Massachusetts: Clark University, 1945. 40 p.

The first 25 years of the Graduate School of Geography of Clark University, founded by Wallace Atwood (1872-1949), professor of geography and president of the University.

666. Balchin, W.G.V. "Geography in the University College of Swansea 1920-1978," pp. 9-27 in Concern for Geography: A Selection of the Work of Professor W.G.V. Balchin (Swansea, Wales: University College of Swansea, 1981).

Geography was taught (by a geologist) from the beginning of the College in 1920. A Department of Geology and Geography was established in 1931 and a separate Department of Geography in 1954.

667. Baschin, Otto. "Das Berliner Geographische Kolloquium (1886-1911)," Zeitschrift der Gesellschaft für Erdkunde zu Berlin, 1911, pp. 570-577.

The first 25 years of the Geographical Colloquium at the University of Berlin. Ferdinand von Richthofen was appointed professor in 1886.

668. Bernleithner, Ernst. "Das Geographische Institut der Universität Wien," Geographischer Jahresbericht aus Osterreich, vol. 25 (1953-1954), 132-145.

Geography in the University of Vienna from 1846 onward. Friedrich Simony was appointed first professor of geography in 1851. Article mostly contains biographical information about the geographers who have taught at the University of Vienna.

669. Bernleithner, Ernst. "Sechshundert Jahre Geographie an der Wiener Universität," pp. 55-125 in Vol. 3 of Studien zur Geschichte der Universität Wien (Graz and Cologne: Verlag Hermann Bohlaus Nachf., 1965).

Geographical instruction at the University of Vienna from the 14th century to the present. Friedrich Simony was the first holder of a chair in geography (1851).

Part One--General and Topical   139

670. Bernleithner, Ernst. "600 Years of Geography at the University of Vienna and Poland," pp. 184-188 in Vol. 4 of *Actes du XIe Congrès International d'Histoire des Sciences* (Warsaw, etc., 1965) (Warsaw, etc.: Ossolineum, 1968).

    University of Vienna founded 1365. Professor designated as "Professor of Geography" for first time in 1774. The first university chair was established 1851, thus introducing geography as an autonomous subject. Essay mostly concerns the University of Vienna. A little about University of Cracow in earlier centuries.

671. Blache, Jules. "La géographie à Harvard et les relations scientifiques franco-américaines," pp. 125-129 in *Harvard et la France* (Le Comité français pour la célébration du troisième centenaire de Harvard) (Paris: La Revue d'histoire moderne, 1936).

    Relations between the geographers of France and those of Harvard University, with emphasis on William Morris Davis. The French looked to Harvard for leadership in physical geography, and the Americans looked to the French for leadership in human geography.

672. Bowen, Emrys G.; Carter, Harold; and Taylor, James A., eds. *Geography at Aberystwyth: Essays Written on the Occasion of the Departmental Jubilee, 1917-18--1967-68* (Cardiff: University of Wales Press, 1968). xxxvi + 276 p.

    See especially pp. xix-xxxvi, "A Retrospect" (anonymous, but presumably written by the editors, especially Bowen). Describes the development of geography in the University of Wales, Aberystwyth, with particular attention to H.J. Fleure (1877-1969).

673. Campbell, John A. *Geography at Queen's: An Historical Survey*. (The Queen's University of Belfast, Department of Geography, Departmental Research Papers, no. 2, 1978). 55 p.

    The first half-century of the Department of Geography at the Queen's University of Belfast, Northern Ireland. "Beginnings" by E. Estyn Evans (pp. 5-15), "Geography at Queen's" by John Campbell (pp. 16-48), and "Fifty Years On" by William Kirk (pp. 49-55).

674. Colby, Charles C. "Narrative of Five Decades," pp. 8-20 in A Half Century of Geography--What Next? (Chicago: University of Chicago, Department of Geography, 1955).

   A history of the first 50 years of the University of Chicago Department of Geography, established in 1903. See also Alice Foster's paper, "The New Department in Its Setting" (pp. 1-7).

675. Davis, William M., and Daly, Reginald A. "Geology and Geography, 1858-1928," pp. 307-328 in The Development of Harvard University since the Inauguration of President Eliot, 1869-1929, ed. Samuel E. Morison (Cambridge: Harvard University Press, 1930).

   Instruction in geology and geography at Harvard from about 1858 onward.

676. Dunbar, Gary S. "Advice to the Still-Born: The Prospects for Geography at Mont Eagle University in 1871," California Geographer, vol. 21 (1981), 95-99.

   Speculation about the possibility of the teaching of geography in Mont Eagle University, an abortive venture of the eccentric San Francisco millionaire Horace Hawes (d. 1871) in 1870-1871.

677. Dunbar, Gary S., ed. Geography in European and American Universities, 1912. Charlottesville: University of Virginia Printing Office, 1965. 38 p.

   A symposium held at the University of Virginia, 12 October 1912, during the Transcontinental Excursion of the American Geographical Society. Essays on geographical instruction in the universities of Germany (J. Partsch), France (L. Gallois), Austria (E. Oberhummer), Great Britain (G. Chisholm), Switzerland (E. Chaix), and the United States (A. Brigham and M. Jefferson). Reprint of original papers published in Proceedings of the Philosophical Society of the University of Virginia, vol. 1 (1915), pp. 99-134, with a new preface by the editor.

678. Dunbar, Gary S. Geography in the University of California (Berkeley and Los Angeles), 1868-1941. Marina del Rey, California: Devorss & Company, 1981. 18 p.

Essay on the growth of geographical instruction in the two largest branches of the University of California before 1941.

679. Engelmann, Gerhard. "Die Geographie an der Universität Leipzig im 19. Jahrhundert," Petermanns Geographische Mitteilungen, vol. 109, no. 1 (March 1965), 32-41.

 Geography at the University of Leipzig in the 19th century. Oscar Peschel was the first holder of a chair in geography (1871). Later Ferdinand von Richthofen and Friedrich Ratzel held the post. Before 1871 lectures on geographical subjects were given by the historian Heinrich Wuttke and the geologist Carl Friedrich Naumann.

680. Fleure, H.J., comp. "Chairs of Geography in British Universities," Geography, vol. 46, part 4 (November 1961), 349-353.

 A list of the chairs (professorships) of geography in British universities, 1903-1961, as well as the early chair at University College, London, held by A. Maconochie, 1833-1836.

681. "T.W.F." [Freeman, T.W.]. "Early Developments in Geography at Manchester University," Geographical Journal, vol. 120, part 1 (March 1954), 118-119.

 Notes on geographical instruction at Manchester University since 1885. The first full-time teacher of geography, H. Yule Oldham, was appointed in 1892.

682. Fuchs, Roland J., and Street, John M., eds. Geography in Asian Universities. Honolulu: The Oriental Publishing Company, n.d. [1976?]. 522 p.

 Current status of geography in 16 countries in East and South Asia. Mostly contemporary description, but some historical information on the beginnings of instruction.

683. Geographisches Institut der ETH. Zur Geschichte des Geographischen Institutes der Eidgenössischen Technischen Hochschule Zürich. Zürich, 1970. 93 p.

 A history of the geographical institute in the Technical University of Zurich, Switzerland. [HB]

684. Gilbert, Edmund William. Geography as a Humane Study. Oxford: Clarendon Press, 1955. 23 p.

History of geographical instruction at the University of Oxford (pp. 3-12).

685. Gilbert, E.W. "Geography at Oxford and Cambridge," The Oxford Magazine, vol. 75, no. 12 (14 February 1957), 274, 276, 278.

The development of geographical instruction at the universities of Oxford and Cambridge, beginning in 1887.

686. Ginkel, Hans J.A. van, and Smidt, Marc de, eds. "75 Years of Human Geography at Utrecht," Tijdschrift voor economische en sociale geografie, vol. 74, no. 5 (1983), 313-406.

Papers concerning the development of human geography at the State University of Utrecht, the Netherlands, since the founding of the Institute of Geography in 1908. Emphasis on Utrecht but considerable information on the history of Dutch geography generally. See especially M.W. Heslinga, "Between German and French Geography: In Search of the Origins of the Utrecht School," pp. 317-334, which treats Dutch geography from the middle of the 19th century to about 1927.

687. Grosjean, Georges. "Aus der Geschichte des geographischen Instituts," Berner Geographische Mitteilungen (1979, published 1980), 9-16.

History of the Geographical Institute at the University of Bern, Switzerland, from its founding in 1886, with preliminary remarks about manifestations of interest in geography in the University from its beginning in 1834.

688. Haefke, Fritz. "150 Jahre Geographie an der Berliner Universität," Wissenschaftliche Zeitschrift der Humboldt-Universität zu Berlin, Mathematisch-Naturwissenschaftliche Reihe, vol. 10, no. 1 (1961), 5-12.

Geographical instruction at the University of Berlin was intiated by J.A. Zeune. Article emphasizes the work of 3 men: Carl Ritter, Ferdinand von Richthofen, and Albrecht Penck.

689. Hartke, Wolfgang. "Geographisches Institut der Technischen Hochschule München" (Festschrift zur 100-Jahrfeier der Geographischen Gesellschaft München 1869-1969, part 1), Mitteilungen der Geographischen Gesellschaft in München, vol. 54 (1969), 51-54.

A brief account of geography in the Technical University of Munich since 1873.

690. Hofmann, Robert. Die Geographie an der Universität Würzburg. (Inaugural-Dissertation.) Dettelbach: Konrad Triltsch, Buch- und Kunstdruckerei, 1912. 92 p.

University-level instruction in geography in Würzburg, 1582-1830.

691. Hurtig, Theodor, and Wegner, Eginhard. "Aus der Geschichte des Geographischen Instituts," pp. 503-515 in Vol. 2 of Festschrift zur 500-Jahrfeier der Universität Greifswald (Greifswald, 1956).

Geographical instruction at the University of Greifswald since mid-18th century. Became a separate discipline in 1881 with the arrival of Rudolf Credner.

692. Johnston, R.J., and Brack, E.V. "Appointment and Promotion in the Academic Labour Market: A Preliminary Survey of British University Departments of Geography, 1933-1982," Transactions of the Institute of British Geographers, n.s. vol. 8, no. 1 (1983), 100-111.

(From Abstract, p. 100) "A brief sketch is provided here of those who have held posts in British university geography departments during the period 1933-1982. Particular attention is paid to where they obtained their first and higher degrees, where those who have obtained readerships and chairs were trained, production rates by department, academic 'inbreeding', and sexual discrimination."

693. Koelsch, William A. "'Better Than Thou': The Rating of Geography Departments in the United States, 1924-1980," Journal of Geography, vol. 80, no. 5 (September-October 1981), 164-169.

Concerns 7 ratings of Ph.D.-granting departments of geography in U.S. universities, 1924-1980.

694. Koelsch, William A. "Terrae Incognitae and Arcana Siwash: Toward a Richer History of Academic Geography," pp. 63-87 in Geographies of the Mind: Essays in Historical Geosophy in Honor of John Kirtland Wright, ed. David Lowenthal and Martyn Bowden (New York: Oxford University Press, 1975).

On the writing of the history of academic geography in the United States.

695. Koelsch, William A. "Wallace Atwood's 'Great Geographical Institute,'" Annals of the Association of American Geographers, vol. 70, no. 4 (December 1980), 567-582.

The plan of the Graduate School of Geography at Clark University (Worcester, Massachusetts), from its inception in 1919 through implementation in 1921 to the end of 1923.

696. Kühn, Arthur. Die Neugestaltung der deutschen Geographie im 18. Jahrhundert: Ein Beitrag zur Geschichte der Geographie an der Georgia Augusta zu Göttingen. ("Quellen und Forschungen zur Geschichte der Geographie und Völkerkunde," vol. 5, ed. Albert Herrmann.) Leipzig: K.F. Koehler Verlag, 1939. 149 p.

Describes the development of geography in 18th-century Germany, with emphasis on the University of Göttingen, from about 1730 onward.

697. Leszczycki, Stanislaw, and Modelska-Strzelecka, Bozena. "Six Centuries of Geography at the Jagiellonian University in Kraków," pp. 276-281 in Vol. 4 of Actes du XIe Congrès International d'Histoire des Sciences (Warsaw, etc., 1965) (Warsaw, etc.: Ossolineum, 1968).

Geographical instruction in the University of Cracow since the 14th century. The holder of the first chair of geography (1849) was Wincenty Pol (1807-1872). Chair abolished 1852 and reinstated 1877.

698. Université de Liège. Cinquantième anniversaire du Seminaire de géographie (1903-1953) et vingt-cinquième anniversaire du Cercle des géographes liégeois (1928-1953). Liège: Imprimerie H. Vaillant-Carmanne, S.A., 1953. 102 p.

See especially pp. 39-48, an historical sketch of geographical instruction in the University of Liège, Belgium, especially in the Geographical Institute (Seminary) founded in 1903, written by O. Tulippe. Pp. 7-24 consist of lists of names of people connected with the Seminary and the Circle, theses, publications, etc.

699. Louis, Herbert. "Die Entwicklung der Geographie an der Universität München bis 1969" (Festschrift zur 100-Jahrfeier der Geographischen Gesellschaft München 1869-1969, part 1), <u>Mitteilungen der Geographischen Gesellschaft in München</u>, vol. 54 (1969), 21-50.

    An historical survey of geography in the University of Munich since the 1870s.

700. Markov, K.K. "Methodological Principles of the Curriculum of a Geography Faculty," <u>Soviet Geography: Review and Translation</u>, vol. 9, no. 5 (May 1968), 358-367.

    Development of geography in the universities of Moscow (founded 1755) and Leningrad (1819). Describes differences between the curricula of the two universities in the 1920s and 1930s. Also describes postwar reforms. Article first published in Russian in 1967.

701. Morawetz, Sieghard. "Hundert Jahre Geographie an der Karl-Franzens-Universität in Graz, 1871-1971," <u>Arbeiten aus dem Geographischen Institut der Universität Graz</u>, no. 15 (1971). 41 p.

    History of geographical instruction in the University of Graz, Austria, with emphasis on the professors. Chair of geography established in 1871 with Robert Rosler (d. 1874) as the first professor.

702. Morris, Rita M.L. "An Examination of Some Factors Related to the Rise and Decline of Geography as a Field of Study at Harvard, 1638-1948," Unpublished Doctor of Education thesis, Graduate School of Education, Harvard University, 1962. 282 p.

    The teaching of geography at Harvard University since its founding in 1638, with emphasis on the modern period, beginning with Nathaniel Southgate Shaler and William Morris Davis.

703. Mücke, E., and Oelke, E. "100 Jahre Geographie in Halle," <u>Petermanns Geographische Mitteilungen</u>, vol. 118, no. 3 (10 June 1974), 204.

Chair of geography at Halle University (Germany) established in 1873, with Alfred Kirchhoff as the first holder. Centenary celebrated 1973.

704. Müller, Theodor. "Die Geschichte der Geographie am Collegium Carolinum zu Braunschweig, 1745-1834," <u>Braunschweiger Jahrbuch</u>, vol. 38 (1957), 75-94.

History of geography in the Collegium Carolinum in Braunschweig, Germany, from 1745 to 1834. The Collegium was founded in 1745, was divided into faculties of arts, commerce, and technology in 1835, became a Technische Hochschule in 1877, and has been known as the Technische Universität Carolo-Wilhelmina zu Braunschweig since 1968. [HB]

705. Nielsen, Niels. "Københavns Universitets Geografiske Institut: Et Bidrag til Dansk Geografis Historie," <u>Geografisk Tidsskrift</u>, vol. 61 (1962), 1-78.

History of geographical instruction at the University of Copenhagen. Pp. 1-53 in Danish; pp. 54-78 in English ("The Copenhagen University Geographical Institute: A Contribution to the History of Danish Geography"). Brief mention of Danish geography before the 17th century. Teaching of geography at the University began around 1635, but geography as a distinct field of teaching and research dates from 1867.

706. Paffen, Karlheinz, and Wenk, Hans-Günther. "Hundert Jahre Lehrstuhl und Institut für Geographie an der Universität Kiel," <u>Kieler Geographische Schriften</u>, vol. 50 (1979), 1-67.

Article treats the history of geography in the University of Kiel, Germany, since 1879, when Theobald Fischer became the holder of the first chair of geography.

707. Parsons, James J., ed. and comp. <u>50 Years of Berkeley Geography, 1923-1973</u>. (Supplement to <u>The Itinerant Geographer</u>.) Berkeley: Department of Geography, University of California, 1973. x + 87 p.

"Selected bibliographies of 104 PhD's granted by the University of California, Berkeley since the establishment of a doctoral program in 1923--with biographical data and a bibliography of Carl Ortwin Sauer."

708. Partsch, Joseph. "Die Geographie an der Universität Breslau von 1702-1901," Festschrift des Geographischen Seminars der Universität Breslau zur Begrüssung des XIII. Deutschen Geographentages, Breslau, 1901, pp. 1-37.

    History of geography in the University of Breslau, Germany, 1702-1901. [HB]

709. Partsch, Joseph. "Das Geographische Seminar," pp. 205-215 in Festschrift zur Feier des 500Jährigen Bestehens der Universität Leipzig, vol. 4, Die Institute und Seminare der Philosophischen Fakultät an der Universität Leipzig, part 1, Die Philologische und die Philosophisch-Historische Sektion (Leipzig: Verlag von S. Hirzel, 1909).

    Geographical instruction at the University of Leipzig since mid-19th century. First professor was Oscar Peschel (1826-1875), appointed in 1871.

710. Passarge, Siegfried. "Das Geographische Seminar des Kolonial-Instituts und der Hansischen Universität: Erinnerungen und Erfahrungen," Mitteilungen der Geographischen Gesellschaft in Hamburg, vol. 46 (1939), 1-104.

    An historical sketch of the Department (Seminar) of Geography at the Colonial Institute in Hamburg from its founding in 1908 to the outbreak of war in 1914 and then its rebirth at the University of Hamburg in 1919. The author played a central role in these developments.

711. Pattison, William D. "Goode's Proposal of 1902: An Interpretation," The Professional Geographer, vol. 30, no. 1 (February 1978), 3-8.

    J. Paul Goode's plan (17 July 1902) for the design of the curriculum for the University of Chicago's Department of Geography (founded 1903).

712. Peel, Ronald. "The Department of Geography, University of Bristol, 1925-75," pp. 411-417 in Processes in Physical and Human Geography: Bristol Essays, eds. Ronald Peel, Michael Chisholm, and Peter Haggett (London: Heinemann Educational Books Ltd., 1975).

An historical sketch of the University of Bristol Department of Geography during its first 50 years. Geography was introduced in 1920 in the Department of Geology, and a separate Department of Geography was created in 1925.

713. Penck, Albrecht. "Die Geographie an der Universität Wien," Foreword, pp. vii-xxii, to Arbeiten des Geographischen Institutes der K.K. Universität Wien (Geographische Abhandlungen, vol. 5, no. 1 1896).

Geographical instruction at the University of Vienna since the 14th century. Friedrich Simony (1851) was first professor of geography in the modern period.

714. Philippson, Alfred. "Geographie," Die Naturwissenschaften, vol. 7 (1919), 561-571.

History of geography at Bonn University since 1818. Part of a series of articles marking the 100th anniversary of Bonn University ("Die Entwicklung der Naturwissenschaften an der Bonner Universität seit ihrer Gründung"). [HB]

715. Université de Rennes. Cinquantième anniversaire du Laboratoire de géographie (1902-1952): Volume jubilaire. Rennes: Nourritures terrestres, 1952. 455 p.

First part, pp. 9-44, is a history of the Laboratory (department) and its personnel since 1902. See especially pp. 47-65, André Meynier's article, "Cinquante ans de géographie française," which treats French geography in the 50-year period beginning in 1902 with the founding of the Laboratory of Geography in the University of Rennes by Emmanuel de Martonne.

716. Ristow, Walter W. "Three Months in the Field," Clark Now, vol. 12, no. 1 (Winter-Spring 1982), 19-24.

Clark University (Worcester, Massachusetts) established a special 3-week field school in 1927. In 1934 there was a 3-month field camp (12 September to 12 December) that covered the eastern United States to Florida and New Orleans.

Part One--General and Topical 149

717. Ryabchikov, A.M. "Geography at Moscow University over the Last 50 Years," Soviet Geography: Review and Translation, vol. 9, no. 5 (May 1968), 343-357.

The development of geography at Moscow University since 1917, with a brief look at earlier developments (beginning with D.N. Anuchin in 1887). Article first published in Russian in 1967.

718. Scargill, D.I. "The RGS and the Foundations of Geography at Oxford," Geographical Journal, vol. 142, part 3 (November 1976), 438-461.

(From author's abstract, p. 438) "With RGS [Royal Geographical Society] help, a Readership was established in 1887.... The Society's contributions were increased when, in 1899, a School of Geography was founded and a diploma course introduced.... In spite of the promptings of the RGS, not until 1932 was a chair of geography established and an Honour School introduced."

719. Schmidt, Gerhard. "50 Jahre Geographisches Institut der Universität Rostock," Wissenschaftliche Zeitschrift der Universität Rostock, Mathematisch-Naturwissenschaftliche Reihe, vol. 10, no. 1 (1961), 127-130.

The Geographical Institute of the University of Rostock, Germany, was founded in 1911. Wille Ule (1861-1940) had been appointed professor (Extraordinarius) in 1907.

720. Schott, Carl, ed. "Hundert Jahre Geographie in Marburg," Marburger Geographische Schriften, no. 71 (1977), 1-237.

Contains two articles on the history of Marburg geography: Gottfried Lange, "Die Marburger Geographie im kosmographischen Zeitalter" (pp. 161-177), which describes geography in Marburg from the founding of the University in 1527 to the establishment of a chair in 1876; and Jürgen Leib, "100 Jahre Lehrstuhl fur Geographie an der Philipps-Universität Marburg" (pp. 179-207), which treats geography at the University of Marburg since the naming of Johannes Justus Rein to the chair in 1876.

721. Steel, Robert W. "Geography at the University of Liverpool," pp. 1-23 in Liverpool Essays in

Geography: A Jubilee Collection, ed. by Robert W. Steel and Richard Lawton (London: Longmans, 1967).

University College, Liverpool, was founded in 1883, and geography was taught as early as 1886. The geographer P.M. Roxby (1880-1947) was appointed lecturer in 1904, and in 1917 he was given the newly created chair of geography.

722. Stein, Harry. "Die Geographie an der Universität Jena: Ein Beitrag zur entwicklung der Geographie als Wissenschaft," Erdkundliches Wissen, no. 29 (Beihefte zur Geographischen Zeitschrift). Wiesbaden: Franz Steiner Verlag GMBH, 1972. 152 p.

Geography at the University of Jena, Germany, from 1786 (Johann Ernst Fabri) onward. As an independent discipline geography dates from 1884 (Fritz Regel). Summary of instructors, 1786-1950, pp. 126-130.

723. Stoddart, David R. "The RGS and the Foundations of Geography at Cambridge," Geographical Journal, vol. 141, part 2 (July 1975), 216-239.

(From author's abstract, p. 216) "The RGS [Royal Geographical Society] approached Cambridge University in 1871 with a proposal to support teaching in geography. The initiative was renewed ... in 1885, and in 1887 agreement was reached for a lectureship financed mainly by the Society.... A tripos leading to an honours degree was approved in 1919 and the first students graduated in 1921. With this success the financial support of the RGS ... came to an end in 1923."

724. Tulippe, Omer. "La Géographie dans les universités allemandes," Bulletin de la Société royale de géographie d'Anvers, vol. 50 (1930), 30-59, 320-355.

A survey of geography in German universities. Mostly the contemporary scene but the article contains some historical material.

725. Uhlig, Harald. "100 Jahre Geographie in Giessen: Das Geographische Institut im Neuen Schloss," Giessener Hochschulblätter, vol. 12, no. 1 (20 March 1965), 14-16.

An historical sketch of geography at the University of Giessen, Germany, with emphasis on the last 100

Part One--General and Topical   151

   years and the professorships of Robert von
   Schlagintweit, Wilhelm Sievers, and Fritz Klute.

726. Uhlig, Harald. "Das Neue Schloss als Geographisches
     Institut. Frühe geographische Vorlesungen. Die
     Giessener Geographen Robert von Schlagintweit und
     Wilhelm Sievers" (Festkolloquium: 100 Jahre
     Geographie in Giessen), Giessener Geographische
     Schriften, no. 6 (1965), 87-103.

     Brief sketch of geographical instruction at Giessen
     University from the Renaissance onward, with emphasis
     on Robert von Schlagintweit and the modern development
     of geography beginning in 1864. Also a few pages on
     Wilhelm Sievers (1860-1921), the first professor
     (Ordinarius) at the University (1903 onward).

727. Vosseler, Paul. "50 Jahre Geographisches Institut der
     Universität Basel," Regio Basiliensis, vol. 3, no. 2
     (April 1962), 205-215.

     History of geographical instruction at the University
     of Basel, Switzerland, from 1912, when Gustav Braun was
     named to the first chair of geography.

728. Wagner, Hermann. "Geographie," Chapter 11, pp. 127-140,
     in Vol. 2 of Die Deutschen Universitäten, ed. by W.
     Lexis (Berlin: Verlag von A. Asher & Co., 1893).

     An historical sketch of geography at German
     universities since the 15th century, with emphasis on
     Carl Ritter and his successors. Abstracted in Scottish
     Geographical Magazine, vol. 9 (1893), 366-371.

729. Wagner, Hermann. "Die Pflege der Geographie an der
     Berliner Universität im ersten Jahrhundert ihres
     Bestehens, 1810-1910," Petermanns Geographische
     Mitteilungen, vol. 56, part 2, no. 4 (October 1910),
     169-176.

     Geographical instruction at the University of Berlin
     began in the founding year of 1810 with August Zeune.
     Emphasis on the long tenure of Carl Ritter (1779-1859),
     who began teaching at the University in 1820. Other
     geographers treated in this article are Heinrich
     Kiepert and Ferdinand von Richthofen.

730. Wenk, Hans-Günther. "Die Geschichte der Geographie und
     der Geographischen Landesforschung an der Universität

Kiel von 1665 bis 1879," <u>Schriften des Geographischen Instituts der Universität Kiel</u>, vol. 24, no. 1 (1966). 253 p.

Geographical instruction and research in the University of Kiel from 1665 (Samuel Reyher) to 1879, when Theobald Fischer was appointed as the first Professor (Ordinarius) of geography.

731. Williams, Juliet. "The First Department," <u>Geographical Magazine</u>, vol. 44, no. 4 (January 1972), 232.

Oxford's School of Geography, established in 1899, "is the oldest autonomous geography department in any British university."

732. Wise, Michael J. "Man and His Environment," pp. 221-243 in <u>Man and the Social Sciences</u>, ed. William S. Robson (Twelve lectures delivered in 1972 at the London School of Economics and Political Science tracing the development of the social sciences during the present century) (Beverly Hills, California: Sage Publications, 1974).

Largely a history of the Joint School of Geography founded in 1922 at the London School of Economics and King's College, London, with emphasis on the work of Halford Mackinder and L. Dudley Stamp.

XII. TEACHING

(BELOW UNIVERSITY LEVEL)

733. Gibbs, David. "The Pedagogy of Geography," <u>Pedagogical Seminary</u> (Worcester, Massachusetts), vol. 14, no. 1 (March 1907), 39-100.

An historical review of geographical teaching methods and textbooks. Sections deal with geography in Europe and all levels of American education (primary and secondary schools, normal schools, colleges, and universities).

734. Gruber, Christian. <u>Die Entwickelung der geographischen Lehrmethoden im XVIII. und XIX. Jahrhundert: Rückblicke</u>

und Ausblicke. Munich and Leipzig: R. Oldenbourg, 1900. viii + 253 p.

The teaching of geography, especially in the schools, in the 18th and 19th centuries, with some attention to present condition and prospects.

735. Himer, Kurt. "Geographie im Schulunterricht des 18. Jahrhunderts," Geographische Zeitschrift, vol. 41, no. 3 (March 1935), 101-111.

Geographical instruction in German schools in the 18th century, with emphasis on Johann Hübner's Kurtze Fragen aus der Alten und Neuen Geographie biss auf gegenwärtige Zeit (various editions, 1696-1755).

736. Hodgson, H.B. "Notes on the History of the Teaching of Geography," Geography, vol. 22, part 1 (March 1937), 44-48.

Brief sketch of the teaching of geography, from ancient times to modern, with emphasis on England from the 16th to the end of the 19th century.

737. Potter, Jefferson R. "History of Methods of Instruction in Geography," Pedagogical Seminary (Worcester, Massachusetts), vol. 1 (1891), 415-424.

A brief sketch of the history of geography and of geographical instruction, with emphasis on the 18th and 19th centuries.

738. Vaughan, J.E. "Aspects of Teaching Geography in England in the early Nineteenth Century," Paedagogica Historica, vol. 12, no. 1 (1972), 128-147.

(Page 128) "Until the middle of the nineteenth century, geography had almost no position in the universities and only a very precarious place in the schools. An examination of some of the textbooks of the period can be used to illustrate some trends and problems of teaching geography before it became fully accepted as part of the school curriculum and established as a respectable academic subject in the universities."

PART TWO

GEOGRAPHY IN VARIOUS COUNTRIES

XIII. THE UNITED STATES

739. Abrahams, Paul P. "Academic Geography in America: An Overview," Reviews in American History, vol. 3, no. 1 (March 1975), 46-52.

A brief survey of academic geography in the United States, beginning with the work of John Wesley Powell and William Morris Davis. Based on books such as Preston James' All Possible Worlds (1972) and Richard Chorley, et al., The History of the Study of Landforms, Volume 2, The Life and Work of William Morris Davis (1973).

740. Allen, John L. "Working the West: Public Land Policy, Exploration, and the Preacademic Evolution of American Geography," pp. 57-68 in The Origins of Academic Geography in the United States, ed. Brian W. Blouet (Hamden, Connecticut: Archon Books, 1981).

Interpretation of the American West, with emphasis on Thomas Jefferson, Thomas Hart Benton, John C. Frémont, and John Wesley Powell.

741. Baker, Marcus. "A Century of Geography in the United States," Philosophical Society of Washington, Bulletin, vol. 13 (1898), 223-239.

Expansion of the United States and the concomitant expansion of transport, postal system, and mapping agencies. Mention of school geographies from the time of Jedidiah Morse.

742. Baker, Marcus. "Geographical Research in the United States," Geographical Journal, vol. 11, no. 1 (January 1898), 52-58.

Review of the work of "the greater geographical agencies of the United States"--General Land Office, Coast and Geodetic Survey, Army Engineers, Geological Survey, Navy, Weather Bureau, etc.

743. Bladen, Wilford A., and Karan, Pradyumna P., eds. The Evolution of Geographic Thought in America: A Kentucky Root. Dubuque, Iowa: Kendall/Hunt Publishing Company, 1983. xviii + 149 p.

The role of Kentucky and Kentuckians in the development of American geography, with emphasis on Nathaniel Southgate Shaler (1841-1906) and Ellen Churchill Semple (1863-1932). Considerable attention is given to Carl Sauer (1889-1975) and to the field station that he established in Kentucky in 1920.

744. Block, Robert H. "The Whitney Survey of California, 1860-74: A Study of Environmental Science and Exploration." Unpublished Ph.D. dissertation, Department of Geography, University of California, Los Angeles, 1982. x + 480 p.

(From Abstract, pp. viii-x) "From 1860 to 1874 American geologist Josiah Dwight Whitney directed the first comprehensive scientific reconnaissance of California.... Known officially as the California State Geological Survey, this organization of nineteenth-century scientists and surveyors searched out, collected, analyzed, described, and mapped the vast array of environmental phenomena found in California.... The Whitney Survey was an incubator for geographical ideas and concepts as well as for field methods and personnel that would later be of crucial importance to the environmental reconnaissance of the greater American West in the closing decades of the Nineteenth Century."

745. Blouet, Brian W., ed. The Origins of Academic Geography in the United States. Hamden, Connecticut: Archon Books, 1981. xii + 342 p.

Emphasis on period from about 1870 to 1946. Papers originally delivered at a conference at the University of Nebraska in April 1979. Papers listed separately under the authors' names in this bibliography.

Part Two--Geography in Various Countries 157

746. Blouet, Brian W., and Stitcher, Teresa L. "Survey of Early Geography Teaching in State Universities and Land Grant Institutions," pp. 327-342 in <u>The Origins of Academic Geography in the United States</u>, ed. Brian W. Blouet (Hamden, Connecticut: Archon Books, 1981).

Alphabetical list of 88 American universities with dates of founding, first teaching of geography, and establishment of geography departments.

747. Bowden, Martyn J. "The Cognitive Renaissance in American Geography: The Intellectual History of a Movement," pp. 65-70 in <u>Les écoles géographiques</u>, ed. Józef Babicz (Warsaw: PWN--Polish Scientific Publishers, 1980).

A survey of studies of geographical cognition (environmental perception) in the United States since 1925.

748. Broek, Jan O.M. "Neuere Strömungen in der amerikanischen Geographie," <u>Geographische Zeitschrift</u>, vol. 44, nos. 7-8 (15 July 1938), 249-258.

Geography in the United States, with emphasis on publications produced in the 1930s.

749. Bushong, Allen D. "Geographers and Their Mentors: A Genealogical View of American Academic Geography," pp. 193-219 in <u>The Origins of Academic Geography in the United States</u>, ed. Brian W. Blouet (Hamden, Connecticut: Archon Books, 1981).

Traces the mentor-student relationship from 1907, when the first Ph.D. degree was earned in an American department of geography, to 1946.

750. Bushong, Allen D. "Geography," pp. 513-516 in <u>The Encyclopedia of Southern History</u>, ed. David C. Roller and Robert W. Twyman (Baton Rouge: Louisiana State University Press, 1979).

Essay covers the history of academic geography in the southern United States as well as the history of the geographical exposition of the region.

751. Cappon, Lester J. "Geographers and Map-makers, British and American, from about 1750 to 1789," <u>Proceedings</u>

of the American Antiquarian Society, vol. 81, part 2 (1972), 243-271.

Geographers and cartographers in Great Britain and British North America (and the new United States) from about 1756 to 1789. Major emphasis on America, little on Britain.

752. Chester, C.M. "Some Early Geographers of the United States," National Geographic Magazine, vol. 15, no. 10 (October 1904), 392-404.

Geographical exploration by the United States Navy, from 1835 onward.

753. Colby, Charles C. "Changing Currents of Geographic Thought in America," Annals of the Association of American Geographers, vol. 26, no. 1 (March 1936), 1-37.

A survey of the geographical work of the state and federal governments, universities, and geographical societies in the United States since 1785, with emphasis on the period since 1852.

754. Cox, Kevin R. "American Geography: Social Science Emergent," Social Science Quarterly, vol. 57, no. 1 (June 1976), 182-207.

Treats the emergence of geography as a social science in the United States since 1900. This was a special issue of the SSQ, devoted to "Social Science in America: The First Two Hundred Years." Other authors dealt with History, Economics, Sociology, Political Science, and Anthropology.

755. Davis, William Morris. Die Erklärende Beschreibung der Landformen. Trans. A. Rühl. Leipzig and Berlin: B.G. Teubner, 1912. xvii + 565 p.

See especially the Foreword (pp. v-xii); Davis' inaugural lecture, "Methods of American Geographical Research," pp. xiii-xvii; and Chapter 1, "The Nature of Geography," pp. 1-21. In these pages Davis gives a succinct overview of American geographical methodology (or at least of his own, which he thought was the same thing). Describes some aspects of human geography as well as landform study. References to contributions of Whitney, Richthofen, Powell, and Gilbert in the

Part Two--Geography in Various Countries 159

American West. Book consists of lectures by Davis at the University of Berlin, 1908-1909.

756. Davis, William Morris. "The Progress of Geography in the United States," Annals of the Association of American Geographers, vol. 14, no. 4 (December 1924), 159-215.

Development of scientific geography in the area that is now the United States, beginning with Lewis Evans in 1755.

757. Dept, G.G. "L'étude et l'enseignement de la géographie aux Etats-Unis," Bulletin de la Société royale belge de géographie, vol. 50 (1926), 119-132.

Growth of geography in American universities and geographical societies, with emphasis on the 20th century. Brief mention of such forerunners as Lewis Evans, Lewis and Clark, and Arnold Guyot.

758. Dryer, Charles R. "A Century of Geographic Education in the United States," Annals of the Association of American Geographers, vol. 14, no. 3 (September 1924), 117-149.

Geography in schools and colleges (emphasis on former) in the United States since about 1800, with particular attention to textbooks.

759. Dunbar, Gary S. "Credentialism and Careerism in American Geography, 1890-1915," pp. 71-88 in The Origins of Academic Geography in the United States, ed. Brian W. Blouet (Hamden, Connecticut: Archon Books, 1981).

The professionalization of geography and its emergence as an academic discipline.

760. Friis, Herman R. "The Role of Geographers and Geography in the Federal Government, 1774-1905," pp. 37-56 in The Origins of Academic Geography in the United States, ed. Brian W. Blouet (Hamden, Connecticut: Archon Books, 1981).

The contributions of geographers associated with the United States government, from George Washington to the founding of the Association of American Geographers.

761. Fuchs, Gerhard. "Das Konzept der Ökologie in der amerikanischen Geographie: Am Beispiel der Wissenschaftstheorie zwischen 1900 und 1930," Erdkunde, vol. 21, no. 2 (June 1967), 81-93.

　　The concept of ecology in American human geography from about 1900 (Davis' ontography) to 1930 (Sauer's cultural morphology and Whittlesey's sequent occupance). Influence of the ecology of biologists (Haeckel, Cowles, etc.) and sociologists (Park and Burgess).

762. Fuchs, Gerhard. "Der Wandel zum anthropogeographischen Denken in der amerikanischen Geographie. Strukturlinien der geographischen Wissenchaftstheorie; dargestellt an den vorliegenden wissenschaftlichen Veröffentlichungen 1900-1930," Marburger Geographische Schriften, no. 32 (1966). 273 p.

　　Human geography in the United States from about 1900 (Davis and Semple) to 1930 (Sauer).

763. Gade, Daniel W. "L'optique culturelle dans la géographie américaine," Annales de géographie, vol. 85, no. 472 (November-December 1976), 672-693.

　　The development of cultural geography in the United States since the 1920s, with emphasis on the "Berkeley School" founded by Carl Sauer of the University of California.

764. Geiser, Samuel Wood. "Geographers of Early Texas: A Bibliographic Note," Texas Geographic Magazine, vol. 7, no. 2 (Autumn 1943), 37-38.

　　(Page 37) "For the period 1820 to 1880, Texas can claim nearly fifty geologists, topographers [excluding Army Engineers and Coast Survey], cartographers, and geographers." Among the most notable were William Bollaert, Ernst Kapp, Edward Belcher, and Julius Froebel.

765. Herbst, Jurgen. "Social Darwinism and the History of American Geography," Proceedings of the American Philosophical Society, vol. 105, no. 6 (15 December 1961), 538-544.

　　Seeks to explain American geography's "lingering sickness" in the 1920s and 1930s and also why European geography showed no similar signs of malaise. Answer is found in the relative strength of Social Darwinism, which was rejected early by the Europeans.

766. Hudson, John C., ed. "Seventy-Five Years of American Geography," <u>Annals of the Association of American Geographers</u>, vol. 69, no. 1 (March 1979), 1-185.

    Essays on American geography in the 20th century by 27 contributors, commemorating the 75th anniversary of the founding of the Association of American Geographers in 1904.

767. James, Preston E. "Continuity and Change in American Geographic Thought," Chapter 1, pp. 3-14, in <u>Problems and Trends in American Geography</u>, ed. Saul B. Cohen (New York: Basic Books, Inc., 1967).

    Brief overview of American geography, from Jedidiah Morse onward. Author seeks to reconcile the disparate branches of geography.

768. James, Preston E. "The Development of Professional Geography in the United States (1885-1940)," pp. 49-63 in <u>Les écoles géographiques</u>, ed. Józef Babicz (Warsaw: PWN--Polish Scientific Publishers, 1980).

    The professionalization of American geography in the period 1885-1940. (Page 50) "In the United States the first opportunity for advanced study in a field called geography was offered at Harvard after 1885."

769. James, Preston E. "Geographical Ideas in America, 1890-1914," pp. 319-326 in <u>The Origins of Academic Geography in the United States</u>, ed. Brian W. Blouet (Hamden, Connecticut: Archon Books, 1981).

    Examines "some of the sources of geographical ideas in America between 1890 and 1914, at a time when professional geography was being formed in this country" (p. 319).

770. James, Preston E. "The Nature and Scope of Geography," <u>Geographical Perspectives</u> (University of Northern Iowa), no. 33 (Spring 1974), 5-19.

    The development of modern geography, from the 1870s onward, with heavy emphasis on the United States.

771. James, Preston E., and Jones, Clarence F., eds. American Geography: Inventory and Prospect. Syracuse, New York: Syracuse University Press, 1954. xii+590 p.

Describes advances in the subfields of geography in the United States in the half-century since the founding of the Association of American Geographers in 1904. Some chapters go back to the beginning of the Republic, and others start with Old World antecedents. Some of the chapters have important historical sections--e.g., Chapter 14, pp. 334-361, "Climatology," by John Leighly, the major part of which is devoted to "The Growth of Climatology in the United States."

772. James, Preston E., and Mather, Cotton. "The Role of Periodic Field Conferences in the Development of Geographical Ideas in the United States," Geographical Review, vol. 67, no. 4 (October 1977), 446-461.

(Page 456) "Field conferences held" in the Middle West by university geographers "between 1923 and 1940 prompted competitive discussion of ideas concerning geographical objectives and methods." The younger geographers held separate "junior conferences" from 1926 to 1932, but the junior and senior conferences were combined beginning in 1935.

773. Joerg, W.L.G. "The Geography of North America: A History of Its Regional Exposition," Geographical Review, vol. 26, no. 4 (October 1936), 640-663.

History of regional geographies of North America, written by Europeans and North Americans, 1877-1934. Emergence of geography in the universities of Europe and North America.

774. Koelsch, William A. "The Enlargement of a World: Harvard Students and Geographical Experience, 1840-1861," Unpublished Ph.D. thesis, Department of History, University of Chicago, 1966.

A study of geographical curiosity in the minds of 19th-century academics and students as revealed in traces of their environmental cognition and behavior. Emphasizes the identification of the textures of educational experience in formal and informal, self-defined learning contexts, and in locales from Harvard

Part Two--Geography in Various Countries 163

Yard and the New England region to Europe and around the world. [WAK]

775. Koelsch, William A. "Terrae Incognitae and Arcana Siwash: Toward a Richer History of Academic Geography," pp. 63-87 in Geographies of the Mind: Essays in Historical Geosophy in Honor of John Kirtland Wright, ed. David Lowenthal and Martyn J. Bowden (New York: Oxford University Press, 1976).

Review of histories of American geography, American geography in college histories, and the geographical experiences of college students (with examples from Harvard in the 1840s and 1850s).

776. Leighly, John. "What Has Happened to Physical Geography?," Annals of the Association of American Geographers, vol. 45, no. 4 (December 1955), 309-318.

An historical survey of physical geography in the United States since 1890.

777. Lewis, G. Malcolm. "Amerindian Antecedents of American Academic Geography," pp. 19-35 in The Origins of Academic Geography in the United States, ed. Brian W. Blouet (Hamden, Connecticut: Archon Books, 1981).

Examines the geographical knowledge of American Indians, as expressed chiefly in maps.

778. Marcus, Melvin G. "Coming Full Circle: Physical Geography in the Twentieth Century," Annals of the Association of American Geographers, vol. 69, no. 4 (December 1979), 521-532.

"A cursory review of American physical geography in this century."

779. Martin, Geoffrey J. "Ontography and Davisian Physiography," pp. 279-289 in The Origins of Academic Geography in the United States, ed. Brian W. Blouet (Hamden, Connecticut: Archon Books, 1981).

W.M. Davis used the word "ontography" from 1902 onward to refer to the study of organic responses to physical controls. Davis himself stuck to geomorphology and left ontography to his disciples.

780. Mikesell, Marvin W. "Continuity and Change," pp. 1-15 in The Origins of Academic Geography in the United States, ed. Brian W. Blouet (Hamden, Connecticut: Archon Books, 1981).

A general essay prefacing the volume. (Page 15) "The history of American geography is best described ... as a gradual, cumulative venture.... Continuity has always been evident in spite of occasional claims of revolution."

781. Parkins, Almon. "The Geography of American Geographers," Journal of Geography, vol. 33, no. 6 (September 1934), 221-230.

Definitions of geography. Survey of more than 40 American geographers. Parkins' correspondence file was deposited in the George Peabody College library and was not to be opened until 1954.

782. Pattison, William D. "The Four Traditions of Geography," Journal of Geography, vol. 63, no. 5 (May 1964), 211-216.

(Page 211) "... the work of American geographers ... has exhibited a broad consistency, and ... this essential unity has been attributable to a small number of distinct but affiliated traditions ... (1) a spatial tradition, (2) an area studies tradition, (3) a man-land tradition and (4) an earth science tradition."

783. Pfeifer, Gottfried. "Entwicklungstendenzen in Theorie und Methode der regionalen Geographie in den Vereinigten Staaten nach dem Kriege," Zeitschrift der Gesellschaft für Erdkunde zu Berlin, 1938, pp. 93-125.

American geography in the 20th century, with emphasis on the 2 decades since World War I and on regional geography. English translation by John Leighly, "Regional Geography in the United States since the War: A Review of Trends in Theory and Method," printed by the American Geographical Society in 1938 (37 pages).

784. Platt, Robert S., ed. "Field Study in American Geography: The Development of Theory and Method Exemplified by Selections," University of Chicago, Department of Geography, Research Paper, No. 61 (July 1959). 405 p.

Part Two--Geography in Various Countries   165

(Page 1) "The project represented by this book is to analyze the development of American field study and recognize the stages through which it has come thus far, by means of representative substantive studies made by American geographers at different times and places [1805-1957]."

785. Platt, Robert S. "The Rise of Cultural Geography in America," pp. 485-490 in International Geographical Union, Proceedings of the Eighth General Assembly and Seventeenth International Congress, Washington, D.C., August 8-15, 1952 (Washington: The United States National Committee of the International Geographical Union, National Academy of Sciences-National Research Council, n.d. [1952]).

Human geography in the United States since 1915.

786. Rugg, Dean S. "The Midwest as a Hearth Area in American Academic Geography," pp. 175-191 in The Origins of Academic Geography in the United States, ed. Brian W. Blouet (Hamden, Connecticut: Archon Books, 1981).

More than any other section of the United States, the Midwest has played a dominant role in the field of geography. This paper explains the importance of the Midwest in the production of geographers and the diffusion of geographical ideas.

787. Sauer, Carl O. "On the Background of Geography in the United States," pp. 59-70 in Heidelberger Studien zur Kulturgeographie: Festgabe zum 65. Geburtstag von Gottfried Pfeifer, ed. Hans Graul and Hermann Overbeck (Heidelberger Geographische Arbeiten, no. 15) (Wiesbaden: Franz Steiner Verlag, 1966).

Geographers and geographical work in the United States from the time of Thomas Jefferson to about 1915.

788. Smith, J. Russell. "American Geography 1900-1904," Professional Geographer, vol. 4, no. 4 (July 1952), 4-7.

Geography in American universities at the beginning of the century.

789. Spencer, Joseph E. "The Evolution of the Discipline of Geography in the Twentieth Century," Geographical

Perspectives (University of Northern Iowa), no. 33 (Spring 1974), 20-36.

The growth of academic geography in the United States in the 20th century.

790. Speth, William W. "Berkeley Geography, 1923-33," pp. 221-244 in *The Origins of Academic Geography in the United States*, ed. Brian W. Blouet (Hamden, Connecticut: Archon Books, 1981).

The Department of Geography at the University of California, Berkeley, in the first decade of Carl Sauer's leadership.

791. Stoddart, David R. "Darwin's Influence on the Development of Geography in the United States, 1859-1914," pp. 265-278 in *The Origins of Academic Geography in the United States*, ed. Brian W. Blouet (Hamden, Connecticut: Archon Books, 1981).

(Page 265) "... the impact of Darwinian views on academic geography in the United States in the half-century following the publication of *The Origin of Species*."

792. Thouez, Jean-Pierre. *L'évolution de la pensée géographique aux Etats-Unis*. Université de Montréal, Département de géographie, Notes et Documents, No. 80-04 (May 1980). 33 p.

A brief overview of some of the major methodological controversies in American geography in the 20th century.

793. Visher, Stephen S. "A Brief History of Geography in Indiana," *Proceedings of the Indiana Academy of Science*, vol. 76 (1966), 95-102.

Contributions to the geography of Indiana by numerous geographers and others, with particular emphasis on work done by people at Indiana University.

794. Visher, Stephen S. "Notable Contributors to American Geography," *Professional Geographer*, vol. 17, no. 3 (May 1965), 25-29; vol. 18, no. 4 (July 1966), 227-229.

Brief biographical sketches of 77 American geographers, all of whom had been members of the Association of American geographers.

Part Two--Geography in Various Countries   167

795. Warntz, William. "Geographia Generalis and the Earliest Development of American Academic Geography," pp. 245-263 in The Origins of Academic Geography in the United States, ed. Brian W. Blouet (Hamden, Connecticut: Archon Books, 1981).

The influence of Varenius' Geographia Generalis (1650) in the colleges of Colonial America.

796. Warntz, William. Geography Now and Then: Some Notes on the History of Academic Geography in the United States. (American Geographical Society, Research Series, No. 25.) New York: American Geographical Society, 1964. 162 p.

(Preface, unpaged) "Survey of the materials related to the history of geography from its colonial beginnings to the present ... with special emphasis on geography's characteristics and its role in the curriculum of the pre-Revolutionary War colleges."

797. Whitbeck, Ray H. "Thirty Years of Geography in the United States," Journal of Geography, vol. 20, no. 4 (April 1921), 121-128.

The vicissitudes of geography in the American educational system, from elementary schools to universities, since 1890.

XIV. GREAT BRITAIN

798. Baker, J.N.L. "Academic Geography in the Seventeenth and Eighteenth Centuries," Scottish Geographical Magazine, vol. 51, no. 3 (May 1935), 129-144.

Paper "is largely concerned with the University of Oxford."

799. Baker, J.N.L. "The Development of Historical Geography in Britain during the Last Hundred Years," Advancement of Science, vol. 8, no. 32 (March 1952), 406-412.

The practice and writing of historical geography in Britain, with emphasis on Oxford and Edinburgh, beginning in 1842 with the publication of James Laurie's System of Universal Geography.

800. Crone, Gerald R. "British Geography in the Twentieth Century," Geographical Journal, vol. 130, part 2 (June 1964), 197-220.

   The development of geography in Great Britain in the first 6 decades of the 20th century. Divided into the following periods: I, 1900-1914; II, 1914-1939; III, 1939-1964.

801. Darby, H.C. "Academic Geography in Britain: 1918-1946," Transactions of the Institute of British Geographers, n.s. vol. 8, no. 1 (1983), 14-26.

   A chronicle of developments in British geography, with emphasis on the universities, between the end of World War I and 1946, when a new era of expansion was beginning. Based in great part on the recollections of the author, former head of the departments of geography at Liverpool, London (University College), and Cambridge.

802. Darby, H.C. "Historical Geography in Britain, 1920-1980: Continuity and Change," Transactions of the Institute of British Geographers, n.s. vol. 8 (1983), 421-428.

   Historical geography and its practitioners in Great Britain from the 1920s to the present, with considerable autobiographical matter concerning the author, who was professor of geography in Liverpool, London (University College), and Cambridge and wrote the monumental Domesday Geography of England (7 vols., 1952-1977).

803. Davies, Gordon L. The Earth in Decay: A History of British Geomorphology, 1578-1878. London: MacDonald & Co. (Publishers) Ltd., 1969. xvi + 390 p.

   The study of landforms in Britain, beginning with William Bourne's Booke Called the Treasure for Traveilers (1578) and ending with T.H. Huxley's Physiography (1877).

804. Dickinson, Robert E. "Die gegenwärtigen Strömungen der britischen Geographie," Geographische Zeitschrift, vol. 44, nos. 7-8 (15 July 1938), 258-269.

   A review of the last half-century of British geography, with emphasis on publications of the 1930s.

Part Two--Geography in Various Countries   169

805. Dury, George H. "Geography and Geomorphology: The Last Fifty Years," <u>Transactions</u> <u>of</u> <u>the</u> <u>Institute</u> <u>of</u> <u>British</u> <u>Geographers</u>, n.s. vol. 8, no. 1 (1983), 90-99.

Geomorphology in Great Britain in the 20th century, with observations on the place of geomorphology vis-à-vis geography and geology and on the differences between geomorphology as practiced in Great Britain, on the one hand, and North America and Australia, on the other.

806. Edwards, K.C. "Sixty Years after Herbertson: The Advance of Geography as a Spatial Science," <u>Geography</u>, vol. 59, part 1 (January 1974), 1-9.

Changes that the author has observed in British geography since the time of A.J. Herbertson (d. 1915).

807. Farmer, B.H. "British Geographers Overseas, 1933-1983," <u>Transactions</u> <u>of</u> <u>the</u> <u>Institute</u> <u>of</u> <u>British</u> <u>Geographers</u>, n.s. vol. 8, no. 1 (1983), 70-79.

(From Abstract, p. 70) "This essay surveys the activities of British professional geographers in what is now known as 'the Third World' during the period since the foundation of the Institute of British Geographers fifty years ago."

808. Freeman, T. Walter. "The British School of Geography," pp. 71-82 in <u>Les</u> <u>écoles</u> <u>géographiques</u>, ed. Józef Babicz (Warsaw: PWN--Polish Scientific Publishers, 1980).

British geography in the 19th and 20th centuries, with emphasis on developments in the universities since 1887.

809. Freeman, T.W. <u>A</u> <u>History</u> <u>of</u> <u>Modern</u> <u>British</u> <u>Geography</u>. London and New York: Longman, 1980. ix + 258 p.

History of geography in Great Britain from 1870 onward. Biographical sketches of nearly 70 British geographers on pp. 205-239.

810. Gilbert, Edmund W. <u>British</u> <u>Pioneers</u> <u>in</u> <u>Geography</u>. Newton Abbot, England: David & Charles, 1972. 271 p.

A collection of 12 essays, mostly previously published, on British geographers and geographical writers from Richard Hakluyt (1552-1616) to the 20th century. Emphasis on Oxford-related geographers.

811. Grady, Alison Dorothy. "The Role of Geographical Societies in the Development of Geography in Britain from 1900-1914." Unpublished Ph.D. thesis in Geography, University of London, Birkbeck College, 1971. 321 p.

(From Abstract, p. 2) "In 1900 the acknowledged and established centres of geographical work were the geographical societies, and it is through their work and influence that the development of geography in the first fourteen years of the century is examined.... After an assessment of the position of geography in 1900, its progress is shown in relation to the various areas of development, namely the universities, the extra-mural departments, the teacher training colleges, the schools and the field of research and publication."

812. Gregory, Stanley. "Quantitative Geography: The British Experience and the Role of the Institute," Transactions of the Institute of British Geographers, n.s. vol. 8, no. 1 (1983), 80-89.

A review of the application of statistical methods to problems in physical and human geography in Great Britain, with emphasis on the last quarter-century.

813. Gribaudi, Pietro. "La geografia in Inghilterra: progressi metodologici e didattici," Rivista Geografica Italiana, vol. 9, no. 5 (May 1902), 324-335.

Recent developments in British geography.

814. Huender, Wilhelmina Johanna. De Engelsche Geographie in de 20ste Eeuw. Utrecht and Amsterdam: Drukkerij J. van Boekhoven, 1934. x + 183 p.

British geography from medieval times onward, with emphasis on the views of 18 geographers in the 20th century.

815. Keltie, John Scott. The Position of Geography in British Universities. (American Geographical Society, Research Series, no. 4.) New York: Oxford University Press, 1921. 33 p.

The history and current status of geography in British universities, following the Royal Geographical

Part Two--Geography in Various Countries   171

   Society's investigation of the status of geography in 1884-1885.

816. Keltie, J. Scott. "Thirty Years Progress in Geographical Education," Geographical Teacher, vol. 7, part 4 (Spring 1914), 215-227.

   Improvement in geographical instruction in British schools and universities since 1884.

817. Lochhead, Elspeth Nora. "The Emergence of Academic Geography in Britain in Its Historical Context." Unpublished Ph.D. thesis in Geography, University of California, Berkeley, 1980. xii + 681 p.

   (Page 1) "This study analyses the emergence of geography in Britain, and the subsequent development of its identity as a separate and distinct discipline, relating this process to the wider trends of events occurring within Britain as a whole, and to the internal philosophical evolution of the discipline." Emphasis is on the period 1870-1914. Extensive bibliography (pp. 617-681).

818. Lochhead, Elspeth N. "Scotland as the Cradle of Modern Academic Geography in Britain," Scottish Geographical Magazine, vol. 97, no. 2 (September 1981), 98-109.

   Describes the role of Scots in the emergence of academic geography in Great Britain in the period 1880-1914. Emphasis on Edinburgh and the Royal Scottish Geographical Society.

819. [Markham, Clements R.]. "Review of British Geographical Work during the Last Hundred Years (1789-1889)," Royal Geographical Society, Supplementary Papers, vol. 3 (1893), 149-199.

   Based on general sketch prepared by Markham and "supplemented with a few additional details by Mr. J. Scott Keltie." Mostly concerned with the work of British explorers since Cook, with a few pages devoted to cartography and marine research.

820. Newbigin, Marion I. "Geography in Scotland since 1889," Scottish Geographical Magazine, vol. 29, no. 9 (September 1913), 471-479.

Progress of geography in the 23 years since A.S. White's report (below). Mapping and natural history surveys in Scotland, geographical education, publishers, work of Royal Scottish Geographical Society, oceanography, and exploration.

821. Robertson, Charles J. "Scottish Geographers: The First Hundred Years," Scottish Geographical Magazine, vol. 89, no. 1 (April 1973), 5-18.

    Scottish topographers (or compilers of geographical information) in the 18th and 19th centuries, "protogeographers" (with emphasis on Patrick Geddes), and university teachers of geography (emphasis on Marion Newbigin, George Chisholm, Alan Ogilvie, John Walter Gregory, Alexander Stevens, John McFarlane, and Andrew O'Dell).

822. Rodrigues, Fátima. "Aspectos da actual geografia inglesa," Finisterra (Lisbon), vol. 14, no. 27 (1979), 5-35.

    Teaching and research in geography in Great Britain since 1960.

823. Rudmose Brown, R.N. "Scotland and Some Trends in Geography: John Murray, Patrick Geddes and Andrew Herbertson," Geography, vol. 33, part 3 (September 1948), 107-120.

    Scottish influences on geography and especially on Andrew Herbertson (1865-1915). Extended discussions of John Murray (1841-1914) and Patrick Geddes (1854-1932).

824. Stoddart, David R. "Geography, Education and Research," Geographical Journal, vol. 147, no. 3 (November 1981), 287-297.

    The author speaks "of the way in which the RGS [Royal Geographical Society] mobilized its resources in the nineteenth century to persuade the universities of the worth of the subject as an academic discipline; of the way in which, after slow beginnings, academic geography has flourished and prospered; and of the dominant figures in our history to whom we owe these developments" (p. 287).

Part Two--Geography in Various Countries   173

825. White, Arthur Silva. "On the Achievements of Scotsmen during the Nineteenth Century in the Fields of Geographical Exploration and Research," <u>Scottish Geographical Magazine</u>, vol. 5, no. 9 (September 1889), 480-497; no. 10 (October), 540-549; and no. 11 (November), 595-605.

The work of Scottish explorers and scientists (geologists, biologists, meteorologists, oceanographers, etc.) in the 19th century. Johnston and Bartholomew in cartography. Founding of Royal Scottish Geographical Society in 1884.

826. Wise, Michael J. "Three Founder Members of the I.B.G.: R. Ogilvie Buchanan, Sir Dudley Stamp, S.W. Wooldridge. A Personal Tribute," <u>Transactions of the Institute of British Geographers</u>, n.s. vol. 8, no. 1 (1983), 41-54.

Author's personal recollections of three founder members of the Institute of British Geographers, all of whom were prominent in the growth of geography in Great Britain, particularly in the post-World War II years: Robert Ogilvie Buchanan (1894-1980), Laurence Dudley Stamp (1898-1966), and Sidney William Wooldridge (1900-1963).

827. Wright, John K. "Some British 'Grandfathers' of American Geography," pp. 144-165 in <u>Geographical Essays in Memory of Alan G. Ogilvie</u>, ed. Ronald Miller and J. Wreford Watson (London, etc.: Thomas Nelson and Sons Ltd., 1959).

Article concerns the British intellectual forbears of Jedidiah Morse (1761-1826), "The Father of American Geography." Included are Patrick Gordon (fl. 1700), Thomas Salmon (1679-1767), William Guthrie (1708-1770), and John Pinkerton (1758-1826).

XV. FRANCE

828. Berdoulay, Vincent R.H. <u>La formation de l'école française de géographie (1870-1914)</u>. (Comité des travaux historiques et scientifiques, Mémoires de la

section de géographie, 11.) Paris: Bibliothèque Nationale, 1981. 245 p.

The emergence of modern academic geography in France, with emphasis on Paul Vidal de la Blache and his students before World War I. Revised version of author's Ph.D. thesis, "The Emergence of the French School of Geography (1870-1914)," University of California, Berkeley, 1974.

829. Berdoulay, Vincent. "The Vidal-Durkheim Debate," Chap. 5, pp. 77-90 in Humanistic Geography: Prospects and Problems, ed. David Ley and Marwyn S. Samuels (Chicago: Maaroufa Press, Inc., 1978).

Concerns "the intellectual debate in France at the turn of the century between Vidalian geography and Durkheimian sociology" (Paul Vidal de la Blache, 1845-1918, and Emile Durkheim, 1858-1917).

830. Broc, Numa. "Les débuts de la géomorphologie en France: le tournant des années 1890," Revue d'histoire des sciences, vol. 28, no. 1 (January 1975), 31-60.

Modern geomorphology in France owes its origins to works such as Les formes du terrain (1888) by E. de Margerie and G.O. de La Noë and Leçons de géographie physique (1896) by A. de Lapparent.

831. Broc, Numa. "L'établissement de la géographie en France: diffusion, institutions, projets (1870-1890)," Annales de géographie, vol. 83, no. 459 (September-October 1974), 545-568.

The growth of geography in France, 1870-1890, with special attention to the work of Ludovic Drapeyron (1839-1901).

832. Broc, Numa. La géographie des philosophes: géographes et voyageurs français au XVIIIe siècle. Paris: Editions Ophrys, 1975. 595 p.

The geographical researches and writings of French travelers and scientists in the 18th century. Divided into two parts: "La fin de l'humanisme en géographie (1700-1765)" and "Nouveaux regards sur le monde (1765-fin du siècle)." Doctoral thesis (Doctorat de l'état), University of Montpellier, 1972. Rich bibliography.

833. Broc, Numa. "La géographie française face à la science allemande (1870-1914)," Annales de géographie, vol. 86, no. 473 (January-February 1977), 71-94.

The development of French geography on the German model in the interwar years.

834. Broc, Numa. "Les grandes missions scientifiques françaises au XIXe siècle (Morée, Algérie, Mexique) et leurs travaux géographiques," Revue d'histoire des sciences et de leurs applications, vol. 34, nos. 3-4 (July-October 1981), 319-358.

Scientific missions that accompanied French military expeditions to Greece, Algeria, and Mexico in the 19th century.

835. Broc, Numa. "L'histoire de la géographie au XVIIIe siècle," L'information historique, vol. 30 (1968), 65-70.

Geographical researches and writings in 18th-century France, with emphasis on physical geography.

836. Broc, Numa. Les montagnes vues par les géographes et les naturalistes de langue française au XVIIIe siècle: Contribution à l'histoire de la géographie. (Ministère de l'éducation nationale, Comité des travaux historiques et scientifiques, Mémoires de la section de géographie, 4.) Paris: Bibliothèque Nationale, 1969. 298 p.

The study of mountains by French geographers and naturalists in the 18th century. Mostly French examples, from the Alps and Pyrenees. Third-cycle doctoral thesis in geography, University of Montpellier, 1966.

837. Broc, Numa. "La pensée géographique en France aux XIXe siècle: continuité ou rupture?," Revue géographique des Pyrénées et du Sud-Ouest, vol. 47, no. 3 (July 1976), 225-247.

Shows the intellectual antecedents of Paul Vidal de la Blache and other French geographers of the late 19th century. Traces the problems of determinism and regionalism.

838. Broc, Numa. "Peut-on parler de géographie humaine au XVIIIe siècle en France," Annales de géographie, vol. 78, no. 425 (January-February 1969), 57-75.

Although it is often said that human geography is a creation of the late 19th century, the author cites geographical passages in the works of Montesquieu, Buffon, Rousseau, Ramond, and Volney.

839. Brouillette, Benoît. "Les instituts de géographie en France," Revue trimestrielle canadienne, vol. 20, no. 77 (March 1934), 73-94.

Describes personnel, programs, and facilities of the geographical establishments of the University of Paris, Collège de France, University of Lyon, and University of Grenoble.

840. Buffin, Frédéric. "Une 'réussite géographique': le Massif central, ou petite histoire de la naissance d'une région géographique," France, Ministère des universités, Comité des travaux historiques et scientifiques, Bulletin de la section de géographie, no. 84 (1979--published 1981), 173-186.

The history of a regional concept--the Massif Central region of France--as studied by historians and geographers in the 19th and 20th centuries.

841. Buttimer, Anne. Society and Milieu in the French Geographic Tradition. (Association of American Geographers, Monograph Series, 6.) Chicago: Rand McNally and Company, 1971. xi + 226 p.

French human geography in the first half of the 20th century, with emphasis on the work of Paul Vidal de la Blache and his students. Outgrowth of author's Ph.D. thesis in Geography, University of Washington, 1965.

842. Chabot, Georges. "Les conceptions françaises de la région géographique," Finisterra (Lisbon), vol. 2 (1967), 5-16.

The development of the regional concept in French geography since the time of Philippe Buache (1700-1773).

843. Chabot, Georges. "Les conceptions françaises de la science géographique," <u>Norsk Geografisk Tidsskrift</u>, vol. 12, nos. 7-8 (1950), 309-321.

Development of modern French geography, with emphasis on Paul Vidal de la Blache and his students.

844. Chabot, Georges; Clozier, René; and Beaujeu-Garnier, Jacqueline, eds. <u>La géographie française au milieu du XXe siècle</u>. ("L'information géographique.") Paris: J.-B. Baillière et fils, éditeurs, 1957. 333 p.

Survey of work done in subfields of geography in France since 1940. 38 collaborators.

845. Chevalier, Michel. "Die französische Schule der Geographie," <u>Wort und Tat</u>, no. 6 (September 1947), 33-53.

French geography since the end of the 19th century, with emphasis on Paul Vidal de la Blache and his students. An article based on Chevalier's essay was published by Carl Troll in <u>Erdkunde</u>, vol. 2, 1948, pp. 344-345; "Die französische Schule der Geographie in den Augen eines französischen Geographen."

846. Claval, Paul. "Le commentaire de cartes et le développement de la géographie française," France, Ministère des universités, Comité des travaux historiques et scientifiques, <u>Bulletin de la section de géographie</u>, no. 84 (1979--published 1981), 163-172.

The place of the analysis of topographic maps in the development of French geography during the last 100 years.

847. Claval, Paul C. "One Hundred Years of Teaching Geography in French Universities," <u>Journal of Geography</u>, vol. 82, no. 3 (May-June 1983), 110-111.

A brief survey of geographical instruction in French universities since the 1870s.

848. Dainville, François de. "Enseignement des 'géographes' et des 'géometres,'" pp. 481-491 in <u>Enseignement et diffusion des sciences en France au XVIIIe siècle</u>, ed. René Taton (Ecole pratique des hautes études,

Sorbonne, "Histoire de la pensée," vol. 11) (Paris: Hermann, 1964).

Practical training in the geographical sciences in the 18th century, with emphasis on surveying and mapping.

849. Ellenberger, François. "De l'influence de l'environnement sur les concepts: l'exemple des théories géodynamiques au XVIIIe siècle en France," <u>Revue d'histoire des sciences et de leurs applications</u>, vol. 33, no. 1 (January 1980), 33-68.

Examines the influence of the environment on the work of the principal French naturalists of the 18th century.

850. Harrison Church, R.J. "The French School of Geography," Chapter 3, pp. 70-90, in <u>Geography in the Twentieth Century</u>, ed. Griffith Taylor, 3rd ed. (New York: Philosophical Library, 1957).

Geography in France since 1870, with emphasis on Paul Vidal de la Blache and his students. Portuguese translation published in <u>Boletim geográfico</u> (Rio de Janeiro), vol. 18, no. 158 (September-October 1960), 784-797.

851. Jaja, Goffredo. "L'Insegnamento della geografia in Francia," <u>Bollettino della Società Geografica Italiana</u>, series 4, vol. 7, no. 11 (November 1906), 1066-1086.

Teaching of geography at the university level in France, with emphasis on the Sorbonne. Little historical depth in the article.

852. Lukermann, Fred. "The 'Calcul des probabilités' and the Ecole française de géographie," <u>The Canadian Geographer</u>, vol. 9, no. 3 (1965), 128-137.

(Page 128) "The specific purpose is to explicate as fully as possible the historical and intellectual context of the period 1884-1927 within which the French School of Geography initially created and expressed its view of the world as a particular geographic foundation." Emphasis on the work of Paul Vidal de la Blache and his colleagues and students, and especially on the theory of probability (<u>contingence</u>) of A.-A. Cournot (1801-1877).

853. Maret, Marie-Paule, and Pinchemel, Philippe. "L'évolution des questions de géographie aux concours d'agrégation des origines à 1914: contribution à l'histoire de la pensée géographique," pp. 77-86 in La pensée géographique française contemporaine: mélanges offerts à André Meynier (Saint-Brieuc: Presses Universitaires de Bretagne, 1972).

Types of geographical questions asked on the agrégation examinations, 1830-1914. Profound change in 1894 with the appearance of the "new geography" of Paul Vidal de la Blache.

854. Margerie, Emmanuel de. "L'oeuvre des géographes français depuis cent ans," Bulletin de la Société royale belge de géographie, vol. 50, nos. 3-4 (1926), 199-215.

Sketch of developments in French geography, cartography, and geology since the founding of the Paris Geographical Society in 1821.

855. Martin, Geoffrey J. "The Region in French Geographic Thought, c. 1900-1930," Papers of the Michigan Academy of Science, Arts, and Letters, vol. 49 (1964), 325-332.

French regional geography in the early 20th century, with emphasis on the writings of Paul Vidal de la Blache and his students.

856. Martonne, Emmanuel de. Geography in France. (American Geographical Society, Research Series, No. 4a.) New York: American Geographical Society, 1924. vii + 70 p.

Geography in France since 1871, with chapters on geographical societies and kindred institutions, geography in the universities (especially Paris), government departments contributing to geography, and the leading geographical journals and series.

857. Martonne, Emmanuel de. "La science géographique," pp. 375-396 in Vol. 1 of La science française (Paris: Librairie Larousse, 1915). New ed. 1933.

Brief history of geography in France since the 17th century.

858. McKay, Donald V. "Colonialism in the French Geographical Movement 1871-1881," Geographical Review, vol. 33, no. 2 (April 1943), 214-232.

French expansionist sentiment in the 1870s, with particular emphasis on the work of the geographical societies in Paris and the provinces and the foundation of the Revue de géographie (1877).

859. Meynier, André. Histoire de la pensée géographique en France (1872-1969). (Collection SUP, "Le géographe," ed. Pierre George.) Paris: Presses Universitaires de France, 1969. 223 p.

History of modern French geography, divided into 3 periods: "Le temps de l'éclosion," 1872-1905; "Le temps de l'intuition," 1905-1939; and "Le temps de craquements," 1939-1969.

860. Moravia, Sergio. "Philosophie et géographie à la fin du XVIIIe siècle," pp. 937-1011 in Vol. 57 of Studies on Voltaire and the Eighteenth Century, ed. Theodore Bestermann (Transactions of the Second International Congress on the Enlightenment, 3) (Geneva: Institut et Musée Voltaire, 1967).

Emphasizes the importance of the ideologues who carried on after the deaths of the great Enlightenment philosophes in the 1770s and 1780s. Especially important were the geographies and travel accounts of such men as Mentelle, J.N. Buache, and Volney. Volney's Voyage en Egypte et en Syrie (1787) is given particular attention.

861. Musset, René. "Der Stand der Geographie und ihre neueren wissenschaftlichen Strömungen in den Ländern französischer Zunge," Geographische Zeitschrift, vol. 44, nos. 7-8 (July 1938), 269-277.

French geography in the 20th century.

862. Ozouf, Marie-Vic. "La Limagne: évolution d'une notion régionale dans la littérature géographique," France, Ministère des universités, Comité des travaux historiques et scientifiques, Bulletin de la section de géographie, no. 84 (1979--published 1981), 187-202.

Evolution of a French region, the Limagne, in the geographical literature of the last 150 years.

863. Perpillou, Aimé. "Geography and Geographical Studies in France during the War and the Occupation," *Geographical Journal*, vol. 107, nos. 1-2 (January-February 1946), 50-57.

    Geography in France during the years 1940-1944.

864. Pinchemel, Philippe, and Pinchemel, Geneviève. "Geographers and the City: A Contribution to the History of Urban Geography in France," pp. 295-318 in *The Expanding City: Essays in Honour of Professor Jean Gottmann*, ed. John Patten (London, etc.: Academic Press, 1983).

    French interest in urban geography from the time of Léon Lalanne (1863) onward.

865. Rhein, Catherine. "La géographie, discipline scolaire et/ou science sociale (1860-1920)?," *Revue française de sociologie*, vol. 23, no. 2 (April-June 1982), 223-251.

    Emergence of geography as a school and university subject in France after 1860. Review of differences between Vidalian geography and Durkheimian sociology in the early 20th century.

XVI. GERMANY

866. Beck, Hanno. "Geographie und Statistik--die Lösung einer Polarität," pp. 269-276 (followed by discussion, pp. 277-281) in *Statistik und Staatsbeschreibung in der Neuzeit vornehmlich im 16.-18. Jahrhundert*, ed. Mohammed Rassem and Justin Stagl ("Quellen und Abhandlungen zur Geschichte der Staatsbeschreibung," vol. 1) (Paderborn, etc.: Ferdinand Schöningh, 1980).

    On the relations between geography and statistics, particularly in the 18th century. German examples.

867. Beck, Hanno. "Zur Geschichte der Geographie, der Pädagogik und des Geographischen Unterrichts" (Auftrag des Hochschulverbandes für Geographie und ihre Didaktik), Geographiedidaktische Forschungen, vol. 8 (1981), 63-83.

History of geography and geographical instruction in German schools and universities, 1799-1980.

868. Beck, Hanno. "Die Streitfälle Fröbel-Ritter und Peschel-Klöden," Petermanns Geographische Mitteilungen, vol. 105, no. 2 (1 June 1961), 105-118.

Interpretation of the methodological controversies between Julius Fröbel and Carl Ritter in 1831 and between Oscar Peschel and Gustav Adolf von Klöden beginning in 1859.

869. Blume, Helmut. "25 Jahre geomorphologische Forschung in der Bundesrepublik Deutschland im Spiegel der Zeitschriften und Schriftenreihen 1950-1974," Geographische Rundschau, vol. 27, no. 9 (September 1975), 361-364.

Review of German geomorphological research published in German journals and monograph series, 1950-1974.

870. Borchert, Günter. "Beiträge von Hamburger Geographen zur Afrikaforschung," Mitteilungen der Geographischen Gesellschaft in Hamburg, vol. 56 (1965), 1-19.

African researches by Hamburg geographers, beginning with Heinrich Barth (1821-1865). This is a Barth memorial volume (209 p.) devoted to articles about African investigations by Hamburg scientists, ethnologists, and missionaries.

871. Braun, Gustav. Zur Methode der Geographie als Wissenschaft. (Zugleich Ergänzungsheft zum 17./38. Jahresbericht der Geographischen Gesellschaft Greifswald.) Greifswald: Verlag von Bruncken & Co., 1925. 24 p.

Essays on geographical methodology, including geography as a science, "Individuelle" and "Spezielle" geography, and the problem of political geography (Staatenkunde). Almost all references are to German works.

Part Two--Geography in Various Countries

872. Drygalski, Erich von. "Die Entwicklung der Geographie seit Gründung des Reiches," Mitteilungen der Geographischen Gesellschaft in Hamburg, vol. 43 (1933), 1-11.

    A review of German geography and exploration since the unification of Germany and the founding of the Hamburg Geographical Society (1873).

873. Genthe, Martha Krug. "German Geographers and German Geography," National Geographic Magazine, vol. 12, no. 9 (September 1901), 324-337.

    German geography from the 15th century onward.

874. Georgi, J. "Deutschland in der Polar-Forschung," Natur und Volk (Frankfurt), vol. 67, no. 9 (1 September 1937), 419-429; no. 10 (1 October 1937), 483-494.

    Role of Germans in exploration and research in polar lands, with emphasis on Arctic.

875. Halkin, Joseph. L'enseignement de la géographie en Allemagne et la réforme de l'enseignement géographique dans les universités belges. (Bibliothèque de la Faculté de philosophie et lettres de l'Université de Liège, 9.) Brussels: Office de publicité/Société belge de librairie, 1900. 171 p.

    Survey of geographical instruction in German gymnasia and universities, with the hope of getting ideas for the stimulation of geography in Belgium.

876. Kettler, J.I. "Ueber die Arbeiten des Geographischen Instituts zu Weimar, 1791-1891: Ein Beitrag zur Geschichte der Geographie," Zeitschrift für wissenschaftliche Geographie, vol. 8 (1891), 316-328, 382-391, 405-440.

    Study apparently never completed, because the journal went out of existence with this volume. These installments seem to deal just with the early work of the Institute and with previous work (before 1791) dealing with mathematical or astronomical geography. Many references to surveyors, cartographers, and geodesists in the 17th and 18th centuries.

877. Kinzel, Hella. "Zur Geschichte des Geographie- unterrichtes in Deutschland bis Mitte des 19.

Jahrhunderts," <u>Wissenschaftliche Zeitschrift der Humboldt-Universität zu Berlin, Gesellschafts- und Sprachwissenschaftliche Reihe</u>, vol. 12, nos. 7-8 (1963), 907-917.

History of geographical instruction in Germany from 1500 to 1850. The following periods form turning-points in science and school teaching: (1) time of humanism, (2) beginning of the 18th century, (3) time of French Revolution, and (4) the first decades of the 19th century.

878. Kolb, Albert. "Deutsche Geographen als Forscher und Lehrer in China," <u>Die Erde</u>, vol. 114, nos. 2-3 (1983), 135-142.

Concerns the research and teaching of 4 German geographers in China in the 1930s (Günther Köhler, Hermann von Wissmann, Wilhelm Credner, and Wolfgang Panzer).

879. Krebs, Norbert. "Der Stand der deutschen Geographie," <u>Geographische Zeitschrift</u>, vol. 44, no. 7-8 (July 1938), 241-249.

Geography in Germany, especially since World War I.

880. Lautensach, Hermann. "L'evoluzione della geografia in Germania nel dopoguerra," <u>Bollettino della Società Geografica Italiana</u>, series 8, vol. 9, no. 1-3 (January-March 1956), 4-13.

Geography in Germany since World War II.

881. Leighly, John. "Methodologic Controversy in Nineteenth Century German Geography," <u>Annals of the Association of American Geographers</u>, vol. 28, no. 4 (December 1938), 238-258.

19th-century German methodological controversies, with special reference to the works of Carl Ritter, Julius Fröbel, Ferdinand von Richthofen, Hermann Wagner, Georg Gerland, and Alfred Hettner.

882. Leser, Hartmut. "Wandel und Bestand methodischer Grundperspektiven der Geomorphologie zwischen den Ansatzen Ferdinand von Richthofens und heute," <u>Die Erde</u>, vol. 114, nos. 2-3 (1983), 103-118.

Part Two--Geography in Various Countries 185

Development of geomorphology in Germany, beginning with the work of Ferdinand von Richthofen (1833-1905), "The Father of Geomorphology." Comparison of Richthofen's inductive and descriptive work with more modern approaches.

883. Lichtenberger, Elisabeth. "The Impact of Political Systems upon Geography: The Case of the Federal German Republic and the German Democratic Republic," Professional Geographer, vol. 31, no. 2 (May 1979), 201-211.

Shows how ideological differences have influenced geographical research in the 2 parts of divided Germany since World War II.

884. Meynen, Emil. Deutscher Geographentag 1881-1963. Gesamtinhaltsverzeichnis der Verhandlungen des 1.-34. Deutschen Geographentages und der aus Anlass der Geographentage Erscheinenen Festschriften. Wiesbaden: Franz Steiner Verlag, 1965. x + 106 p.

Statistical and bibliographical summary of the first 34 conferences (now biennial) of German-speaking geographers.

885. Meynen, Emil, ed. Institut für Landeskunde. 25 Jahre Amtliche Landeskunde. Bad Godesberg: Bundesanstalt fur Landeskunde und Raumforschung, 1967.

See especially pp. 1-62, Meynen's article on the activities of the first quarter-century of the Institute.

886. Oberhummer, Eugen. "Deutschland und die Erdkunde," Mitteilungen der Geographische Gesellschaft in Wien, vol. 83, nos. 4-6 (15 June 1940), 102-105.

Brief sketch of German role in promoting geography and exploration, with emphasis on the 19th century.

887. Oertel, Karl Otto. Die Naturschilderung bei den Deutschen Geographischen Reisebeschreibern des 18. Jahrhunderts. (Inaugural-Dissertation, Leipzig.) [Leipzig, 1898.] 90 p.

German geographical description in the 18th century. Alexander von Humboldt and his 18th-century precursors, such as J.P. Kolb, J.G. Gmelin, G.W. Steller, and the Forsters, Reinhold and Georg.

888. Overbeck, Hermann. "Die Entwicklung der Anthropogeographie (inbesondere in Deutschland) seit der Jahrhundertwende und ihre Bedeutung für die geschichtliche Landesforschung," Blätter für Deutsche Landesgeschichte, vol. 91 (1954), 182-244.

The development of human geography (especially in Germany) since 1900 and its importance for research in regional history. [HB]

889. Overbeck, Hermann. "Ritter--Riehl--Ratzel. Die grossen Anreger zu einer historischen Landschafts- und Länderkunde Deutschlands im XIX. Jahrhundert," Die Erde, vol. 3, no. 3-4 (May 1952), 197-210.

Carl Ritter, Wilhelm Heinrich Riehl, and Friedrich Ratzel, the leaders in the historical geography of Germany in the 19th century.

890. Partsch, Joseph. "Die geographische Arbeit des 19. Jahrhunderts," pp. 35-45 in Aus fünfzig Jahren, Verlorene Schriften, ed. H. Waldbaur (Breslau: Ferdinand Hirt, 1927).

Geographical achievements of the 19th century (exploration, survey, mountain measurements, university research and teaching), with almost exclusive emphasis on Germans. This was Partsch's inaugural address as Rector of the University of Breslau, 15 October 1899, and was published at that time in the Schlesische Zeitung. See pp. 7-34 for an anonymous biographical sketch of Partsch and pp. 170-184 for his bibliography.

891. Penck, Albrecht. "Neuere Geographie," pp. 31-56 in Sonderband zur Hundertjahrfeier der Gesellschaft, Zeitschrift der Gesellschaft für Erdkunde zu Berlin, 1828-1928 (Berlin, 1928).

Penck's views of the changes in German geography during the past 100 years, beginning with the "classical" geography of Humboldt and Ritter.

892. Richter, Bernhard. "Die Entwicklung der Naturschilderung in den deutschen geographischen Reisebeschreibungen, mit besonderer Berücksichtigung der Naturschilderung in der ersten Hälfte des 19. Jahrhunderts" (Inaugural-

Dissertation, Leipzig, 1900), Euphorion: Zeitschrift für Litteraturgeschichte, Erganzungsheft, 5 (1901), 1-93.

Nature sentiment and description in German works, with emphasis on the Humboldtian era (first half of the 19th century).

893. Schott, Gerhard. Deutschlands Anteil an der geographischen Erforschung der Meere. (Beiheft zur Marine Rundschau, July Number 1907.) Berlin: Ernst Siegfried Mittler und Sohn, 1907. 24 p.

    History of German interest in oceanography. [HB]

894. Schulte-Althoff, Franz-Josef. Studien zur politischen Wissenschaftsgeschichte der deutschen Geographie im Zeitalter des Imperialismus. (Bochumer Geographische Arbeiten, no. 9.) Paderborn: Ferdinand Schöningh, 1971. 251 p.

    History of German geography in relation to German imperialism in the period 1870-1914.

895. Schultz, Hans-Dietrich. Die deutschsprachige Geographie von 1800 bis 1970: Ein Beitrag zur Geschichte ihrer Methodologie. (Abhandlungen des Geographischen Instituts--Anthropogeographie, vol. 29.) Berlin: Geographische Institut der Freien Universität, 1980. 478 p.

    A history of German geographical methodology, 1800-1970. Considerable discussion of the concept of "landschaft."

896. Schultz, Hans-Dietrich. "Die Situation der Geographie nach dem Ersten Weltkrieg. Eine unbekannte Umfrage aus dem Jahre 1919, historisch kommentiert," Die Erde, vol. 108, nos. 1-2 (1977), 75-102.

    Explication of an 8-page survey of geography published in Leipzig in 1919: Neue Aussprüche von 15 Hochschulprofessoren über Wesen/Wert und Methode der Erdkunde (reproduced here on pp. 95-102). (From English abstract, p. 75) "Not one of the ideas contributed by these scientists to the methodological discussion has become outdated. New problems and new solutions have simply been added to the old ones." Author discusses development of geography in relation to the concepts of Thomas Kuhn and Karl Popper.

897. Specklin, Robert. "Les régions françaises dans la littérature allemande (1870-1945)," France, Ministère des universités, Comité des travaux historiques et scientifiques, Bulletin de la section de géographie, no. 84 (1979--pub. 1981), 117-161.

Description of France and its regions in the German geographical literature from 1870 to 1945. Bibliography of 408 items (pp. 138-161).

898. Troll, Carl. "Die deutsche geographische Japan-Forschung vor und nach der Meiji-Restauration. Zum Gedenken von Johannes Justus Rein (1835-1918), Philipp Franz von Siebold (1796-1866) und Engelbert Kaempfer (1651-1716)," Erdkunde, vol. 22, no. 1 (March 1968), 7-13.

German geographical research on Japan, before and after the Meiji Restoration of 1868, with emphasis on the 3 researchers named in the title. Accompanied by a brief review of work published since the time of J.J. Rein.

899. Troll, Carl. "Die geographische Wissenschaft in Deutschland in den Jahren 1933 bis 1945: Eine Kritik und Rechtfertigung," Erdkunde, vol. 1, nos. 1-3 (May 1947), 3-48.

Vicissitudes of geography in Hitler's Germany. Partial translation by Eric Fischer appeared in Annals of the Association of American Geographers, vol. 39 (1949), 99-137.

900. Van Valkenburg, Samuel. "The German School of Geography," pp. 91-115 in Geography in the Twentieth Century, ed. Griffith Taylor, 3rd ed. (New York: Philosophical Library, 1957).

German geography from about 1900 to 1945.

901. Wagner, Hermann. Festrede in Namen der Georg-Augusts-Universität zur Akademischen Preisverteilung am 4. Juni 1890. Göttingen: Druck der Dieterichschen Univ.-Buchdruckerei, n.d. (1890). 30 p.

Development of geography in Germany from the middle of the 18th century, with particular emphasis on Göttingen. Discussion of geography ends on p. 25.

Part Two--Geography in Various Countries   189

902. Wagner, Hermann. "Gothas Bedeutung für die Pflege der Astronomie und Geographie," pp. 146-167 in <u>Gotha und sein Gymnasium</u>, ed. Heinrich Anz (Gotha/Stuttgart: Verlag Friedrich Andreas Perthes A.-G., 1924).

Concerns early geographical instruction in the Gotha Gymnasium, beginning with Johann Galletti (1750-1828), but mostly emphasizes the geographical and cartographic activities of the Justus Perthes Anstalt. Geography is discussed on pp. 152-167, astronomy on pp. 146-152.

903. "W.W." [Wolkenhauer, Wilhelm]. "Zeittafel zur Geschichte der Pflege und Forderung der Geographie in Bremen," <u>Deutsche Geographische Blätter</u>, vol. 18, nos. 1-2 (1895), 12-13.

Chronological table of significant events in Bremen beginning with the arrival of Adam of Bremen, "the first German geographer," in 1067. Next date is 1758 and majority are in 19th century.

XVII. ITALY

904. Almagià, Roberto. <u>La geografia</u>. ("Le Guide ICS" [Profili Bibliografici de <u>L'Italia Che Scrive</u>].) Rome: Istituto per la Propaganda della Cultura Italiana, 1919. viii + 109 p.

After a few introductory pages on "The Historical Evolution of Geography as a Science," the text is devoted to Italian geography since the middle of the 19th century. Lengthy bibliography on geography in Italy (pp. 69-109).

905. Almagià, Roberto. "La geografia in Italia dal 1860 al 1960," <u>L'Universo</u>, vol. 41, no. 3 (May-June 1961), 419-432.

Geography in Italy from 1860 to 1960, with emphasis on the work of academic geographers such as Giuseppe Dalla Vedova and the Marinellis (Giovanni and Olinto), and of institutions such as the Istituto Geografico Militare.

906. Arena, Gabriella. "Il pensiero geografico in Italia (bibliografia tematica)," Università di Roma, Facoltà

di Lettere e Filosofia, Pubblicazioni dell'Istituto di Geografia, series C, no. 4 (1973). 38 p.

An annotated bibliography of works by Italian geographers on the history and methodology of geography, 1857-1973.

907. Baldacci, Osvaldo. "Storia della geografia," pp. 469-506 in Un sessantennio di ricerca geografica italiana (Rome: Società Geografica Italiana, 1964).

A survey of Italian writings on the history of geography, the history of exploration, and the historical geography of Antiquity and the Middle Ages. Useful bibliography of 211 items. [LG]

908. Bertacchi, Cosimo. Conversazione geografiche: per la storia della geografia in Italia. ("Piccola Biblioteca di Scienze Moderne," no. 310.) Turin: Fratelli Bocca, Editori, 1925. 262 p.

Several papers on the history of geography in Italy. See "Per la storia della geografia in Italia nell'epoca contemporanea" (pp. 3-51), which concerns geography in Italy since the time of Vico (1668-1744), with emphasis on the period following the founding of the Società Geografica Italiana (1867); "L'Italia e il suo mare" (pp. 53-101), which concerns geographical and cartographic study of the Mediterranean from ancient times onward; "Per il VII Congresso Geografico Italiano" (pp. 103-130), which treats the Italian Geographical Congresses (held every 3 years, beginning 1892) and also the work of contemporary Italian geographers; "La geografia nella scienza, nella scuola e nella vita sociale" (pp. 183-214), an historical and methodological essay on the nature of geography and its place among the sciences; and "Antonio Cecchi e la politica coloniale dell'Italia" (pp. 215-262), which concerns Antonio Cecchi (1849-1896) and Italian exploration and colonialism in Africa.

909. Bertacchi, Cosimo. Geografi ed esploratori italiani contemporanei. Milan: S.A. Prof. Giovanni de Agostini, 1929. 452 p.

Italian geographers and explorers (including missionaries and promoters of exploration). Introductory section (pp. 9-42) deals with geographers and travelers from the late Middle Ages to the early 19th

Fig. 8. Italian geographers on a field excursion (Passo Sella) in June 1931. From left, Pietro Gribaudi, Luigi Filippo De Magistris, Attilio Mori, Antonio Renato Toniolo, Roberto Almagià, and Leonardo Ricci. Reproduced through the courtesy of Professor Elio Migliorini and Dr. Claudia Mancini.

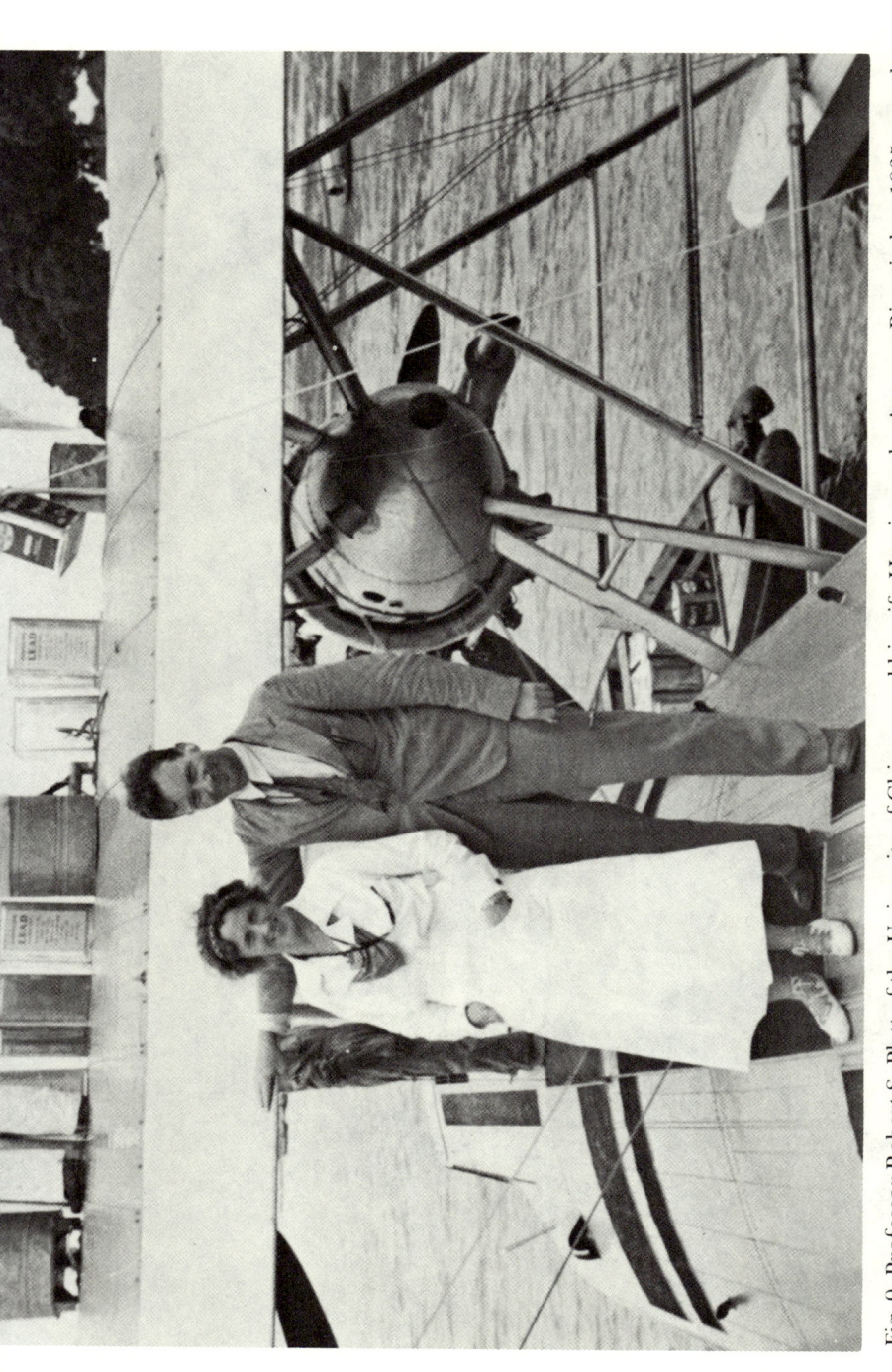

Fig. 9. Professor Robert S. Platt of the University of Chicago and his wife Harriet on the Amazon River in late 1935 or early 1936. On the wing are mechanics refueling the plane with five-gallon gasoline cans. Reproduced through the courtesy of Mrs. Nancy Platt Ravfield and Professor Richard Hartshorne.

Part Two--Geography in Various Countries   191

century, and the rest of the book treats geographers and explorers in the 19th and 20th centuries.

910. Caraci, Giuseppe. "Storia della geografia, geografia storica, toponomastica," pp. 541-556 in Vol. 2 of <u>Un secolo di progresso scientifico italiano, 1839-1939</u> (Rome: Società Italiana per Il Progresso delle Scienze, 1939).

Bibliographical essay on Italian writings on the history of geography, historical geography, and place-name studies during the previous century. Bibliography (pp. 551-556) lists works published after 1920. Earlier works were cited in the <u>Guida bibliografica</u> (2nd ed. 1922) of the Fondazione Leonardo.

911. Corna-Pellegrini, Giacomo, and Brusa, Carlo, eds. <u>La ricerca geografica in Italia 1960-1980</u>. (Convegno sullo Stato della Ricerca Geografica in Italia 1960-1980 Svoltosi sotto gli Auspici del Consiglio Nazionale delle Ricerche ed Organizzato dall'Istituto de Geografia Umana della Facoltà di Lettere e Filosofia dell'Università degli Studi di Milano, held at Varese 31 March-2 April 1980.) Varese: Ask Edizioni, 1980. xvi + 1073 p.

Contributions of 115 authors on recent developments in geography in Italy.

912. Gambi, Lucio. "Uno schizzo di storia della geografia in Italia," pp. 3-37 in <u>Una geografia per la storia</u> by Gambi ("Piccola Biblioteca Einaudi," no. 211) (Turin: Giulio Einaudi editore, 1973).

History of geography in Italy since 1680, with emphasis on the period since 1830. Also published as "Esquissse d'une histoire de la géographie en Italie," pp. 9-37 in <u>Travaux de géographie fondamentale</u> (Cahiers de géographie de Besançon, no. 23) (Paris: Les Belles Lettres, 1974).

913. Gribaudi, Dino. "Per una storia italiana delle conoscenze e delle dottrine geografiche," <u>Atti del XIV Congresso Geografico Italiano</u>, Bologna, 1947, pp. 256-262.

Laments the lack of an Italian work that treats the history of geographical doctrines and knowledge in an organized way and indicates the principles on which such a work should be based. Also treats the matter of periodization. [LG]

914. Gribaudi, Pietro. "La geografia nel secolo XIX specialmente in Italia," Rivista di Fisica, Matematica e Scienze Naturali (Pavia), vol. 1, no. 1 (January 1900), 47-51; no. 2 (February 1900), 133-139; no. 4 (April 1900), 325-333; vol. 2, no. 7 (July 1900), 49-62; no. 12 (December 1900), 508-534.

    A survey of geography from ancient times onward, with emphasis on the 19th century and on Italy. Long discussions of German, English, and French geography. Reprinted in Pietro Gribaudi, Scritti di varia geografia (Università di Torino, Facoltà di Economia e Commercio, Laboratorio di Geografia Economica "Pietro Gribaudi," no. 9) (Turin: G. Giappichelli Editore, 1955), pp. 199-229. The 1955 work begins with a biography and bibliography of Gribaudi.

915. Ippolito, Guglielmo. "Cartografia," pp. 505-539 in Vol. 2 of Un secolo di progresso scientifico italiano 1839-1939 (Rome: Società Italiana per Il Progresso delle Scienze, 1939).

    Official and private cartography in Italy since 1839, with emphasis on topographic mapping.

916. Malesani, Emilio. "Geografia dell'Italia," pp. 601-612 in Vol. 2 of Un secolo di progresso scientifico italiano 1839-1939 (Rome: Società Italiana per Il Progresso delle Scienze, 1939).

    A bibliographical essay on studies in the geography of Italy, with emphasis on the works of Italian geographers.

917. Migliorini, Elio. "Die heutigen neuen Strömungen in der italienischen Geographie," Geographische Zeitschrift, vol. 44, nos. 7-8 (July 1938), 277-283.

    Italian geography in the 19th and 20th centuries, with emphasis on the recent past.

918. Migliorini, Elio, and Bernardy, Amy A. Geografia e viaggi--arti e tradizioni popolari. ("Bibliografie del Ventennio.") Rome: Istituto Nazionale per le Relazioni Culturali con L'Estero, 1943. xvi + 239 p.

Part Two--Geography in Various Countries 193

Elio Migliorini, "Geografia e viaggi" (xvi, 1-206), includes a subsection, "Storia della geografia e geografia storica" (pp. 11-29), that treats the history of geography and also historical geography. Italian sources only. Heavy emphasis on exploration.

919. Milanini Kemeny, Anna. La Società d'Esplorazione Commerciale in Africa e la politica coloniale (1879-1914). (Pubblicazioni della Facoltà di Lettere e Filosofia dell'Università di Milano, no. 67, Pubblicazioni dell'Istituto di Geografia Umana, 3.) Florence: La Nuova Italia Editrice, 1973. xi + 258 p.

The Society for Commercial Exploration in Africa from its beginning until the death of its active president, Giuseppe Vigoni, in 1914. Society was merged with Istituto Coloniale Fascista in 1928.

920. Mongini, Giovanni Maria. "La geografia nelle riunioni degli scienziati italiani, 1839-1871," Rome, Istituto universitario pareggiato di Magistero "Maria SS. Assunta," Istituto di geografia, Pubblicazioni, no. 4 (1975). 84 p.

A fairly detailed summary of the geographical contributions presented at the Italian Science Congresses in the years 1839-1847, 1862, 1873, and 1875. [LG]

921. Natali, Giovanni. "La geografia in Italia nella prima metà del secolo XIX," Rivista d'Italia, 18th year, vol. 2, no. 8 (August 1915), 255-284.

Geography in Italy in the first half of the 19th century. Special mention of geographical publications in Antologia (Florence, 1821-) and Progresso delle Scienze, Lettere ed Arti (Naples, 1832-) and in the annual scientific congresses (1839-1847).

922. Riccardi, Riccardo. "Antropogeografia," pp. 571-584 in Vol. 2 of Un secolo di progresso scientifico italiano 1839-1939 (Rome: Società Italiana per Il Progresso delle Scienze, 1939).

A bibliographical essay on human (or anthropo-) geography, with emphasis on Italian geographers of the 20th century. Includes political geography.

923. Sestini, Aldo. "Geografia fisica," pp. 557-570 in Vol. 2 of <u>Un secolo di progresso scientifico italiano 1839-1939</u> (Rome: Società Italiana per Il Progresso delle Scienze, 1939).

   A bibliographical essay on studies in physical geography in Italy since 1839, with emphasis on the 20th century.

924. Toniolo, Antonio Renato. "Introduzione [allo studio dei progressi raggiunti nel campo della geografia]," pp. 501-504 in Vol. 2 of <u>Un secolo di progresso scientifico italiano 1839-1939</u> (Rome: Società Italiana per Il Progresso delle Scienze, 1939).

   Brief overview of developments in geography in Italy since 1839.

925. Toschi, Umberto. "La geografia economica," pp. 585-599 in Vol. 2 of <u>Un secolo di progresso scientifico italiano 1839-1939</u> (Rome: Società Italiana per Il Progresso delle Scienze, 1939).

   A bibliographical essay on Italian studies in economic geography, beginning with the <u>Filosofia della statistica</u> (1826) of Melchiorre Gioia (1767-1829).

XVIII. OTHER EUROPEAN COUNTRIES (INCLUDING THE USSR)

926. Académie Royale de Belgique, Classe des Sciences. <u>Florilège des sciences en Belgique pendant le XIXe siècle et le début du XXe</u>. Brussels: Académie Royale de Belgique, 1967. 1067 p.

   Chapters on cartography (pp. 525-549) and geography (pp. 551-592). See especially Omer Tulippe, "Quelques événements importants relativement à l'effort géographique aux XIXe et XXe siècles en Belgique" (pp. 555-559) and the chapters that follow on the various branches of geography.

927. Ahlmann, Hans W., and Friberg, Nils. "Neue Strömungen in der nordischen geographischen Forschung," <u>Geographische Zeitschrift</u>, vol. 44, nos. 7-8 (15 July 1938), 307-315.

Part Two--Geography in Various Countries 195

Geography in Sweden, Norway, and Denmark in the 20th century, with emphasis on the 1920s and 1930s.

928. Amorim Girão, Aristides de. "L'evoluzione degli studi geografici in Portogallo," Bollettino della Società Geografica Italiana, series 8, vol. 9, nos. 9-10 (September-October 1956), 396-400.

Teaching of geography at the university level in Portugal began in the 16th century at the University of Coimbra. Author credits Bernardino Barros Gomes with the inauguration of scientific geography in 1878. Article deals chiefly with the work of the Lisbon Geographical Society and the universities of Coimbra and Lisbon in the 20th century.

929. Babicz, Józef. "Trois étapes de recherches sur l'histoire de la géographie en Pologne," pp. 24-36 in Studia z dziejów geografii i kartografii / Etudes d'histoire de la géographie et de la cartographie, ed. Józef Babicz (Warsaw: Polish Academy of Science, 1973).

History of geography in Poland, beginning with Joachim Lelewel (1785-1861). Article also printed in Polish on pp. 11-23.

930. Beck, Hanno. "Die Geschichte des Geographie in Polen," Erdkunde, vol. 21, no. 3 (August 1967), 240-242.

A review of works on the history of geography in Poland, with emphasis on the writings of Boleslaw Olszewicz and Józef Babicz.

931. Beck, Hanno. "Wiener Forschungen zur Geschichte der Geographie und der Reisen," Geographische Zeitschrift, vol. 65, no. 1 (First Quarter 1977), 49-52.

Recent Viennese researches on the history of geography and exploration.

932. Becker, Jeronimo. Los Estudios Geográficos en España (Ensayo de una Historia de la Geografía). (Publicaciones de la Real Sociedad Geográfica.) Madrid: Establecimiento Tipográfico de Jaime Ratés, 1917. 366 p.

Spanish geographical work from the Roman period onward, including exploration, survey, and cartography.

933. Becker, M.A. "Zur Geschichte der Geographie in Oesterreich seit 1750," Mittheilungen der Geographischen Gesellschaft in Wien, vol. 16 (1873), 193-213.

    Historical sketch of the study and practice of geography in Austria since 1750.

934. Berg, Lev Semenovich. Geschichte der russischen geographischen Entdeckungen. Trans. Rolf Ulbrich. Leipzig: VEB Bibliographisches Institut, 1954. 281 p.

    Geography in Russia since the 16th century. Emphasis on exploration, but there is also much on geography and geographers, such as Voiekov, Dokuchaev, and D.N. Anuchin. First published in Russian in 1946.

935. Bernleithner, Ernst. "Die Entwicklung der österreichischen Länderkunde von ihren Anfängen bis zur Errichtung der ersten Lehrkanzel für Geographie in Wien (1851)," Mitteilungen der Geographischen Gesellschaft Wien, vol. 97, no. 2 (1955), 111-127.

    Development of Austrian cartography and regional geography (local or topographical studies) from the 15th century to 1851, when Friedrich Simony took up the first chair of geography in the University of Vienna.

936. Blüthgen, Joachim. "Entwicklung, Stand und Aufgaben der Geographie in Schweden (Sammelreferat)," Zeitschrift für Erdkunde, vol. 9, no. 3-4 (February 1941), 65-88.

    Bibliographical essay on recent researches (since about 1930) in Sweden by Swedes and others. 255 references.

937. Blüthgen, Joachim. "Die Geographie in Dänemark," Zeitschrift für Erdkunde, vol. 10, no. 3 (March 1942), 129-144.

    Bibliographical essay on geographical researches on Denmark by Danes and others. Mostly recent materials (from about 1930 onward). Not much on development of geography in Denmark. 147 references.

Part Two--Geography in Various Countries 197

938. Dussart, Frans. La géographie aux Pays-Bas. (Travaux du Cercle des géographes liégeois, Fascicule 35, and Travaux du Seminaire de géographie de l'Université de Liège, Fasc. 55.) (Reprinted from the Bulletin de la Société royale belge de géographie, 1937) Liège: J. Wyckmans, Editeur, 1937. 30 p.

Survey of geography in the Netherlands, from Varenius to the present.

939. Evers, W. "Stand und Aufgaben der Geographie in Norwegen," Zeitschrift für Erdkunde, vol. 7, no. 17 (3 September 1939), 689-706.

Bibliographical essay (270 items) on geographical researches in Norway. Of greater historical interest than Blüthgen's essays (above) on Denmark and Sweden because it contains brief sections on geography in the university, societies and journals, etc.

940. Fiechter, Alfredo. "L'Istituto Geografico Militare di Vienna: Origine, Attività, Fine," L'Universo, vol. 5, no. 11 (November 1924), 793-808.

The mapping activities of the Military Geographical Institute of Vienna, 1839-1918, and earlier work in Lombardy beginning in 1773.

941. Galon, Rajmund. "Die Geographie in Polen, ihre Fortschritte und Ziele," Geographische Zeitschrift, vol. 44, nos. 7-8 (July 1938), 297-306.

Recent trends in Polish geography.

942. George, Pierre. "Aperçu de l'histoire de la géographie dans l'empire russe et en Union soviétique depuis deux cents ans," Comité des travaux historiques et scientifiques, Bulletin de la section de géographie, vols. 61-62 (1946-1948 [pub. 1953]), 57-101.

A recounting of the great events and the great names in Russian and Soviet geography and exploration since the time of Peter the Great. Treats the methodology of geography in the Soviet period, including the views of A.A. Grigoryev (1883-1968). [PP]

943. Gerasimov, I.P. "Fifty Years of Development of Soviet Geographic Thought," Soviet Geography: Review and Translation, vol. 9, no. 4 (April 1968), 238-252.

Trends in Soviet geography since the Revolution. Article first published in Russian in 1967. The remaining 9 articles in this issue of Soviet Geography deal with the development of geography in several of the Soviet republics and in the Mongolian People's Republic since 1917 (pp. 253-332).

944. Gerasimov, Innokenty P., ed. <u>A Short History of Geographical Science in the Soviet Union</u>. Trans. John Williams. Moscow: Progress Publishers, 1976. 178 p.

Russian geography, cartography, and exploration from the 15th century onward. Part 1 (pp. 19-104), "Russian Geography before the Revolution"; Part 2 (pp. 105-178), "The Development of Soviet Geographical Science."

945. Gottschalk, M.K.E. "Some Aspects of the Development of Historical Geography in the Netherlands," <u>Tijdschrift van het Koninklijk Nederlandsch Genootschap</u>, vol. 77, no. 3 (July 1960), 319-323.

(Page 319) "The civil-engineer J.C. Ramaer and the engineer-officer A.A. Beekman are to be accounted the pioneers of historical geography in the Netherlands. Their activities in this field already started at the end of the last century...." Numerous examples of 20th-century work, especially the geographers such as J.P. Bakker and H.J. Keuning and the historians B.H. Slicher van Bath and S.J. Fockema Andreae.

946. Hägerstrand, Torsten. "Proclamations about Geography from the Pioneering Years in Sweden," <u>Geografiska Annaler</u> (Series B: Human Geography), vol. 64B, no. 2 (1982), 119-125.

Contrasting approaches to geography by the Swedish geographers Helge Nelson (1882-1966) and Sten De Geer (1886-1933).

947. Harris, Chauncy D., ed. <u>Soviet Geography: Accomplishments and Tasks</u>. (American Geographical Society, Occasional Publication, No. 1.) New York: American Geographical Society, 1962. xviii + 409 p.

English translations of papers by 56 Soviet geographers. See especially Chapter 2, pp. 9-13, "Russian Geography," by A.A. Grigor'yev, which is a

Part Two--Geography in Various Countries   199

   brief sketch of Russian geography and exploration from
   the medieval period to 1917; and Chapters 3-8, pp. 15-
   60, which concern the history and present state of
   Soviet geography. Chapter 8, pp. 53-60, "Research in
   the History of Geographic Knowledge," by D.M. Lebedev,
   discusses Soviet research in the history of geography
   and exploration.

948. Hassinger, Hugo. Osterreichs Anteil an der Erforschung
     der Erde: Ein Beitrag zur Kulturgeschichte
     Osterreichs. Vienna: Verlag Adolf Holzhausens Nfg.,
     1949. 194 p.

   History of Austrian geography, cartography, and
   exploration from the 12th century to the present.

949. Helmfrid, Staffan. "Hundra år svensk geografi," Ymer
     (1976/77 Yearbook), pp. 360-372.

   Geography in Sweden since the founding of the Swedish
   Society for Anthropology and Geography in 1877.

950. Hooson, David J.M. "The Development of Geography in
     Pre-Soviet Russia," Annals of the Association of
     American Geographers, vol. 58, no. 2 (June 1968),
     250-272.

   Geography in Russia from the time of Peter the Great
   to 1917. Describes the carry-over into the Soviet
   period as well.

951. "D.H." [Hooson, David]. "Geography," pp. 265-266 in
     The Cambridge Encyclopedia of Russia and the Soviet
     Union, ed. Archie Brown, John Fennell, Michael Kaser,
     and H.T. Willetts (Cambridge: Cambridge University
     Press, 1982).

   Geography in Russia and the Soviet Union from Peter
   the Great to the present. See also adjacent article on
   "Exploration" (pp. 264-265) by "J.P." (Judith Pallot).

952. Hooson, David J.M. "Phases in the 20th Century
     Development of Russian and American Geography: A
     Comparison," Proceedings of the Association of
     American Geographers, vol. 1 (1969), 66-69.

   Comparative study of geographical thought and
   practice in the U.S. and U.S.S.R. in the 20th century.
   Convergence in the 1920s, followed by a sharp

divergence in the Stalin era, and now, "in the last dozen years or so," another convergence. Emphasis in this brief paper is strongly on the Russian side, with very little about American geography.

953. Howe, G. Melvyn. "Geography in the Soviet Universities," Geographical Journal, vol. 124, part 1 (March 1958), 80-84.

   The author's personal review of geographical instruction in Soviet universities in 1956, with brief mention of some methodological changes since 1927.

954. Isachenko, A.G. "Fifty Years of Soviet Landscape Science," Soviet Geography: Review and Translation, vol. 9, no. 5 (May 1968), 402-407.

   Study of natural regions in the USSR since the Revolution. Paper first published in Russian in 1967.

955. Kalesnik, S.V. "The Development of General Earth Science in the USSR," Soviet Geography: Review and Translation, vol. 9, no. 5 (May 1968), 393-402.

   Development of general physical geography in the USSR since 1917. Paper first published in Russian in 1967.

956. Keuning, H.J. "Un siècle de géographie aux Pays Bas," Tijdschrift voor Economische en Sociale Geografie, vol. 47, nos. 6-7 (June-July 1956), 141-144.

   Traces the development of geography in the Netherlands, beginning in 1873 with the founding of the Royal Dutch Geographical Society. Mostly concerns the growth of human geography and sociografie.

957. Kielczewska-Zaleska, Maria. "Gli studi geografici in Polonia dopo la seconda guerra mondiale," Bollettino della Società Geografica Italiana, series 8, vol. 12, nos. 1-3 (January-March 1959), 1-12.

   After brief mention of the state of geography in Poland between the wars, the article concentrates on the period since World War II.

958. Köhler, Franz. "Zu einigen Aufgaben der Geschichte der Geographie und ihre Fortschritte in sowjetischen Arbeiten," Petermanns Geographische Mitteilungen, vol. 122, no. 2 (1978), 127-133.

Part Two--Geography in Various Countries 201

Analysis of Soviet and East German publications on the history of geography since 1971.

959. Kondracki, Jerzy. "The West Slav Geographers, Part II.--The Development of Geography in Poland," pp. 122-127 in Geography in the Twentieth Century, ed. Griffith Taylor, 3rd ed. (New York: Philosophical Library, 1957).

The development of geography in Poland, with emphasis on the period from the middle of the 19th century to just after World War II.

960. Konstantinov, D.A. "Economic Geography in the USSR on the 50th Anniversary of Soviet Power," Soviet Geography: Review and Translation, vol. 9, no. 5 (May 1968), 417-424.

Economic geography in the Soviet Union since the Revolution. Discusses controversies of the 1920s and reorientation along Marxist-Leninist lines in the 1930s. (Page 423) "Economic geography is the most widespread geographic discipline in institutions of higher learning in the USSR."

961. Kostrowicki, Jerzy. "Geography in Poland since the War," Geographical Journal, vol. 122, part 4 (December 1956), 441-450.

After a review of Polish geography since the 15th century, article emphasizes period since World War II.

962. Král, Jiří. "The West Slav Geographers, Part I.--Czechoslovak Geography in the Twentieth Century," pp. 116-121 in Geography in the Twentieth Century, ed. Griffith Taylor, 3rd ed. (New York: Philosophical Library, 1957).

Geography in Czechoslovakia from 1902 to just after World War II.

963. Lefèvre, Marguerite. "Historique de l'évolution de la géographie en Belgique," Acta geographica lovaniensa (Louvain, Belgium), vol. 5 (1967), 117-137.

First treats cartography since the 16th century and then geography proper ("proprement dite"), beginning with Philippe Vander Maelen (1795-1869) in the 19th century.

964. Liss, Carl-Christoph. "Die Entwicklung der Geographie in Spanien in den letzten vier Jahrzehnten," Erdkunde, vol. 34, no. 4 (December 1980), 293-297.

A survey of geography in Spain, with emphasis on the period since the end of the Civil War in 1939.

965. Matley, Ian M. "The Marxist Approach to the Geographical Environment," Annals of the Association of American Geographers, vol. 56, no. 1 (March 1966), 97-111.

Environmentalism in Russian and Soviet geography in the 19th and 20th centuries.

966. Mead, William R. "The Eighteenth Century Military Reconnaissance of Finland: A Neglected Chapter in the History of Finnish Geography," Acta Geographica (Helsinki), vol. 20 (1968), 255-271.

Origins, nature, and significance of the late 18th-century military reconnaissance manuscript maps of Finland that are deposited in the Kungliga Krigsarkivet in Stockholm.

967. Mead, W.R. The Geographical Tradition in Finland. London: H.K. Lewis & Co., Ltd., 1963. 18 p.

Early geographical studies in Finland, with emphasis on the period 1740-1860 and the work of Zachris Topelius (1818-1898).

968. Mead, W.R. "Luminaries of the North: A Reappraisal of the Achievements and Influence of Six Scandinavian Geographers," Transactions of the Institute of British Geographers, no. 57 (November 1972), 1-13.

Development of geography in Scandinavia, as mirrored in the biographies of Conrad Malte-Brun (1775-1826), Adolf Erik Nordenskiöld (1832-1901), Fridtjof Nansen (1861-1930), Zachris Topelius (1818-1898), Rudolf Kjellén (1864-1922), and Sven Hedin (1865-1952).

969. Melón, Amando. "España en la historia de la geografía," Estudios Geográficos (Madrid), vol. 4, no. 11 (May 1943), 195-232.

The 15th and 16th centuries are the Spanish Period in the history of geography, the 17th century is the Dutch Period, the 18th is the French Period, and the 19th is the German Period. Article emphasizes cartography and exploration in medieval and Renaissance Spain. Reprinted in Estudios Geográficos, vol. 38 (1977), 65-106.

970. Melón, Amando, and Gordejuela, Ruiz de. "Esquema sobre los modeladores de la moderna ciencia geográfica," Estudios Geográficos, vol. 6, no. 20-21 (August-November 1945), 393-442.

    Modern geography from Varenius through the disciples of Vidal de la Blache. Article reprinted in Estudios Geográficos, vol. 38 (1977), 317-368.

971. Nikitin, N.P. "A History of Economic Geography in Prerevolutionary Russia," Soviet Geography: Review and Translation, vol. 7, no. 9 (November 1966), 3-37.

    Economic geography in Russia from the 18th century to 1917. Paper originally published in Russian in 1965.

972. Oberhummer, Eugen. "Die Entwicklung der Erdkunde in Osterreich seit der Mitte des 19. Jahrhunderts," Mitteilungen der K. K. Geographischen Gesellschaft in Wien, vol. 51, nos. 11-12 (November-December 1908), 433-452.

    Geography in Austria since mid-19th century, with a brief glance at earlier developments.

973. Oestreich, Karl. "Die neueren Strömungen in der niederlandischen Geographie," Geographische Zeitschrift, vol. 44, nos. 7-8 (July 1938), 289-297.

    Geography in the Netherlands in the 19th and 20th centuries.

974. Pethybridge, Roger. "British Contributions to the Study of Russian Geography in the Nineteenth Century," Scottish Geographical Magazine, vol. 87, no. 1 (April 1971), 58-65.

    Geographical observations of British visitors to Russia in the 19th century.

975. Pokshishevskiy, V.V. "Relationships and Contacts between Prerevolutionary Russian and Soviet Geography and Foreign Geography," Soviet Geography: Review and Translation, vol. 7, no. 9 (November 1966), 56-76.

Russian contacts with other geographers (chiefly European) and their publications from the early 18th century to the present. Paper originally published in Russian in 1965.

976. Portmann, J.-P. "La géographie à Neuchâtel: notice historique," Bulletin de la Société neuchâteloise de géographie, series 2, no. 23 (1978), 3-14.

Geographers, cartographers, and travelers who originated in Neuchâtel, Switzerland, or lived there, from the 18th century to 1978.

977. Prillinger, Ferdinand. "Landeskundliche Forscherpersönlichkeiten in Salzburg im 18. und 19. Jahrhundert," Mitteilungen der Osterreichischen Geographischen Gesellschaft, vol. 109, nos. 1-3 (1967), 207-226.

German and Austrian contributors to the geographical description of the Salzburg area in the 18th and 19th centuries, with a brief postscript on the 20th century.

978. Reparaz Ruiz, Gonzalo de. "Les études scientifiques et la géographie en Espagne au XVIIIe siècle," Bulletin hispanique (Bordeaux, etc.), vol. 44, nos. 2-4 (April-December 1942), 103-153; vol. 45, no. 1 (January-June 1943), 10-25.

Geography and cartography in 18th-century Spain, including the life of Isidoro de Antillón (1778-1814).

979. Rio, Maria Isabel del. "La geografía en España desde 1940 a 1972, a través de las principales revistas geográficas," Estudios Geográficos, vol. 36, nos. 140-141 (August-November 1975), 1031-1046.

Development of geography in Spain, 1940-1972, as reflected in journal articles.

980. Sanchez, Francisca. "El acceso al profesorado en la geografía española (1940-1979)," Geo-Critica (Barcelona), no. 32 (March 1981). 51 p.

Geography in Spanish universities since the Civil War, with particular emphasis on access to teaching positions.

Part Two--Geography in Various Countries   205

981. Şandru, Ion, and Cucu, V. "The Development of Geographical Studies in Rumania," <u>Geographical Journal</u>, vol. 132, part 1 (March 1966), 43-48.

   Geography in Romania from earliest times to the present. The first University Department of Geography was created at Bucharest in 1900.

982. Saushkin, Yu. G. "A History of Soviet Economic Geography," <u>Soviet Geography: Review and Translation</u>, vol. 7, no. 8 (October 1966), 3-104.

   Soviet economic geography since 1917. Article originally published in Russian in 1965.

983. Saushkin, Yu. G., Kosmachev, K.P., and Bykov, V.I. "The Scientific School of Baransky-Kolosovsky and Its Role in the Development of Soviet Geography," pp. 83-89 in <u>Les écoles géographiques</u>, ed. Józef Babicz (Warsaw: PWN--Polish Scientific Publishers, 1980).

   The development of economic geography in the Soviet Union by N.N. Baransky (1881-1963) and N.N. Kolosovsky (1891-1954).

984. Schmidt, Erwin. "German-Russian Scientific Contacts in Geography," pp. 290-295 in Vol. 4 of <u>Actes du XIe Congrès International d'Histoire des Sciences</u> (Warsaw, etc., 1965) (Warsaw, etc.: Ossolineum, 1968).

   German travelers in Russia, particularly in the 18th and 19th centuries.

985. Shnitnikov, A.V. "Geographical Problems in Climatology, Hydrology and Glaciology over the Last 50 Years," <u>Soviet Geography: Review and Translation</u>, vol. 9, no. 5 (May 1968), 407-416.

   Russian work in climatology, hydrology, and glaciology since the Revolution, with brief mention of earlier developments.

986. Ström, Ernst Ture. "Svensk kulturgeografisk forskning med särskild hänsyn till tiden efter 1940," <u>Gothia</u>, series 2, no. 8 (1958), 38-75.

   Study of Swedish geographical research, with emphasis on human geography, after 1940. Substantial introduction (pp. 38-49) covers developments in geography (physical as well as human) in Sweden,

Germany, France, United States, and Great Britain from 1800 to 1940.

987. Umlauft, Friedrich. "Geographischen Unterricht in Osterreich-Ungarn 1848-1898," Zeitschrift für Schulgeographie (Vienna), vol. 20 (1899), 353-369.

   History of geographical instruction in Austria-Hungary from 1848 to 1898. [HB]

988. Umlauft, Friedrich, ed. Die Pflege der Erdkunde in Oesterreich, 1848-1898. (Festschrift der K.K. Geographischen Gesellschaft aus Anlass des fünfzigjährigen Regierungs-Jubiläums Sv. Majestät des Kaisers Franz Joseph I.) (Special issue of Mitteilungen der K. K. Geographischen Gesellschaft in Wien, vol. 41, 1898.) Vienna: R. Lechner (Wilh. Müller), 1898. xix + 317 p.

   27 chapters on the development of geography (plus geology, cartography, and exploration) in Austria, 1848-1898.

989. Yli-Jokopii, Pentti. "Trends in Finnish Geography in 1920-1979 in the Light of the Journals of the Period," Fennia, vol. 160, no. 1 (1982), 95-193.

   Analysis of Finnish geographical research during the period since Independence. Brief sketch of geography in Finland since 1640. Trends in geographical research as shown in the periodical literature of the last 6 decades. Concerns articles by Finns and non-Finns on Finnish geography as well as articles by Finns on topics outside Finland.

990. Zonneveld, J.I.S. "Physical Geography in the Netherlands," Erdkunde, vol. 33, no. 1 (March 1979), 1-10.

   Dutch research in physical geography since 1930.

XIX. OTHER COUNTRIES

991. Chatterjee, Shiba Prasad. Fifty Years of Science in India: Progress of Geography. Calcutta: Indian Science Congress Association, 1964. iv + 277 p.

The progress of geography in India during the half-century ending in 1964. Mostly deals with personalities and publications in the various subfields of geography in recent years. The first postgraduate offerings in geography were initiated at the Aligarh Muslim University in 1931. Supplement issued in 1968: Progress of Geography in India (1964-1968) (Calcutta: Indian Science Congress Association, 1968), vi + 82 p.

992. Dean, William G., et al. "Canadian Geography, 1967," The Canadian Geographer, vol. 11, no. 4 (1967), 195-371.

The centennial issue of The Canadian Geographer marking the 100th anniversary of Canadian Confederation. (Page 195) "... the papers herein explore many of the major developments in geography in Canada up to 1967 and the prospects for the future." See especially J. Lewis Robinson, "Growth and Trends in Geography in Canadian Universities" (pp. 216-229), for the teaching of geography in the 19th and early 20th centuries as well as the present scene. Most of the other chapters concern recent research and publication in the various branches of geography and give little attention to the development of the subject. An exception is J.T. Parry's "Geomorphology in Canada" (pp. 280-311), which provides considerable background material, not only on geomorphology but also on the early development of geography in the universities (thus overlapping Robinson's article).

993. Geiger, Pedro P. "The Development of Geography in Brazil," East Lakes Geographer, vol. 6 (1970), 56-62.

(Page 56) "Modern scientific geography first appeared in Brazil in the 1930's. This was the period when the first Faculties of Philosophy were established in the universities, in which geography was first taught as a separate course of study." Describes influence of several foreign geographers, mostly Europeans, in the 1940s and 1950s.

994. Hamelin, Louis-Edmond. "Petite histoire de la géographie dans le Québec et à l'Université Laval," Cahiers de géographie de Québec, no. 13 (October 1962-March 1963), 137-152.

Société de géographie de Québec, the oldest geographical society in Canada, was founded in 1877. The first French-Canadian professional geographers were Benoît Brouillette and Pierre Dagenais (1930s). Visits of French geographers began in 1928 with Raoul Blanchard. Most of the developments in the universities began after World War II.

995. Inouye, Syuzi. "Die japanische Geographie der letzten zehn Jahre," Geographische Zeitschrift, vol. 44, nos. 7-8 (July 1938), 284-289.

   The work of Japanese geographers during the previous 10 years.

996. Jeon, Sang-woon. "Geography and Cartography," pp. 273-315 in Science and Technology in Korea: Traditional Instruments and Techniques (The M.I.T. East Asian Science Series, 4) (Cambridge, Massachusetts: The MIT Press, 1974).

   Geography and cartography (emphasis on latter) in Korea from the 4th century to 1866. Based largely on book previously published in Korean in 1966.

997. Nakano, T. "Some Prevailing Trends of the Historical Development of Geosciences in the Far East," Geoforum, vol. 1, no. 3 (1970), 63-80.

   Geological and physical geographical studies in the Far East (Japan, China, and Southeast Asia) since the 16th century.

998. Pinchemel, Philippe. "L'histoire de la géographie japonaise," L'espace géographique, vol. 9, no. 2 (April-June 1980), 165-171.

   Japanese geography since the 8th century, with emphasis on the 20th century.

999. Reichman, Shalom, and Gerson, Ran. "Uniqueness and Generality in Israeli Geography," pp. 24-35 in Geography in Israel, ed. D.H.K. Amiran and Y. Ben-Arieh (Jerusalem: The Israel National Committee of the International Geographical Union, 1976).

   Trends in Israeli geography in the last 2 generations, with emphasis on works produced in the 1960s and '70s.

Fig. 10. The Italian geographer Giotto Dainelli at his field camp near Lake Tana in Ethiopia, 1937 or 1938. Reproduced through the courtesy of the Società Geografica Italiana and Dr. Claudia Mancini.

Fig. 11. An interior view of the Villa Celimontana in Rome, the seat of the Italian Geographical Society (Società Geografica Italiana), one of the grand old geographical societies that were prominent in the 19th and early 20th centuries and in the latter half of the 20th century. Reproduced through the courtesy of the Società Geografica Italiana

Part Two--Geography in Various Countries 209

1000. Robinson, J. Lewis. "The Development and Status of Geography in Universities and Government in Canada," Yearbook of the Association of Pacific Coast Geographers, vol. 13 (1951), 3-13.

Emergence of professional geography in Canada in the 20th century. Although geography was first taught at the University of New Brunswick in 1800, the real development of geography as an academic discipline did not come until the 20th century, especially after 1935, when "the first full department of geography in Canada was established at [the University of] Toronto."

1001. Roh, Do-yang. "History of Korean Geography," Korea Journal, vol. 13, no. 12 (December 1973), 19-30.

History of geography in Korea from ancient times (Samje) to the present. Four periods: (1) Ancient (Samje), (2) Medieval (Geomantic), (3) Recent (Topographical), and (4) Modern (Scientific) (from the time of the Japanese colonial rule to the present).

1002. Spate, O.H.K., and Jennings, J.N. "Australian Geography 1951-1971," Australian Geographical Studies, vol. 9 (October 1972), 197-224.

Sections 1 and 2, "The Rise of Australian Geography" and "Developments in Human Geography," written by Spate; Section 3, "Gaps and Emphases in Physical Geography," by Jennings. Rise of geography in Australia since 1951, with emphasis on the expansion of the subject in universities. In 1951 the "full-time fully-recognised university geographers in Australia" numbered only 6, but by 1971 "there were altogether over 140 Lecturers and above, including some 20 Professors."

1003. Suizu, Ichiro, et al. Geographical Languages in Different Times and Places: Japanese Contributions to the History of Geographical Thoughts [sic]. Kyoto: Geography Institute, Kyoto University, 1980. 130 p.

A miscellany of essays on the history of geography and cartography, with emphasis on the history of geography in Japan. See especially the 9 chapters in Part IV, "Encounter of Japan and Europe." Parts III and IV concern the geographical "languages" of Japan, China, and India--i.e., the various geographical concepts of the peoples of East and South Asia.

1004. Takeuchi, Keiichi. "Some Remarks on the History of Regional Description and the Tradition of Regionalism in Japan," *Progress in Human Geography*, vol. 4, no. 2 (June 1980), 238-248.

   Geography in Japan since 1868, with emphasis on local and regional studies.

1005. Tümertekin, Erol. "The Development of Human Geography in Turkey," pp. 6-18 in *Turkey: Geographic and Social Perspectives*, eds. Peter Benedict, Erol Tümertekin, and Fatma Mansur ("Social, Economic and Political Studies of the Middle East," vol. 9) (Leiden: E. J. Brill, 1974).

   Emergence of academic geography in Turkey from the 1870s onward, with emphasis on human geography. French influences predominated at first, but in 1915 an Institute of Geography was created with German help in the University of Istanbul.

PART THREE

BIOGRAPHICAL WORKS

XX. THE UNITED STATES

Unlike the previous sections of the bibliography, the biographical sections are arranged by the name of the subject. Works by particular authors can easily be found through the use of the author index.

ATWOOD, Wallace Walter

1006. Koelsch, William A. "Wallace Walter Atwood, 1872-1949," <u>Geographers: Biobibliographical Studies</u>, vol. 3 (1979), 13-18.

Wallace Atwood (1872-1949) was President of Clark University, 1920-1946, and founded the Clark Graduate School of Geography in 1921.

BARRETT, Robert LeMoyne

1007. Martin, Geoffrey J. "Robert LeMoyne Barrett, 1871-1969: Last of the Founding Members of the Association of American Geographers," <u>Professional Geographer</u>, vol. 24, no. 1 (February 1972), 29-31.

The American geographer and traveler Robert Barrett (1871-1969) was the longest-lived charter member of the Association of American Geographers (founded 1904).

BINGHAM, Millicent Todd

1008. Berman, Mildred. "Millicent Todd Bingham: Human Geographer and Literary Scholar," <u>Professional Geographer</u>, vol. 32, no. 2 (May 1980), 199-204.

Millicent Todd Bingham (1880-1968) was the only woman to earn a Ph.D. in geography from Harvard University (1923). After 1929 she left geography to take up her mother's work in editing the poems and letters of Emily Dickinson.

BLODGET, Lorin

1009. Dunbar, Gary S. "Lorin Blodget, 1823-1901," Geographers: Biobibliographical Studies, vol. 5 (1981), 9-12.

Life of the American climatologist and statistician Lorin Blodget (1823-1901), "The Father of American Climatology."

BOAS, Franz

1010. Benison, Saul. "Geography and the Early Career of Franz Boas," American Anthropologist, vol. 51, no. 3 (July-September 1949), 523-526.

The German-American anthropologist Franz Boas (1858-1942) and his early interests in the field of geography.

1011. Speth, William W. "The Anthropogeographic Theory of Franz Boas," Anthropos, vol. 73, nos. 1-2 (1978), 1-31.

The German-American anthropologist Franz Boas (1858-1942), professor of anthropology at Columbia University (1896-1936), was trained in geography and taught the subject at the University of Berlin (1886) but left geography for anthropology in 1888.

1012. Trindell, Roger T. "Franz Boas and American Geography," Professional Geographer, vol. 21, no. 5 (September 1969), 328-332.

The German-American anthropologist Franz Boas (1858-1942) was trained in geography and might have continued in that field if he had had proper encouragement when he came to the United States in 1887.

BOWMAN, Isaiah

1013. Coughlan, Robert. "Isaiah Bowman," Life, vol. 19, no. 17 (22 October 1945), 118-120, 123-126, 129.

Biographical sketch of the American geographer Isaiah Bowman (1878-1950).

1014. Knadler, George A. "Isaiah Bowman," pp. 137-138 in Vol. 2 of International Encyclopedia of the Social Sciences (New York: The Macmillan Company and The Free Press, 1968).

A biographical sketch of Isaiah Bowman (1878-1950).

1015. Martin, Geoffrey J. "Isaiah Bowman," Geographers: Biobibliographical Studies, vol. 1 (1977), 9-18.

A biographical sketch of Isaiah Bowman (1878-1950).

1016. Martin, Geoffrey J. The Life and Thought of Isaiah Bowman. Hamden, Connecticut: Archon Books, 1980. xvi + 272 p.

Biography of the American geographer Isaiah Bowman (1878-1950), Director of the American Geographical Society (1915-1935), and President of Johns Hopkins University (1935-1948).

1017. Smith, Neil. "Political Geographers of the Past. Isaiah Bowman: Political Geography and Geopolitics," Political Geography Quarterly, vol. 3, no. 1 (January 1984), 69-76.

The American geographer Isaiah Bowman (1878-1950), the author of The New World (1921), took pains to distinguish his "scientific" political geography from the non-scientific geopolitics of the Nazis, but the present author shows how Bowman actually mixed science and politics "in fine geopolitical style."

1018. Wright, John K., and Carter, George F. "Isaiah Bowman," pp. 39-64 in Vol. 33 of Biographical Memoirs, National Academy of Sciences (New York: Columbia University Press, 1959).

A biographical memoir of the American geographer Isaiah Bowman (1878-1950).

BRIGHAM, Albert Perry

1019. James, Preston E. "Albert Perry Brigham, 1855-1932," Geographers: Biobibliographical Studies, vol. 2 (1978), 13-19.

A biographical sketch of Albert Perry Brigham (1855-1932), who taught geography and geology at Colgate University from 1892 to 1925.

BROOKS, Alfred Hulse

1020. Sherwood, Morgan B. "Alfred Hulse Brooks, 1871-1924," Geographers: Biobibliographical Studies, vol. 1 (1977), 19-23.

Alfred Brooks (1871-1924) was a geologist and geographer who served in the U.S. Geological Survey from 1894 (Alaska from 1898). Chief of the Alaska unit of the USGS from 1903.

BROWN, Ralph Hall

1021. Darby, H. Clifford. "Ralph H. Brown," pp. 155-156 in Vol. 2 of International Encyclopedia of the Social Sciences (New York: The Macmillan Company and the Free Press, 1968).

A biographical sketch of the American geographer Ralph H. Brown (1898-1948).

1022. Miles, Linda Jeanne. "Ralph Hall Brown: Gentlescholar [sic] of American Geography," Unpublished Ph.D. thesis in Geography, University of Oklahoma, 1982. vii + 170 p.

Biography of Ralph Hall Brown (1898-1948), professor of geography at the University of Minnesota, 1929-1948.

COLBY, Charles Carlyle

1023. Calef, Wesley. "Charles Carlyle Colby, 1884-1965," Geographers: Biobibliographical Studies, vol. 6 (1982), 17-22.

Biographical sketch of the American geographer Charles Colby (1885-1965), who taught at the University of Chicago from 1916 to 1949.

COOK, O.F.

1024. Gade, Daniel W. "The Contributions of O.F. Cook to Cultural Geography," Professional Geographer, vol. 22, no. 4 (July 1970), 206-209.

O.F. Cook (1867-1949) was a biologist with the United States Department of Agriculture from 1898 to 1937.

Part Three--Biographical Works

This paper concerns Cook's work on human modification of natural vegetation and transoceanic diffusion.

CRESSEY, George Babcock

1025. James, Preston E., and Perejda, Andrew D. "George Babcock Cressey, 1896-1963," *Geographers: Biobibliographical Studies*, vol. 5 (1981), 21-25.

The American geographer George Cressey (1896-1963), professor of geography at Syracuse University, 1931-1963; known for his textbooks on Asia and the Soviet Union.

DALY, Charles Patrick

1026. Hammond, Harold E. *A Commoner's Judge: The Life and Times of Charles Patrick Daly*. Boston: The Christopher Publishing House, 1954. 456 p.

A biography of the New York judge Charles Daly (1816-1899), who was President of the American Geographical Society from 1864 to 1899.

DARBY, William

1027. Kennedy, J. Gerald. *The Astonished Traveler: William Darby, Frontier Geographer and Man of Letters*. Baton Rouge: Louisiana State University Press, 1981. xiii + 238 p.

A biography of the American geographer William Darby (1775-1854), author of *A Geographical Description of the State of Louisiana* (1816) and various other geographies and gazetteers.

DAVIDSON, George

1028. Davenport, Charles B. "Biographical Memoir of George Davidson, 1825-1911," National Academy of Sciences, *Biographical Memoirs*, vol. 18, no. 9 (1938), 189-217.

Biography of the American scientist George Davidson (1825-1911), who served in the Coast Survey (renamed Coast and Geodetic Survey in 1878) from 1845 to 1895 and subsequently as professor of geography at the University of California.

1029. Dunbar, Gary S. "George Davidson, 1825-1911," *Geographers: Biobibliographical Studies*, vol. 2 (1978), 33-37.

A biographical sketch of the American geodesist and geographer George Davidson (1825-1911).

1030. King, William F. "George Davidson: Pacific Coast Scientist for the U.S. Coast and Geodetic Survey, 1845-1895," Unpublished Ph.D. thesis in History, Claremont Graduate School, Claremont, California, 1973. vii + 306 p.

George Davidson's (1825-1911) 50-year career with the U.S. Coast Survey (renamed Coast and Geodetic Survey in 1878).

1031. Lewis, Oscar. George Davidson, Pioneer West Coast Scientist. Berkeley and Los Angeles: University of California Press, 1954. viii + 146 p.

Biography of the American scientist George Davidson (1825-1911), who served with the U.S. Coast Survey for 50 years before becoming professor of geography at the University of California in 1898.

1032. Sherwood, Morgan B. "A Pioneer Scientist in the Far North: George Davidson and the Development of Alaska," Pacific Northwest Quarterly, vol. 53, no. 2 (April 1962), 77-80.

The Alaska experiences of the American geographer George Davidson (1825-1911) in the period 1867-1869 and his subsequent interest in the development of that area.

1033. Wagner, Henry R. "George Davidson, Geographer of the Northwest Coast of America," Quarterly of the California Historical Society, vol. 11, no. 4 (December 1932), 299-320.

The experiences of George Davidson (1825-1911) on the west coast of North America, from California northward.

DAVIS, William Morris

1034. Baulig, Henri. "L'oeuvre de William Morris Davis," L'information géographique, vol. 12, no. 3 (May-June 1948), 101-108.

A sketch of the life and work of the American geographer William Morris Davis (1850-1934).

Part Three--Biographical Works

1035. Beckinsale, Robert P. "W.M. Davis and American Geography: 1880-1934," pp. 107-122 in The Origins of Academic Geography in the United States, ed. Brian W. Blouet (Hamden, Connecticut: Archon Books, 1981).

Davis' leadership in the "New Geography" in the United States.

1036. Beckinsale, Robert P., and Chorley, Richard J. "William Morris Davis, 1850-1934," Geographers: Biobibliographical Studies, vol. 5 (1981), 27-33.

A biographical sketch of the American geographer William Morris Davis (1850-1934), who taught at Harvard University from 1878 to 1912.

1037. Broc, Numa. "Davis et la France," Bulletin de la Société languedocienne de géographie, vol. 8, no. 1 (1974), 87-95.

Explores the connections between the American geographer William Morris Davis (1850-1934) and France, which was the European country where his ideas found their warmest reception.

1038. Chorley, Richard J.; Beckinsale, Robert P.; and Dunn, Antony J. The History of the Study of Landforms, or The Development of Geomorphology, Vol. 2, The Life and Work of William Morris Davis. London: Methuen & Co. Ltd., 1973. xxii + 874 p.

A biography of the American geomorphologist William Morris Davis (1850-1934), who taught at Harvard University from 1878 to 1912 and founded the Association of American Geographers in 1904.

1039. Daly, Reginald A. "Biographical Memoir of William Morris Davis, 1850-1934," National Academy of Sciences, Biographical Memoirs, vol. 23, no. 11 (1945), 263-303.

Biography of the American geographer William Morris Davis (1850-1934).

1040. Hartshorne, Richard. "William Morris Davis--The Course of Development of His Concept of Geography," pp. 139-149 in The Origins of Academic Geography in the United States, ed. Brian W. Blouet (Hamden, Connecticut: Archon Books, 1981).

Davis' changing view of geography, from the 1870s to 1923.

1041. Huntington, Ellsworth. "William Morris Davis, Geographer," Bulletin of the Geographical Society of Philadelphia, vol. 10, no. 4 (October 1912), 224-234.

Contributions of the American geographer William Morris Davis (1850-1934), particularly in the field of geomorphology.

1042. Judson, Sheldon. "William Morris Davis--an Appraisal," Zeitschrift für Geomorphologie, vol. 4, nos. 3-4 (December 1960), 193-201.

Life and work of the Harvard geographer and geomorphologist William Morris Davis (1850-1934), with contemporary and posthumous reactions by geographers and geologists.

1043. Martin, Lawrence. "William Morris Davis: Investigator, Teacher, and Leader in Geomorphology," Annals of the Association of American Geographers, vol. 40, no. 3 (September 1950), 172-177.

Appreciation of the life and work of the Harvard geographer William Morris Davis (1850-1934). Followed by remarks by O.D. von Engeln, Roderick Peattie, Wellington Jones, and J.K. Wright.

DE BRAHM, William Gerard

1044. De Vorsey, Louis. "William Gerard De Brahm: Eccentric Genius of Southeastern Geography," Southeastern Geographer, vol. 10, no. 1 (April 1970), 21-29.

The work of the German-born W.G. De Brahm (1717-1799) in the survey of the southern colonies (later southeastern United States).

DWIGHT, Timothy

1045. Whitford, Kathryn, and Whitford, Philip. "Timothy Dwight's Place in Eighteenth-Century Science," Proceedings of the American Philosophical Society, vol. 114, no. 1 (16 February 1970), 60-71.

The American scientist, historian, and college president (Yale) Timothy Dwight (1752-1817), whose Travels in New England and New York (4 vols., 1821-

Part Three--Biographical Works 219

1822) is here compared with Thomas Jefferson's Notes on the State of Virginia.

EVANS, Lewis

1046. Klinefelter, Walter. Lewis Evans and His Maps. Transactions of the American Philosophical Society, n.s. vol. 61, part 7 (July 1971). 65 p.

The Welsh-born American geographer and cartographer Lewis Evans (c. 1699-1756) of Pennyslvania, whose best-known map was "A general Map of the Middle British Colonies" (1755).

FAIRBANKS, Harold

1047. Pattison, William. "Harold W. Fairbanks, California Geographer," Journal of Geography, vol. 59, no. 8 (November 1960), 351-357.

Harold Fairbanks (1860-1952) was a geographer, geologist, and author of numerous schoolbooks on geography.

GILBERT, Grove Karl

1048. Baulig, Henri. "La leçon de Grove Karl Gilbert," Annales de géographie, vol. 67, no. 362 (July-August 1958), 289-307.

Contributions of the American geomorphologist Grove Karl Gilbert (1843-1918).

1049. Davis, William Morris. "Biographical Memoir, Grove Karl Gilbert, 1843-1918," National Academy of Sciences, Memoirs, vol. 21, no. 5 (1927). v + 303 p.

Biography of the American geomorphologist Grove Karl Gilbert (1843-1918).

1050. James, Preston E. "Grove Karl Gilbert, 1843-1918," Geographers: Biobibliographical Studies, vol. 1 (1977), 25-33.

Grove Karl Gilbert (1843-1918) was a geologist and physical geographer who served with the U.S. Geological Survey from its founding in 1879.

1051. Penck, Albrecht. "Grove Karl Gilbert," Zeitschrift der Gesellschaft für Erdkunde zu Berlin, 1929, nos. 7-8, pp. 265-278.

An appreciation of the life and work of the American geologist and geomorphologist G.K. Gilbert (1843-1918).

1052. Pyne, Stephen J. *Grove Karl Gilbert: A Great Engine of Research*. Austin: University of Texas Press, 1980. xiv + 306 p.

Biography of the American geologist and geomorphologist Grove Karl Gilbert (1843-1918), who served with the U.S. Geological Survey from its founding in 1879 (Chief Geologist, 1889-1892).

1053. Yochelson, Ellis L., ed. "The Scientific Ideas of G.K. Gilbert," Geological Society of America, *Special Paper* 183 (1980). viii + 148 p.

14 papers (by 15 authors) on the work of the American geologist and geomorphologist Grove Karl Gilbert (1843-1918).

GILMAN, Daniel Coit

1054. Wright, John K. "Daniel Coit Gilman, Geographer and Historian," *Geographical Review*, vol. 51, no. 3 (July 1961), 381-399.

The American scholar Daniel Coit Gilman (1831-1908) taught geography at Yale University from 1863 to 1872, after which he served as president of the University of California and Johns Hopkins University.

GUYOT, Arnold Henri

1055. Anstey, Robert L. "Arnold Guyot, Teacher of Geography," *Journal of Geography*, vol. 57, no. 9 (December 1958), 441-449.

The Swiss-American geographer Arnold Guyot (1807-1884) taught geography and geology at the College of New Jersey (Princeton University) from 1854.

1056. Anstey, Robert L., comp. "A List of the Known Maps, Books, Articles, and Lectures by Arnold Guyot," Special Libraries Association, Geography and Map Division, *Bulletin*, 39 (February 1960), 7-18.

Chronological list of the publications of the Princeton geographer Arnold Guyot.

Part Three--Biographical Works 221

1057. Anstey, Robert L. "The Search for Guyot's Geographical Writings," <u>Professional Geographer</u>, vol. 4, no. 4 (July 1952), 8-11.

   Writings of the Swiss-American geographer Arnold Guyot. Search resulted in the "List" (above).

1058. Ferrell, Edith H. "Arnold Henry Guyot, 1807-1884," <u>Geographers: Biobibliographical Studies</u>, vol. 5 (1981), 63-71.

   The Swiss-American geographer Arnold Guyot (1807-1884), professor of physical geography and geology at the College of New Jersey (Princeton).

1059. Jones, Leonard C. <u>Arnold Guyot et Princeton</u>. Université de Neuchâtel, Faculté des Lettres, Recueil de Travaux, no. 14 (1929). 125 p.

   The Swiss-American geographer and geologist Arnold Guyot (1807-1884) taught at the College of New Jersey (renamed Princeton University in 1896) from 1854 onward.

1060. Jones, Leonard C. "Arnold Henry Guyot," <u>Union College Bulletin</u>, vol. 23, no. 2 (<u>Faculty Papers of Union College</u>, vol. 1, no. 1) (January 1930), 31-65.

   Biography of A.H. Guyot (1807-1884). Based largely on Jones' Neuchatel monograph (above).

HILGARD, Eugene Woldemar

1061. Jenny, Hans. <u>E.W. Hilgard and the Birth of Modern Soil Science</u>. (Collana della Rivista "Agrochimica," 3.) Pisa, Italy: Agrochimica, 1961. 144 p.

   Biography of Eugene Woldemar Hilgard (1833-1916), Professor of Agricultural Botany, University of California, Berkeley, 1875-1906. Closely paralleled Russian work in soil science. Hilgard "was an early discoverer of the soil profile and the climatic principle in soil formation" (p. 76).

HILLARD, George Stillman

1062. Manheim, Frank J. "George Stillman Hillard--An Early American Apostle of Human Geography," <u>Bulletin of the Geographical Society of Philadelphia</u>, vol. 35, no. 1 (January 1937), 1-7.

George Hillard (1808-1879) "was the first American who definitely recognized and publicized the intimate relation between history and geography" and "aided in making the Guyot-Ritter revolution possible in this country." Article mostly concerns Hillard's 1845 lecture, "The Connection between Geography and History."

HUNTINGTON, Ellsworth

1063. Martin, Geoffrey J. <u>Ellsworth Huntington, His Life and Work</u>. Hamden, Connecticut: Archon Books, 1973. xx + 315 p.

Biography of the American geographer Ellsworth Huntington (1876-1947).

1064. Spate, O.H.K. "Ellsworth Huntington," pp. 26-27 in Vol. 7 of <u>International Encyclopedia of the Social Sciences</u> (New York: The Macmillan Company and The Free Press, 1968).

"Ellsworth Huntington (1876-1947), American geographer, was the most notable exponent of environmentalism in the English-speaking world in the twentieth century."

JAMES, Preston E.

1065. Robinson, David J., ed. <u>Studying Latin America: Essays in Honor of Preston E. James</u>. (Dellplain Latin American Studies, 4) Ann Arbor, Michigan: University Microfilms International, 1980. xiii + 273 p.

A festschrift in honor of Preston James (b. 1899), professor emeritus of geography at Syracuse University, with emphasis on his Latin American interests. Contains 8 essays by as many authors, including a biographical sketch of James by the editor (pp. 1-101) and a paper on 20th-century Brazilian geography by Kempton E. Webb (pp. 177-194).

JEFFERSON, Mark S.W.

1066. Martin, Geoffrey J. "'The Law of Primate Cities' Re-examined," <u>Journal of Geography</u>, vol. 60, no. 4 (April 1961), 165-172.

The contributions of the American geographer Mark Jefferson (1863-1949) to urban geography, with special attention to "The Law of Primate Cities" (1938,

Part Three--Biographical Works 223

published in 1939 as "The Law of the Primate City"). Jefferson's 1938 data compared with 1959.

1067. Martin, Geoffrey J. <u>Mark Jefferson: Geographer</u>. Ypsilanti: Eastern Michigan University Press, 1968. x + 370 p.

Biography of the American geographer Mark S.W. Jefferson (1863-1949), who taught in the Michigan State Normal College, Ypsilanti, 1901-1939.

JEFFERSON, Thomas

1068. Clark, Austin H. "Thomas Jefferson and Science," <u>Journal of the Washington Academy of Sciences</u>, vol. 33, no. 7 (15 July 1943), 193-203.

The scientific (including the geographical) observations and activities of the American President Thomas Jefferson (1743-1826).

1069. Dunbar, Gary S. "Thomas Jefferson, Geographer," Special Libraries Association, Geography and Map Division, <u>Bulletin</u>, no. 40 (April 1960), 11-16.

The geographical interests of the American President Thomas Jefferson (1743-1826).

1070. Greely, A.W. "Jefferson as a Geographer," <u>National Geographic Magazine</u>, vol. 7, no. 8 (August 1896), 269-271.

The geographical interests of Thomas Jefferson (1743-1826), third President of the United States; "... of all our Presidents Jefferson is the only one of whom we can say, 'He was a geographer.'" Paper reprinted on pp. i-vii of Vol. 13 of <u>The Writings of Thomas Jefferson</u>, ed. Andrew A. Lipscomb (Washington: The Thomas Jefferson Memorial Association, 1905).

1071. Martin, Edwin T. <u>Thomas Jefferson: Scientist</u>. New York: Henry Schuman, 1952. x + 289 p.

The scientific (including geographical) interests of the American President Thomas Jefferson (1743-1826).

1072. Oliver, John W. "Thomas Jefferson--Scientist," <u>Scientific Monthly</u>, vol. 56, no. 5 (May 1943), 460-467.

The scientific (including geographical) observations and activities of the American President Thomas Jefferson (1743-1826).

1073. Surface, George T. "Thomas Jefferson: A Pioneer Student of American Geography," <u>Bulletin of the American Geographical Society</u>, vol. 41, no. 12 (December 1909), 743-750.

The geographical interests of the American President Thomas Jefferson (1743-1826).

JOHNSON, Douglas Wilson

1074. Bucher, Walter H. "Biographical Memoir of Douglas Wilson Johnson, 1878-1944," National Academy of Sciences, <u>Biographical Memoirs</u>, vol. 23, no. 5 (1947), 197-230.

Biography of the American geomorphologist D.W. Johnson (1878-1944).

MARSH, George Perkins

1075. Ekirch, Arthur A., Jr. "George Perkins Marsh: Pioneer," pp. 70-80 in <u>Man and Nature in America</u> by Ekirch (New York: Columbia University Press, 1963).

The American geographer and diplomat G.P. Marsh (1801-1882) in his role as a pioneer conservationist.

1076. Faybusovich, E.L. "Centennial of the Russian Edition of George Perkins Marsh's <u>Man and Nature</u>," <u>Soviet Geography: Review and Translation</u>, vol. 9, no. 10 (December 1968), 886-890.

"In 1866, two years after publication in the United States, there appeared a Russian translation of a book by the American geographer George Perkins Marsh titled <u>Man and Nature</u>." Marsh was largely ignored in the West. "Only in Russia were the progressive ideas of Marsh promptly accepted." Translation of Russian paper originally published in 1967.

1077. Lowenthal, David. "George Perkins Marsh," pp. 23-25 in Vol. 10 of <u>International Encyclopedia of the Social Sciences</u> (New York: The Macmillan Company and The Free Press, 1968).

"George Perkins Marsh (1801-1882), an American geographer, is known today primarily as the founding father of the conservation movement."

1078. Lowenthal, David. "George Perkins Marsh and the American Geographical Tradition," Geographical Review, vol. 43, no. 2 (April 1953), 207-213.

Geographical interests and views of the American geographer, conservationist, and diplomat George Perkins Marsh (1801-1882).

1079. Lowenthal, David. "George Perkins Marsh and the Nature and Purpose of Geography," Geographical Journal, vol. 126, part 4 (December 1960), 413-417.

Hitherto unpublished manuscript that G.P. Marsh (1801-1882) had written in 1869 to serve as an introduction to an American edition of Elisée Reclus' La terre. (Page 413) "American geographers before the twentieth century devoted little attention to the history and methodology of the field.... The dearth of methodological records lends the more significance to [Marsh's essay].... Although Marsh never considered himself a professional geographer, his book 'Man and Nature' [1864] was the most important geographical work of the nineteenth century.... [This statement] is the only one in which Marsh specifically concerned himself with the nature and purpose of geography."

1080. Lowenthal, David. George Perkins Marsh, Versatile Vermonter. New York: Columbia University Press, 1958. 441 p.

Biography of the American geographer, conservationist, and diplomat George Perkins Marsh (1801-1882).

MAURY, Matthew Fontaine

1081. Brown, Ralph Minthorne. "Bibliography of Commander Matthew Fontaine Maury, including a Biographical Sketch," Bulletin of the Virginia Polytechnic Institute, vol. 24, no. 2 (1930). 61 p.

Biobibliographical work on the American naval officer and physical geographer Matthew Fontaine Maury (1806-1873), author of Physical Geography of the Sea (1855).

1082. Corbin, Diana Fontaine Maury. A Life of Matthew Fontaine Maury. London: Sampson Low, Marston, Searle, & Rivington, 1888. vi + 326 p.

Life of oceanographer and naval officer M.F. Maury (1806-1873) by his daughter.

1083. Cotter, Charles H. "Matthew Fontaine Maury (1806-1873): 'Pathfinder of the Seas,'" Journal of Navigation, vol. 32, no. 1 (January 1979), 75-83.

Biographical sketch of the American pioneer of oceanography M.F. Maury (1806-1873).

1084. Du Val, Miles. "Matthew Fontaine Maury: Benefactor of Mankind," Explorers Journal, vol. 43, no. 4 (December 1965), 203-215.

Life of American naval officer M.F. Maury (1806-1873).

1085. Hansen, Robert C. "The Cartographic Contributions of Matthew Fontaine Maury," Special Libraries Association, Geography and Map Division, Bulletin, no. 119 (March 1980), 25-29.

The cartographic work of the American naval officer M.F. Maury (1806-1873). (Page 30) "His cartographic contributions, particularly his wind and current charts, remain his most important achievement."

1086. Klapp, Orrin E. "Matthew Fontaine Maury, Naval Scientist," United States Naval Institute Proceedings, vol. 71, no. 11 (November 1945), 1315-1325.

Biographical sketch of the American naval officer M.F. Maury (1806-1873), author of The Physical Geography of the Sea (1855).

1087. Leighly, John. "Matthew Fontaine Maury, 1806-1873," Geographers: Biobibliographical Studies, vol. 1 (1977), 59-63.

Biographical sketch of the American naval officer and physical geographer M.F. Maury (1806-1873).

1088. Lewis, Charles L. Matthew Fontaine Maury, the Pathfinder of the Seas. Annapolis, Maryland: The United States Naval Institute, 1927. xvii + 264 p.

The American naval officer and geographer M.F. Maury (1806-1873), author of Physical Geography of the Sea (1855).

1089. Wayland, John W. The Pathfinder of the Seas: The Life of Matthew Fontaine Maury. Richmond, Virginia: Garrett & Massie, Inc., 1930. xiii + 191 p.

Biography of the American naval officer and oceanographer M.F. Maury (1806-1873).

1090. Williams, Frances Leigh. Matthew Fontaine Maury, Scientist of the Sea. New Brunswick, New Jersey: Rutgers University Press, 1963. xx + 720 p.

Biography of the American naval officer M.F. Maury (1806-1873), author of Physical Geography of the Sea (1855).

MAY, Jacques

1091. Thouez, Jean-Pierre. "Jacques M. May, 1896-1975," Geographers: Biobibliographical Studies, vol. 7 (1983), 85-88.

The French-born American physician Jacques May (1896-1975) conducted research in medical geography at the American Geographical Society from 1948.

MORSE, Jedidiah

1092. Brown, Ralph Hall. "The American Geographies of Jedidiah Morse," Annals of the Association of American Geographers, vol. 31, no. 3 (September 1941), 145-217.

Jedidiah Morse (1761-1826) is sometimes called "The Father of American Geography" because he was a prominent writer of school geographies in the Federal Period.

1093. Brown, Ralph Hall. "A Letter to the Reverend Jedidiah Morse, Author of The American Universal Geography," Annals of the Association of American Geographers, vol. 41, no. 3 (September 1951), 188-198.

An appreciation of the geographical works of the American writer and clergyman Jedidiah Morse (1761-1826).

MUIR, John

1094. Leighly, John. "John Muir's Image of the West," Annals of the Association of American Geographers, vol. 48, no. 4 (December 1958), 309-318.

Geographical observations of the American naturalist John Muir (1838-1914) in the Sierra Nevada of California.

PARK, Robert

1095. Entrikin, John Nicholas. "Robert Park's Human Ecology and Human Geography," Annals of the Association of American Geographers, vol. 70, no. 1 (March 1980), 43-58.

The geographical views of the American sociologist Robert Park (1864-1944), who had been exposed to the German geographers Georg Gerland and Alfred Hettner in his student days.

PLATT, Robert Swanton

1096. Thoman, Richard S. "Robert Swanton Platt, 1891-1964," Geographers: Biobibliographical Studies, vol. 3 (1979), 107-116.

Robert S. Platt (1891-1964) taught geography at the University of Chicago, 1920-1957.

POWELL, John Wesley

1097. Darrah, William C. Powell of the Colorado. Princeton: Princeton University Press, 1951. ix + 426 p.

A biography of the American geographer, geologist, and ethnologist J.W. Powell (1834-1902).

1098. James, Preston E. "John Wesley Powell, 1834-1902," Geographers: Biobibliographical Studies, vol. 3 (1979), 117-124.

John Wesley Powell (1834-1902) was a geographer, geologist, and ethnologist who served as Director of the Bureau of Ethnology (Smithsonian) from 1879 and Director of the U.S. Geological Survey (1881-1894).

Part Three--Biographical Works   229

1099. Stegner, Wallace. Beyond the Hundredth Meridian: John
       Wesley Powell and the Second Opening of the West.
       Boston: Houghton Mifflin Company, 1954. xxiii + 438 p.

       The American geographer and geologist J.W. Powell
       (1834-1902) and his work in the scientific exploration
       of the western United States from 1868 onward.

RAISZ, Erwin Josephus

1100. Yacher, Leon.   "Erwin Josephus Raisz, 1893-1968,"
       Geographers: Biobibliographical Studies, vol. 6
       (1982), 93-97.

       The Hungarian-born American cartographer Erwin Raisz
       (1893-1968) worked at the Institute of Geographical
       Exploration at Harvard University from 1931 to 1950.

ROYCE, Josiah

1101. Entrikin, John Nicholas. "Royce's 'Provincialism': A
       Metaphysician's Social Geography," pp. 208-226 in
       Geography, Ideology and Social Concern, ed. David
       Stoddart (Oxford: Basil Blackwell, 1981).

       (Page 208) "The writings of America's foremost
       absolute idealist philosopher, Josiah Royce (1855-
       1916), provide an illustration of the similarity of
       many of the contemporary themes of humanist social
       geographers and the themes of idealist social
       philosophers of the nineteenth century."

RUSSELL, Richard Joel

1102. Walker, H. Jesse. "Richard Joel Russell, 1895-1971,"
       Geographers: Biobibliographical Studies, vol. 4
       (1980), 127-138.

       Richard J. Russell (1895-1971) was a physical
       geographer who taught at Louisiana State University
       from 1928 onward.

SALISBURY, Rollin

1103. Pattison, William D.   "Rollin D Salisbury, 1858-
       1922," Geographers: Biobibliographical Studies, vol.
       6 (1982), 105-113.

       The American geologist Rollin Salisbury (1858-1922)
       was Professor of Geology at the University of Chicago

from 1892 and founded the Department of Geography in 1903.

1104. Pattison, William D. "Rollin Salisbury and the Establishment of Geography at the University of Chicago," pp. 151-163 in The Origins of Academic Geography in the United States, ed. Brian W. Blouet (Hamden, Connecticut: Archon Books, 1981).

The work of the geologist Rollin Salisbury in establishing the Department of Geography at the University of Chicago in 1903.

1105. Visher, Stephen S. "Rollin D. Salisbury and Geography," Annals of the Association of American Geographers, vol. 43, no. 1 (March 1953), 4-11.

The work of the geologist and physiographer Rollin Salisbury (1858-1922) in establishing and directing the Department of Geography at the University of Chicago.

SAUER, Carl Ortwin

1106. Hooson, David J.M. "Carl O. Sauer," pp. 165-174 in The Origins of Academic Geography in the United States, ed. Brian W. Blouet (Hamden, Connecticut: Archon Books, 1981).

The legacy of Carl Sauer (1889-1975) and the "Berkeley School of Geography."

1107. Leighly, John. "Carl Ortwin Sauer, 1889-1975," Geographers: Biobibliographical Studies, vol. 2 (1978), 99-108.

Carl Sauer (1889-1975) was Professor of Geography at the University of California, Berkeley, 1923-1957.

1108. Parsons, James J. "Carl O. Sauer," pp. 17-19 in Vol. 14 of International Encyclopedia of the Social Sciences (New York: The Macmillan Company and The Free Press, 1968).

A biographical sketch of Carl Sauer (1889-1975), professor of geography at the University of California.

1109. Pfeifer, Gottfried. "Carl Ortwin Sauer zum 75. Geburtstage am 24.XII.1964," Geographische Zeitschrift, vol. 53, no. 1 (February 1965), 1-9.

Part Three--Biographical Works   231

An appreciation of the life and work of Carl Sauer (1889-1975), professor of geography at the University of California, Berkeley, from 1923. Followed by list of doctoral dissertations supervised by Sauer, 1927-1964: "Die Berkeleyer geographische Schule im Spiegel der unter Leitung und auf Anregung von C.O. Sauer hervorgegangenen Dissertationen 1927-1964," pp. 74-77.

1110. Speth, William W. "Carl Ortwin Sauer on Destructive Exploitation," *Biological Conservation*, vol. 11, no. 2 (February 1977), 145-160.

(From Abstract, p. 145) "A persistent theme in the work of Carl O. Sauer (1889-1975) is destructive exploitation, the process of economic devastation and its aftermath associated with Occidental commercial culture."

1111. West, Robert C. *Carl Sauer's Fieldwork in Latin America*. (Dellplain Latin American Studies, 3.) Ann Arbor, Michigan: University Microfilms International, 1979. xviii + 165 p.

The field experiences of Carl Sauer (1889-1975) in Latin America, 1926-1967, with emphasis on Mexico. Considerable biographical information included.

1112. Williams, Michael. "'The Apple of My Eye': Carl Sauer and Historical Geography," *Journal of Historical Geography*, vol. 9, no. 1 (January 1983), 1-28.

Concerns the interest of the American geographer Carl Sauer (1889-1975) in the time element in geography, as well as his views on academic and intellectual matters generally.

SEMPLE, Ellen Churchill

1113. Berman, Mildred. "Sex Discrimination and Geography: The Case of Ellen Churchill Semple," *Professional Geographer*, vol. 26, no. 1 (February 1974), 8-11.

The American geographer Ellen Churchill Semple (1863-1932) taught at the University of Chicago and Clark University. In her last will she revoked a gift of $1000 to Clark University, partly because she had discovered that the trustees had fixed her salary at $500 a year less than that of male professors.

1114. Broc, Numa. "Les classiques de Miss Semple: essai sur les sources des 'Influences of Geographic Environment,' 1911," Annales de géographie, vol. 90, no. 497 (January-February 1981), 87-102.

Sources used by the American geographer Ellen Churchill Semple (1863-1932) in writing her book Influences of Geographic Environment (1911).

1115. Bronson, Judith Conoyer. "Ellen Semple: Contributions to the History of American Geography," Unpublished Ph.D. thesis in History, Saint Louis University, 1973. vi + 263 p.

Biography of Ellen Churchill Semple (1863-1932), professor of geography at Chicago and Clark universities, and analysis of her work, preceded by a long chapter (pp. 14-80) on "American Geography before Ellen Semple."

1116. Bushong, Allen D. "Women as Geographers: Some Thoughts of Ellen Churchill Semple," Southeastern Geographer, vol. 15, no. 2 (November 1975), 102-109.

The attempts of Ellen Churchill Semple (1863-1932) to interest women in careers in geography.

1117. Gelfand, Lawrence. "Ellen Churchill Semple: Her Geographical Approach to American History," Journal of Geography, vol. 53, no. 1 (January 1954), 30-41.

The life and work of the American geographer Ellen Churchill Semple (1863-1932), with emphasis on her book American History and Its Geographic Conditions (1903).

1118. Hawley, Arthur J. "Environmental Perception: Nature and Ellen Churchill Semple," Southeastern Geographer, vol. 8 (1968), 54-59.

The uses of the term "nature" by the American geographer Ellen Churchill Semple (1863-1932).

1119. Sauer, Carl. "Ellen Churchill Semple," pp. 661-662 in vol. 13 of Encyclopaedia of the Social Sciences (London: Macmillan and Co. Ltd., 1934).

Ellen Churchill Semple (1863-1932) was a human geographer who taught at the University of Chicago and Clark University.

Part Three--Biographical Works 233

SHALER, Nathaniel Southgate

1120. Bacon, H. Philip. "Fireworks in the Classroom: Nathaniel Southgate Shaler," Journal of Geography, vol. 54, no. 7 (October 1955), 349-353.

   The American geologist and geographer Nathaniel Shaler (1841-1906) taught at Harvard University from 1864. "Shaler was an exceedingly entertaining lecturer," and his "lectures were commonly described as 'the fireworks.'"

1121. Koelsch, William A. "Nathaniel Southgate Shaler, 1841-1906," Geographers: Biobibliographical Studies, vol. 3 (1979), 133-139.

   Nathaniel S. Shaler (1841-1906) taught at Harvard University from 1864.

1122. Livingstone, David N. "Environment and Inheritance: Nathaniel Southgate Shaler and the American Frontier," pp. 123-138 in The Origins of Academic Geography in the United States, ed. Brian W. Blouet (Hamden, Connecticut: Archon Books, 1981).

   The views of Harvard professor N.S. Shaler (1841-1906) on the American West.

1123. Livingstone, David N. "Nature and Man in America: Nathaniel Southgate Shaler and the Conservation of Natural Resources," Institute of British Geographers, Transactions, n.s. vol. 5, no. 3 (1980), 369-382.

   Contributions of the Harvard geologist and geographer N.S. Shaler (1841-1906) to the conservation of natural resources.

1124. Shaler, Nathaniel. The Autobiography of Nathaniel Southgate Shaler, with a Supplementary Memoir by His Wife. Boston and New York: Houghton Mifflin Company, 1909. viii + 481 p.

   The autobiography of Harvard geologist and geographer N.S. Shaler (1841-1906).

SMITH, Joseph Russell

1125. Rowley, Virginia M. J. Russell Smith: Geographer, Educator, and Conservationist. Philadelphia: University of Pennsylvania Press, 1964. 247 p.

Biography of Joseph Russell Smith (1874-1966), professor of geography at the University of Pennsylvania and Columbia University. Appendix, pp. 202-217, "The Development of Modern Geography in Europe and America."

STEVENS, Isaac

1126. Meinig, Donald W. "Isaac Stevens: Practical Geographer of the Early Northwest," Geographical Review, vol. 45, no. 4 (October 1955), 542-558.

The geographical activities of Isaac Stevens (1818-1862), first governor of Washington Territory and leader of the Pacific Railroad Survey (northern route).

TURNER, Frederick Jackson

1127. Billington, Ray Allen. Frederick Jackson Turner: Historian, Scholar, Teacher. New York: Oxford University Press, 1973. x + 599 p.

Biography of the Harvard historian and historical geographer Frederick Jackson Turner (1861-1932).

1128. Block, Robert H. "Frederick Jackson Turner and American Geography," Annals of the Association of American Geographers, vol. 70, no. 1 (March 1980), 31-42.

The historian F.J. Turner (1861-1932) and his connections with geography and geographers.

ULLMAN, Edward

1129. Eyre, John D. "Edward Ullman: A Career Profile," pp. 1-15 in A Man for All Regions: The Contributions of Edward L. Ullman to Geography, ed. J.D. Eyre (Papers of the Fourth Carolina Geographical Symposium, 1977) (University of North Carolina at Chapel Hill, Department of Geography, Studies in Geography, no. 11, 1978).

A biographical sketch of the American geographer Edward Ullman (1912-1976), professor of geography at the University of Washington from 1951. Ullman's biographical data and bibliography are listed in an appendix (pp. 144-158).

Part Three--Biographical Works

WARD, Robert DeCourcy

1130. Koelsch, W.A. "Robert DeCourcy Ward, 1867-1931," Geographers: Biobibliographical Studies, vol. 7 (1983), 145-150.

The American climatologist Robert Ward (1867-1931) taught at Harvard University from 1895 onward (professor from 1910).

WISSLER, Clark

1131. Speth, William W. "Clark Wissler, 1870-1947," Geographers: Biobibliographical Studies, vol. 7 (1983), 151-154.

A biographical sketch of the American anthropologist Clark Wissler (1870-1947), curator of anthropology at the American Museum of Natural History in New York, with emphasis on his geographical work.

WRIGHT, John Kirtland

1132. Kahn, E.J., Jr. "Profiles: Big Geographer," The New Yorker, vol. 17, no. 24 (26 July 1941), 20-24, 26-28.

A biographical sketch of the American geographer John K. Wright (1891-1969) and a description of the activities of the American Geographical Society.

XXI. EUROPE

ABBADIE, Antoine d'

1133. Pastoureau, Mireille. "Antoine d'Abbadie, 1810-1897," Geographers: Biobibliographical Studies, vol. 3 (1979), 29-33.

Antoine d'Abbadie (1810-1897) was a French-Irish explorer in Ethiopia.

AJO, Reino

1134. Gould, Peter. "The Trail Blazer: Reino Ajo and Contemporary Geography," Terra, vol. 84, no. 2 (1972), 64-66.

Contributions of the Finnish geographer Reino Ajo (1902-1974).

AMUNDSEN, Roald

1135. Steinmetzler, Johannes. "Roald Amundsen (1872-1928)," Geographisches Taschenbuch 1960/61 (Wiesbaden: Franz Steiner Verlag GMBH, 1960), pp. 446-453.

Biographical sketch of the Norwegian polar explorer Roald Amundsen (1872-1928).

ANCEL, Jacques

1136. Specklin, Robert. "Jacques Ancel, 1882-1943," Geographers: Biobibliographical Studies, vol. 3 (1979), 1-6.

Jacques Ancel (1882-1943) was a French political geographer who taught at the Institut des hautes études internationales of the University of Paris from 1926.

ANDREE, Karl

1137. Plewe, Ernst. "Karl Theodor Andree (1808-1875)," Geographisches Taschenbuch 1977/78 (Wiesbaden: Franz Steiner Verlag GMBH, 1977), pp. 165-175.

The German economic geographer and writer Karl Andree (1808-1875), who founded Globus, a geographical and ethnological journal, in 1861.

ANUCHIN, Dmitry Nikolaevich

1138. Esakov, Vasily Alexeyevich. "Dmitry Nikolaevich Anuchin, 1843-1923," Geographers: Biobibliographical Studies, vol. 2 (1978), 1-8.

D.N. Anuchin (1843-1923) was "an eminent Russian geographer, anthropologist, ethnographer and archaeologist ... and the founder of the Russian school of geography at the University of Moscow" (p. 1).

ARBOS, Philippe

1139. Derruau, Max. "Philippe Arbos, 1882-1956," Geographers: Biobibliographical Studies, vol. 3 (1979), 7-12.

Philippe Arbos (1881-1956) was a French geographer who taught at the University of Clermont-Ferrand, 1919-1952.

Part Three--Biographical Works

ARQUÉ, Paul

1140. Papy, Louis. "Cavaillès, Arqué and Revert: Three Geographers of Bordeaux," Geographers: Biobibliographical Studies, vol. 7 (1983), 5-9. (Arque on pp. 7-8)

The French geographer Paul Arqué (1887-1970), who taught at the University of Bordeaux from 1945 to 1957.

AUERBACH, Bertrand

1141. Broc, Numa. "Bertrand Auerbach (1856-1942): un pionnier de la géographie en Lorraine," Revue géographique de l'Est, vol. 14, nos. 3-4 (July-December 1974), 411-415.

The French geographer Bertrand Auerbach (1856-1942) taught geography at the University of Nancy from 1885.

BAER, Karl Ernst von

1142. Heydenreich, Adolf. Karl Ernst von Baer als Geograph. (Inaugural-Dissertation, Munich [Technische Hochschule].) Munich: Theodor Ackermann, 1908. 87 p.

The Russian scientist Karl Ernst von Baer (1792-1876) was primarily a zoologist, but he produced much work in geography and geology.

1143. Raikov, Boris E. Karl Ernst von Baer, 1792-1876: Sein Leben und sein Werk. Acta Historica Leopoldina (Leipzig), no. 5 (1968). 516 p.

Biography of the Russian naturalist K.E. von Baer (1792-1876).

BALBI, Adriano

1144. Errera, Carlo. "Adriano Balbi," p. 406 in Vol. 2 of Encyclopaedia of the Social Sciences (London: Macmillan and Co. Ltd., 1930).

The Italian geographer Adriano Balbi (1782-1848) lived in Paris, 1821-1832, and published under the name Adrien Balbi.

BANKS, Joseph

1145. Cameron, Hector C. Sir Joseph Banks, the Autocrat of the Philosophers. London: The Batchworth Press, 1952. xx + 341 p.

Biography of the English scientist Joseph Banks (1743-1820).

BANSE, Ewald

1146. Banse, Ewald. Die Seele der Geographie: Geschichte einer Entwicklung. Braunschweig and Hamburg: Georg Westermann, 1924. 96 p.

Autobiography of German geographer Ewald Banse (1883-1953), stressing his views on geography and his role in the construction of the "New Geography."

BARANSKII, Nikolai N.

1147. Harris, Chauncy D. "Nikolai N. Baranskii," pp. 10-12 in Vol. 2 of International Encyclopedia of the Social Sciences (New York: The Macmillan Company and The Free Press, 1968).

Nikolai N. Baranskii (1881-1963), the founder of Soviet economic geography, was a professor at the Moscow State University.

1148. Hönsch, Fritz, and Lawrow, Sergej B. "Der Einfluss von N.N. Baranskii auf die Herausbildung der sowjetischen Geographie," Geographische Berichte, vol. 26, no. 3 (March 1981), 133-142.

The Soviet geographer N.N. Baranskii (1881-1963) and his influence on the development of Soviet geography and cartography.

BARTH, Heinrich

1149. Plewe, Ernst. "Heinrich Barth und Carl Ritter: Briefe und Urkunden," Die Erde, vol. 96, no. 4 (1965), 245-278.

Letters and documents illustrating the relations between the German geographers Carl Ritter (1779-1859) and Heinrich Barth (1821-1865) in the years 1848-1858.

1150. Plewe, Ernst. "Heinrich Barths Habilitation im Urteil von Carl Ritter und August Boeckh," Die Erde, vol. 94, no. 1 (1963), 5-12.

The opinions of Carl Ritter and August Boeckh on Heinrich Barth's habilitation thesis at the University of Berlin (1848). This paper demonstrates that, contrary to popular opinion, Ritter did indeed "interest scientists of the younger generation in his conception of geography."

Part Three--Biographical Works 239

1151. Schiffers, Heinrich, ed. Heinrich Barth, ein Forscher in Afrika: Leben--Werk--Leistung. Wiesbaden: Franz Steiner GmbH, 1967. x + 540 + unpaged section of 39 illustrations.

20 chapters on the life and work of the German African explorer and geographer Heinrich Barth (1821-1865). 6 chapters by Schiffers, including Chap. 3 (pp. 68-92) on Barth's contributions to geography. Also chapters by geographers Gerhard Engelmann, Hanno Beck, and Mansell Prothero.

1152. [Schiffers, Heinrich]. "Souvenirs de H. Barth (1821-1865)," Acta Geographica, nos. 69-70 (4th quarter 1967), 1-41.

Issue contains several papers, most of them by Schiffers, on the German geographer and explorer Heinrich Barth (1821-1865).

1153. Schubert, Gustav von. Heinrich Barth, der Bahnbrecher der deutschen Afrikaforschung. Berlin: Verlag von Dietrich Reimer, 1897. x + 184 p.

The life and work of the German geographer and African explorer Heinrich Barth (1821-1865).

BARTHOLOMEW, John George

1154. Allan, Douglas A. "John George Bartholomew, A Centenary," Scottish Geographical Magazine, vol. 76, no. 2 (September 1960), 85-88.

John George Bartholomew (1860-1920), Scottish cartographer and publisher, whose contributions to geography included support of the Royal Scottish Geographical Society and the first lectureship in geography at the University of Edinburgh.

BASTIAN, Adolf

1155. Achelis, Thomas. Adolf Bastian. (Sammlung gemeinverständlicher wissenschaftlicher Vorträge, neue folge, 6th series, no. 128.) Hamburg: Verlagsanstalt und Druckerei A.-G., 1891. 36 p.

Life and work of the German ethnographer Adolf Bastian (1826-1905), who was influential with geographers.

BAULIG, Henri

1156. Juillard, Etienne. "Henri Baulig (1877-1962)," pp. 119-131 in "Les géographes français," Secrétariat d'état aux universités, Comité des travaux historiques et scientifiques, Bulletin de la section de géographie, vol. 81 (For the years 1968-1974) (Paris: Bibliothèque Nationale, 1975).

Biographical sketch of the French geographer Henri Baulig (1877-1962), professor of geography at the University of Strasbourg from 1919. Baulig was a co-founder, with Emmanuel de Martonne, of modern French geomorphology.

1157. Juillard, Etienne, and Klein, Claude. "Henri Baulig, 1877-1962," Geographers: Biobibliographical Studies, vol. 4 (1980), 7-17.

Henri Baulig (1877-1962) was a French geomorphologist who taught at the University of Strasbourg, 1919-1947.

1158. Klein, Claude. "La leçon de Henri Baulig," Revue géographique de l'Est, vol. 17, nos. 3-4 (July-December 1977), 191-201.

An appraisal of the work of the French physical geographer Henri Baulig (1877-1962).

BENNETT, Arnold

1159. Hudson, Brian J. "The Geographical Imagination of Arnold Bennett," Institute of British Geographers, Transactions, n.s. vol. 7, no. 3 (1982), 365-379.

Geographical ideas of the English novelist Arnold Bennett (1867-1931).

BERG, Lev Semyonovich

1160. Gerasimov, I.P., et al. "The Centennial of L.S. Berg," Soviet Geography: Review and Translation, vol. 18, no. 1 (January 1977), 1-32.

Translations of 5 Russian articles marking the 100th anniversary of the birth of the Soviet geographer L.S. Berg (1876-1950).

1161. Murzayev, E.M. "Lev Semenovich Berg, 1876-1950," Geographers: Biobibliographical Studies, vol. 5 (1981), 1-7.

Life of the Soviet geographer L.S. Berg (1876-1950), professor and head of the Department of Physical Geography at the Geographical Institute, later at the University of Leningrad (1916-1950).

BERGHAUS, Heinrich

1162. Beck, Hanno. "Heinrich Berghaus und Alexander von Humboldt," Petermanns Geographische Mitteilungen, vol. 100, no. 1 (15 February 1956), 4-16.

Relations between the German cartographer Heinrich Berghaus (1797-1884) and the geographer Alexander von Humboldt (1769-1859) during the period 1815-1859. Emphasis on Berghaus and the plan (1848) for a geography textbook for schools in India.

1163. Engelmann, Gerhard. Heinrich Berghaus: Der Kartograph von Potsdam. Deutsche Akademie der Naturforscher Leopoldina (Halle/Saale), Acta Historica Leopoldina, no. 10 (1977). 411 p.

Life of the German cartographer Heinrich Berghaus (1797-1884). Text ends on p. 187, and the rest of the book is made up of notes and bibliography.

1164. Engelmann, Gerhard. "Heinrich Berghaus 1797-1884," Geographisches Taschenbuch 1979/80 (Wiesbaden: Franz Steiner Verlag GmbH, 1979), pp. 62-71.

Life of the German cartographer Heinrich Berghaus (1797-1884).

BERLIOUX, Etienne-Félix

1165. Broc, Numa. "Il y a un siècle: Etienne-Félix Berlioux," Revue de géographie de Lyon, vol. 50, no. 2 (1975), 167-170.

The French historian Etienne-Félix Berlioux (1828-1910) was named professor of geography at the University of Lyon in 1874.

BERNARD, Augustin

1166. Larnaude, Marcel. "Emile-Félix Gautier (1864-1940) et Augustin Bernard (1864-1947)," pp. 107-118 in "Les géographes français," Secrétariat d'état aux universités, Comité des travaux historiques et scientifiques, Bulletin de la section de géographie, vol. 81 (For the years 1968-1974) (Paris: Bibliothèque Nationale, 1975).

Concerning the French geographer Augustin Bernard (1865-1947), professor of geography and of the colonization of North Africa at the Sorbonne from 1902 to 1933.

1167. Sutton, Keith. "Augustin Bernard, 1865-1947," Geographers: Biobibliographical Studies, vol. 3 (1979), 19-27.

Augustin Bernard (1865-1947) was a French geographer who occupied the chair of geography and North African civilization at the University of Paris from 1902 onward.

BIASUTTI, Renato

1168. Cerulli, Ernesta. "Il contributo di Renato Biasutti agli studi etnologici," Rivista Geografica Italiana, vol. 87, no. 3 (September 1980), 305-312.

The ethnological work of the Italian geographer Renato Biasutti (1878-1965).

1169. Sestini, Aldo. "Renato Biasutti e gli inizi degli studi antropogeografici in Italia," Rivista Geografica Italiana, vol. 87, no. 3 (September 1980), 313-323.

The study of human geography in Italy, 1894-1914, with emphasis on the work of Renato Biasutti (1878-1965).

BLACHE, Jules

1170. Nicod, Jean. "Jules Blache, 1893-1970," Geographers: Biobibliographical Studies, vol. 1 (1977), 1-8.

Jules Blache (1893-1970) was a French geographer who specialized in glacial geomorphology and mountain geography. Taught at the universities of Grenoble and Nancy and served as rector of the Academy of Aix-Marseille.

BLANCHARD, Raoul

1171. Guichonnet, Paul, and Masseport, Jean. "Raoul Blanchard (1877-1965)," pp. 133-144 in "Les géographes français," Secrétariat d'état aux universités, Comité des travaux historiques et scientifiques, Bulletin de la section de géographie, vol. 81 (For the years 1968-1974) (Paris: Bibliothèque Nationale, 1975).

Biographical essay on the French geographer Raoul Blanchard (1877-1965), professor of geography at the University of Grenoble from 1905 to 1948. He wrote especially on Alpine and Canadian topics.

1172. Hamelin, Louis-Edmond. "La géographie de Raoul Blanchard," Canadian Geographer, vol. 5, no. 1 (Spring 1961), 1-9.

The French geographer Raoul Blanchard (1877-1965), with emphasis on his connections with French Canada.

BLUMENBACH, Johann Friedrich

1173. Plischke, Hans. Johann Friedrich Blumenbachs Einfluss auf die Entdeckungsreisenden seiner Zeit. Abhandlungen der Gesellschaft der Wissenschaften zu Göttingen, Philologisch-Historische Klasse, 3rd series, no. 20 (1937). ix + 107 p.

The German scientist J.F. Blumenbach (1752-1840) and his influence on the travelers and expeditions of his time.

BONPLAND, Aimé

1174. Bouvier, René, and Maynial, Edouard. Aimé Bonpland, explorateur de l'Amazonie, botaniste de la Malmaison, planteur en Argentine, 1773-1858. Paris: Société d'édition d'enseignement supérieur, 1950. 190 p.

Biography of Aimé Bonpland (1773-1858), the French botanist who accompanied Humboldt to South America.

1175. Castellanos, A. "Bonpland en los países del Plata," Revista de la Academia Colombiana de Ciencias Exactas, Físicas y Naturales, vol. 12, no. 45 (November 1963), 57-86.

Aimé Bonpland (1773-1858) was the French botanist who went to South America with Alexander von Humboldt in 1799. Article concerns Bonpland's activities in the Rio de la Plata area.

1176. Schulz, Wilhelm. "Aimé Bonpland, Alexander von Humboldts Begleiter auf der Amerikareise 1799-1804: Sein Leben und Wirken, besonders nach 1817 in Argentinien," Akademie der Wissenschaften und der Literatur in Mainz, Abhandlungen der Mathematisch-Naturwissenschaftlichen Klasse, 1960, no. 9, pp. [583]-633.

Aimé Bonpland (1773-1858), the French naturalist who accompanied Humboldt to South America in 1799, with emphasis on his life in Argentina after 1817.

BRATESCU, Constantin

1177. Nimigeanu, George. "Constantin Bratescu, 1882-1945," Geographers: Biobibliographical Studies, vol. 4 (1980), 19-24.

Constantin Bratescu (1882-1945) was a Romanian geomorphologist who taught at the universities of Cernauti and Bucharest.

BRÜCKNER, Eduard

1178. Kinzl, Hans. "Eduard Brückner: Ein führender Gletscher- und Eiszeitforscher," Geographisches Taschenbuch 1970/72 (Wiesbaden: Franz Steiner Verlag GMBH, 1970), pp. 262-265.

The glaciologist and physical geographer Eduard Brückner (1862-1927) was professor of geography at the University of Vienna from 1906.

BRUNHES, Jean

1179. Deffontaines, Pierre. "La aportación geográfica de Jean Brunhes," Revista de geografía (Barcelona), vol. 2 (July-December 1968), 161-167.

The contributions of the French geographer Jean Brunhes (1869-1930), professor of geography at the Collège de France from 1912.

1180. Delamarre, Mariel Jean-Brunhes. "Jean Brunhes (1869-1930)," pp. 49-80 in "Les géographes français," Sécretariat d'état aux universités, Comité des travaux historiques et scientifiques, Bulletin de la section de géographie, vol. 81 (For the years 1968-1974) (Paris: Bibliothèque Nationale, 1975).

Part Three--Biographical Works 245

An appreciation of the French geographer Jean Brunhes (1869-1930), professor of geography at the Collège de France from 1912, by his daughter Mariel, herself a distinguished geographer.

BUACHE, Philippe

1181. Broc, Numa. "Un géographe dans son siècle: Philippe Buache (1700-1773)," Dix-huitième siècle, vol. 3 (1971), 223-235.

The French geographer Philippe Buache (1700-1773) and his attempts at systematic physical description of the earth. Believed that chains of mountains or hills separated river basins, even in the case of the Seine-Loire divide.

1182. Drapeyron, Ludovic. "Les deux Buache, ou l'origine de l'enseignement géographique par versants et par bassins," Revue de géographie, vol. 21 (July 1887), 6-16.

Concerns 2 French geographers, Philippe Buache (1700-1773) and his nephew Jean-Nicolas Buache de Neuville (1741-1825). Influence of P. Buache's Essai de géographie physique (1752).

BUCHANAN, John Young

1183. Stoddart, David R. "Buchanan--The Forgotten Apostle," Geographical Magazine, vol. 44, no. 12 (September 1972), 858-862.

The British oceanographer John Young Buchanan (1844-1925) was a lecturer in geography at the University of Cambridge, 1889-1893. He had been a chemist on the Challenger Expedition (1872-1876).

BÜSCHING, Anton Friedrich

1184. Büttner, Manfred, and Jakel, Reinhard. "Anton Friedrich Büsching, 1724-1793," Geographers: Biobibliographical Studies, vol. 6 (1982), 7-15.

Anton Büsching (1724-1793) was a German polyhistor (geographer, philosopher, theologian, and historian) who wrote Erdbeschreibung (1754-1792) and issued the Magazin für Historie und Geographie (1767-1793).

1185. Plewe, Ernst. "D. Anton Friedrich Büsching. Das Leben eines deutschen Geographen in der zweiten Halfte des 18. Jahrhunderts" (Lautensach-Festschrift), <u>Stuttgarter Geographischen Studien</u>, vol. 69 (1957), 107-120.

The life of the German geographer Anton Friedrich Büsching (1724-1793) and a reevaluation of his contributions. [HB]

BUFFON, Georges-Louis Leclerc, Comte de

1186. Badey, Lucien. "Buffon, précurseur de la science démographique," <u>Annales de géographie</u>, vol. 38, no. 213 (15 May 1929), 206-220.

The writings of the French naturalist Buffon (Georges-Louis Leclerc, Comte de Buffon, 1707-1788) that concern population data, including mortality and average lifespan (but no discussion of population density).

1187. Glacken, Clarence J. "Count Buffon on Cultural Changes of the Physical Environment," <u>Annals of the Association of American Geographers</u>, vol. 50, no. 1 (March 1960), 1-21.

The views of the French naturalist Georges-Louis Leclerc, later Comte de Buffon (1707-1788), on human modification of the natural landscape, as expressed in his <u>Histoire naturelle</u>.

CAMENA D'ALMEIDA, Pierre

1188. Papy, Louis. "Pierre Camena d'Almeida, 1865-1943," <u>Geographers: Biobibliographical Studies</u>, vol. 7 (1983), 1-4.

The French geographer Pierre Camena d'Almeida (1865-1943), professor of geography at the University of Bordeaux, 1899-1935.

CAPOT-REY, Robert

1189. Bisson, Jean. "Robert Capot-Rey, 1897-1977," <u>Geographers: Biobibliographical Studies</u>, vol. 5 (1981), 13-19.

The French geographer Robert Capot-Rey (1897-1977), who devoted some 30 years to Saharan research.

Part Three--Biographical Works

CARTER, Clement Cyril

1190. Jay, L.J. "Two Eminent Gentlemen," Geographical Magazine, vol. 50, no. 3 (December 1977), 215-216.

The English schoolmasters Basil Bentham Dickinson (1863-1941) of Rugby School and C.C. Carter (1875-1949) played leading roles in the early years of the British teachers' organization, the Geographical Association (founded 1893).

CAVAILLÈS, Henri

1191. Papy, Louis. "Cavaillès, Arqué and Revert: Three Geographers of Bordeaux," Geographers: Biobibliographical Studies, vol. 7 (1983), 5-9. (Cavailles on pp. 5-7)

The French geographer Henri Cavaillès (1870-1951), who taught at the University of Bordeaux from 1922 to 1941.

CHARPENTIER, Jean de

1192. Teller, James T. "Jean de Charpentier, 1786-1855," Geographers: Biobibliographical Studies, vol. 7 (1983), 17-22.

The German-born Swiss naturalist Jean de Charpentier (1786-1855), who "can be credited with bringing about the initial conversion of scientists to the theory of glaciation."

CHEKHOV, Anton

1193. Matley, Ian M. "Chekhov and Geography," Russian Review, vol. 31, no. 4 (October 1972), 376-382.

The Russian writer Anton Chekhov (1860-1904) and his interest in geography, as shown in his plays The Wood Demon and Uncle Vanya. Article explores sources of Chekhov's geographical knowledge, including the works of the Russian geographers D.N. Anuchin, A.I. Voiekov, and Petr Semenov Tian-Shanskii.

CHESNEY, Francis Rawdon

1194. Marshall-Cornwall, James. "Three Soldier-Geographers," Geographical Journal, vol. 131, part 3 (September 1965), 357-365.

The geographical activities of 3 British army officers--William Martin Leake, Edward Sabine, and Francis Rawdon Chesney (1789-1872).

CHISHOLM, George G.

1195. Maclean, Kenneth. "George G. Chisholm: His Influence on University and School Geography," Scottish Geographical Magazine, vol. 91, no. 2 (September 1975), 70-78.

The British geographer George G. Chisholm (1850-1930), who taught at the University of Edinburgh from 1908 to 1923.

1196. Wise, Michael J. "A University Teacher of Geography," Institute of British Geographers, Transactions, Publication no. 66 (November 1975), 1-16.

Biographical sketch of the British geographer George Chisholm (1850-1930), who taught at the University of Edinburgh from 1908.

CHOLLEY, André

1197. Gras, Jacques. "André Cholley (1886-1968)," pp. 153-171 in "Les géographes français," Secrétariat d'état aux universités, Comité des travaux historiques et scientifiques, Bulletin de la section de géographie, vol. 81 (For the years 1968-1974) (Paris: Bibliothèque Nationale, 1975).

A sketch of the life and work of the French geographer André Cholley (1886-1968), professor of geography at the Sorbonne from 1927 (Dean of the Sorbonne, 1945-1953, and director of the Institute of Geography, 1944-1956).

CHRISTALLER, Walter

1198. Hottes, Karlheinz; Hottes, Ruth; and Schöller, Peter. "Walter Christaller, 1893-1969," Geographers: Biobibliographical Studies, vol. 7 (1983), 11-16.

The German geographer Walter Christaller (1893-1969), well known for his work in central place theory.

1199. Hottes, Karlheinz, and Schöller, Peter. "Werk und Wirkung Walter Christallers," Geographische Zeitschrift, vol. 56, no. 2 (June 1968), 81-84.

Part Three--Biographical Works 249

Homage to the German geographer Walter Christaller (1893-1969), who produced the seminal work in central place theory, Die zentralen Orte in Süddeutschland (1933). Followed by a bibliography of Christaller's writings, 1933-1967, compiled by Hanns J. Buchholz (pp. 85-87) and an autobiographical essay by Christaller telling how he came to produce the work on central place theory ("Wie Ich der Theorie der Zentralen Orte Gekommen Bin," pp. 88-101).

1200. Hottes, Ruth. "Walter Christaller (1893-1969). Un aperçu de sa vie," Mosella, vol. 9, nos. 3-4 (July-December 1979), 97-107.

A biographical sketch of the German geographer Walter Christaller (1893-1969), best known for his work in central place theory.

1201. Hottes, Ruth. "Walter Christaller: Ein Uberblick über Leben und Werk," Geographisches Taschenbuch 1981/1982 (Wiesbaden: Franz Steiner Verlag GmbH, 1981), pp. 59-70.

Biographical sketch of the German geographer Walter Christaller (1893-1969), who is best known for his pioneering work in central place theory, Die zentralen Orte in Süddeutschland (1933). Abridged English translation by Guido G. Weigend appeared in Annals of the Association of American Geographers, vol. 73, no. 1 (March 1983), 51-54.

1202. Wirth, Eugen. "Fünfzig Jahre Theorie der zentralen Orte," Geographische Zeitschrift, vol. 70, no. 4 (Fourth Quarter 1982), 293-297.

Describes Walter Christaller's doctoral examination at the University of Erlangen on 2 November 1932. His thesis on central places of southern Germany is called "the most famous geographical dissertation of this century." Includes Christaller's autobiographical sketch, 1891-1932, and the opinion of his mentor, Robert Gradmann, on the thesis.

COELLO, Francisco

1203. Gómez Pérez, José. "El geógrafo don Francisco Coello de Portugal y Quesada," Estudios geográficos, vol. 27, no. 103 (May 1966), 249-308.

The role of the Spanish geographer Francisco Coello (1822-1898) in the topographic mapping of Spain and in the establishment of the Madrid Geographical Society.

COLE, Grenville Arthur James

1204. Davies, Gordon L. "The Making of Irish Geography, II: Grenville Arthur James Cole (1859-1924)," Irish Geography, vol. 10 (1977), 90-94.

The geographical work of G.A.J. Cole (1859-1924), professor of geology in the Royal College of Science, Dublin, from 1890, and director of the Geological Survey of Ireland from 1905.

COOK, James

1205. Beaglehole, J.C. The Life of Captain James Cook. Stanford, California: Stanford University Press, 1974. xi + 760 p.

Biography of the English navigator and explorer James Cook (1728-1778).

CORNISH, Vaughan

1206. Gilbert, Edmund W. Vaughan Cornish (1862-1948) and the Advancement of Knowledge Relating to the Beauty of Scenery in Town and Country. Oxford: Oxford Preservation Trust, 1965. 21 p.

The life of the English geographer Vaughan Cornish (1862-1948), with particular attention to his interest in landscape esthetics.

1207. Goudie, Andrew S. "Vaughan Cornish: Geographer," Institute of British Geographers, Transactions, Publication no. 55 (March 1972), 1-16.

The life of the English geographer Vaughan Cornish (1862-1948) and his contributions to "the study of wave forms, the study of frontiers and capitals in historical and political geography and the study of aesthetic geography."

CORTAMBERT, Eugene

1208. Broc, Numa. "Eugène Cortambert et la Place de la géographie dans la classification des connaissances humaines (1852)," Revue d'histoire des sciences et de leurs applications, vol. 29, no. 4 (October 1976), 337-345.

(From English abstract, p. 337) "[The French geographer] Eugène Cortambert (1805-1881), known mainly as a teacher and popularizer, tried to turn geography, which until his time had been mere nomenclature, into a genuine science (Physiographie, 1836). He argued in favour of the autonomy of geography in the face of two encroaching neighbours: geology and history. In 1852, he defined geography's place in the classification of the sciences, a problem which Ampère, Comte, and Cournot had avoided. He proposed the creation of a new category of disciplines, the 'physico-moral sciences,' in which geography would fit quite naturally."

1209. Broc, Numa. "Eugène Cortambert, 1805-1881," Geographers: Biobibliographical Studies, vol. 2 (1978), 21-25.

The French savant Eugène Cortambert (1805-1881) was a geographer, textbook writer, librarian in geographical section of the Bibliothèque Nationale, and teacher in a girls' school.

CURZON, George Nathaniel

1210. Goudie, Andrew S. "George Nathaniel Curzon--Superior Geographer," Geographical Journal, vol. 146, part 2 (July 1980), 203-209.

The geographical interests of the English statesman George Nathaniel Curzon (1859-1925), one-time Viceroy of India and President of the Royal Geographical Society.

CVIJIĆ, Jovan

1211. Vasovic, Milorad. "Jovan Cvijić, 1865-1927," Geographers: Biobibliographical Studies, vol. 4 (1980), 25-32.

Jovan Cvijić (1865-1927) was a Yugoslav geographer who was professor of geography at Belgrade University, 1893-1927.

DAINVILLE, François de

1212. Grivot, Françoise. "Le Père François de Dainville (1909-1971)," pp. 197-198 in "Les géographes français,"

Secrétariat d'état aux universités, Comité des travaux historiques et scientifiques, <u>Bulletin de la section de géographie</u>, vol. 81 (For the years 1968-1974) (Paris: Bibliothèque Nationale, 1975).

Sketch of the French geographer, cartographer, and Jesuit priest François de Dainville (1909-1971), who taught the history of cartography in various institutions in Paris from 1959 to 1970.

DALLA VEDOVA, Giuseppe

1213. Porena, Filippo. "L'opera di Giuseppe Dalla Vedova," pp. ix-xxi in <u>Scritti di geografia e di storia della geografia concernenti l'Italia pubblicati in onore di Giuseppe Dalla Vedova</u> (Florence: Tip. M. Ricci, 1908).

Life and work of the Italian geographer Giuseppe Dalla Vedova (1834-1919). Festschrift contains 15 other articles by Italian geographers on various aspects of the geography of Italy, including history of geography and cartography (mostly pre-1750). Includes a paper by Goffredo Jaja on the geographical concepts of the Italian economist Melchiorre Gioia (1767-1829). [LG]

DARBY, Henry Clifford

1214. Perry, Peter J. "H.C. Darby and Historical Geography: A Survey and Review," <u>Geographische Zeitschrift</u>, vol. 57, no. 3 (September 1969), 161-176.

The contributions of the British geographer H.C. Darby (b. 1909), professor of geography at the universities of Liverpool, London (University College), and Cambridge. Author and editor of <u>The Domesday Geography of England</u> (7 vols., 1952-1977).

DARWIN, Charles

1215. Stoddart, David R. "Darwin's Impact on Geography," <u>Annals of the Association of American Geographers</u>, vol. 56, no. 4 (December 1966), 683-698.

Influence of biological, and specifically Darwinian, ideas on geography during the past century.

Part Three--Biographical Works

1216. Ule, Willi. "Darwins Bedeutung in der Geographie," Deutsche Rundschau für Geographie und Statistik, vol. 31, no. 10 (July 1909), 433-443.

An essay on the significance to geography of the work of the British naturalist Charles Darwin (1809-1882).

1217. Vallaux, Camille. "Deux précurseurs de la géographie humaine: Volney et Charles Darwin," Revue de synthèse, vol. 15, no. 2 (June 1938), 81-93.

If human geography is a product of the late 19th century with such geographers as Friedrich Ratzel, there were numerous earlier workers who were actually doing human geography before it was defined. Among these precursors were the French savant Constantin François Chasse-Boeuf, Comte de Volney (1757-1820), and the English naturalist Charles Darwin (1809-1882).

DAVID, Mihai

1218. Gugiuman, Ion. "Mihai David, 1886-1954," Geographers: Biobibliographical Studies, vol. 6 (1982), 31-33.

The Romanian geographer Mihai David (1886-1954) taught in Iaşi University from 1922 to 1945.

DEMANGEON, Albert

1219. Perpillou, Aimé. "Albert Demangeon (1872-1940)," pp. 81-106 in "Les géographes français," Secrétariat d'état aux universités, Comité des travaux historiques et scientifiques, Bulletin de la section de géographie, vol. 81 (for the years 1968-1974) (Paris: Bibliothèque Nationale, 1975).

An appreciation of the French geographer Albert Demangeon (1872-1940), professor of geography at the Sorbonne from 1911 to 1940, by his son-in-law, Aimé Perpillou (1902-1976), himself a distinguished Sorbonne professor.

DICKINSON, Basil Bentham

1220. Jay, L.J. "Two Eminent Gentlemen," Geographical Magazine, vol. 50, no. 3 (December 1977), 215-216.

The English schoolmasters B.B. Dickinson (1863-1941) of Rugby School and C.C. Carter played leading roles in the early years of the British teachers' organization, the Geographical Association (founded 1893).

DIETRICH, Philippe-Frédéric de

1221. Kugler, Ernst. "Philipp Friedrich von Dietrich: Ein Beitrag zur Geschichte der Vulkanologie," Münchener Geographische Studien, no. 7 (1899). 88 p.

Studies in vulcanology from the 17th to the 19th centuries, with particular attention to the work of the French mineralogist and political figure Philippe-Frédéric de Dietrich.

DILTHEY, Wilhelm

1222. Rose, Courtice. "Wilhelm Dilthey's Philosophy of Historical Understanding: A Neglected Heritage of Contemporary Humanistic Geography," pp. 99-133 in Geography, Ideology and Social Concern, ed. D.R. Stoddart (Oxford: Basil Blackwell, 1981).

The German philosopher Wilhelm Dilthey (1833-1911) and his relevance to modern geography.

DIMITRESCU-ALDEM, Alexandre

1223. Cotet, Petre. "Alexandre Dimitrescu-Aldem, 1880-1917," Geographers: Biobibliographical Studies, vol. 3 (1979), 35-37.

Alexandre Dimitrescu-Aldem (1880-1917) was a Romanian geographer who specialized in geomorphology.

DÖRRIES, Hans

1224. Müller-Wille, Wilhelm. "Hans Dörries als Geograph und Landesforscher," Berichte zur deutschen Landeskunde, vol. 14, no. 1 (January 1955), 1-11.

Biographical sketch of the German geographer Hans Dörries (1897-1945).

DOKUCHAEV, Vasily Vasilyevich

1225. Esakov, Vasily Alexeyevich. "Vasily Vasilyevich Dokuchaev, 1846-1903," Geographers: Biobibliographical Studies, vol. 4 (1980), 33-42.

(Page 33) "Vasily Dokuchaev, the great Russian naturalist, gained renown for his work in geography, soil science, and geology. He was the founder of modern genetic soil science...."

Part Three--Biographical Works 255

1226. Gerassimow (Gerasimov), I.P. "Alexander von Humboldt und Wassilij Wassiljewitsch Dokutschajew," pp. 43-48 in Alexander von Humboldt: Vorträge und Aufsätze Anlässlich der 100. Wiederkehr seines Todestages am 6. Mai 1959, ed. Johannes F. Gellert (Geographische Gesellschaft der Deutschen Demokratischen Republik, Wissenschaftliche Abhandlungen, Vol. 2) (Berlin: VEB Deutscher Verlag der Wissenschaften, 1960).

A description of the works of Alexander von Humboldt (1769-1859) and those of V.V. Dokuchaev (1846-1903), a Russian geographer of a later era.

DRAPEYRON, Ludovic

1227. Broc, Numa. "Ludovic Drapeyron, 1839-1901," Geographers: Biobibliographical Studies, vol. 6 (1982), 35-38.

The French geographer Ludovic Drapeyron (1839-1901) taught at the Lycée Charlemagne in Paris and founded the Revue de géographie (1877-).

DRYGALSKI, Erich von

1228. Tiggesbäumker, Günter. "Erich von Drygalski, 1865-1949," Geographers: Biobibliographical Studies, vol. 7 (1983), 23-29.

German geographer and polar explorer Erich von Drygalski (1865-1949), who taught at the universities of Berlin (1899-1906) and Munich (1906-1935).

DUBOIS, Marcel

1229. Broc, Numa. "Nationalisme, colonialisme et géographie: Marcel Dubois (1856-1916)," Annales de géographie, vol. 87, no. 481 (May-June 1978), 326-333.

The French geographer Marcel Dubois (1856-1916) taught at the Sorbonne from 1885 and collaborated with Vidal de la Blache in the founding of the Annales de géographie in 1891.

EDELMAN, C.H.

1230. Heslinga, M.W., and Wiggers, A.J. "Over de betekenis van C.H. Edelman voor de geografie," Tijdschrift van het Koninklijk Nederlandsch Aardrijkskundig Genootschap, 2nd series, vol. 83, no. 1 (January 1966), 4-14.

The contributions of the Dutch soil scientist C.H. Edelman (d. 1964) to geography. Edelman held the chair of pedology at the State Agricultural University in Wageningen.

ESCHWEGE, Wilhelm Ludwig von

1231. Beck, Hanno. "Wilhelm Ludwig von Eschwege und die klassische deutsche Geographie," Erdkunde, vol. 9, no. 2 (May 1955), 89-92.

Wilhelm Ludwig von Eschwege (1777-1855), "The Father of Brazilian Geology," had contacts with both Humboldt and Goethe.

EVEREST, George

1232. Heaney, G.F. "Sir George Everest," Geographical Journal, vol. 133, part 2 (June 1967), 209-211.

The Englishman George Everest (1790-1866) served as Surveyor General of India and Superintendent of the Great Trigonometrical Survey. His name was given to the world's highest mountain.

FAUCHER, Daniel

1233. Taillefer, François. "Daniel Faucher (1882-1970)," pp. 173-183 in "Les géographes français," Secrétariat d'état aux universités, Comité des travaux historiques et scientifiques, Bulletin de la section de géographie, vol. 81 (For the years 1968-1974) (Paris: Bibliothèque Nationale, 1975).

The life and work of the French geographer Daniel Faucher (1882-1974), professor of geography at the University of Toulouse, 1926-1952.

FAWCETT, Charles Bungay

1234. Freeman, T.W. "Charles Bungay Fawcett, 1883-1952," Geographers: Biobibliographical Studies, vol. 6 (1982), 39-46.

The English geographer Charles Fawcett (1883-1952) was Professor of Geography at University College, London, from 1928 to 1949.

Part Three--Biographical Works 257

FILCHNER, Wilhelm

1235. Beck, Hanno. "Wilhelm Filchner (1877-1957),"
Geographisches Taschenbuch 1966/69 (Wiesbaden: Franz
Steiner Verlag GMBH, 1968), pp. 227-238.

Biographical sketch of Wilhelm Filchner (1877-1957),
German explorer in Asia and Antarctica.

FLEURE, Herbert John

1236. Campbell, John A. Some Sources of the Humanism of H.
J. Fleure. University of Oxford, School of
Geography, Research Papers, no. 2 (1972). 45 p.

Essay on the British geographer and anthropologist
H.J. Fleure (1877-1969) with emphasis on the humanistic
concerns that enlivened his geography.

1237. Evans, E. Estyn. "H.J. Fleure," pp. 494-495 in Vol.
5 of International Encyclopedia of the Social
Sciences (New York: The Macmillan Company and The
Free Press, 1968).

Biographical sketch of the British geographer and
anthropologist Herbert John Fleure (1877-1969),
professor of geography at the universities of
Aberystwyth and Manchester.

FLINDERS, Matthew

1238. Ritchie, G.S. "Matthew Flinders, Hydrographer,"
Journal of Navigation, vol. 27, no. 3 (July 1974),
283-297.

The English navigator and chartmaker Matthew Flinders
(1774-1814), explorer of the Australian coastline.

FONCIN, Pierre

1239. Broc, Numa. "Patriotisme, régionalisme et
géographie: Pierre Foncin (1841-1916)," L'information
historique, vol. 38, no. 1 (January-February 1976),
30-32.

The French geographer Pierre Foncin (1841-1916) was
the first holder of the chair of geography at the
Faculté des Lettres de Bordeaux (1877). He was active
in the Alliance Française and served as Inspector-
General of Public Instruction (1882-1911).

FORBES, James David

1240. Cunningham, Frank F. "James David Forbes, 1809-1868," Geographers: Biobibliographical Studies, vol. 7 (1983), 31-37.

Scottish glaciologist James David Forbes (1809-1868), professor of natural philosophy at the University of Edinburgh from 1833 to 1860 and principal of the University of St. Andrews from 1860 to 1867.

FORMOZOV, Alexander Nikolayevich

1241. Rakhilin, V.K. "Alexander Nikolayevich Formozov, 1899-1973," Geographers: Biobibliographical Studies, vol. 7 (1983), 39-46.

The Soviet biogeographer A.N. Formozov (1899-1973), professor of soil science at Moscow University, 1935-1956. Worked in the biogeography department of the Institute of Geography of the USSR Academy of Sciences, 1942-1962.

FORSTER, Georg

1242. Beck, Hanno. "Georg Forster--Geograph, Weltumsegler und Revolutionär (1754-1794)," pp. 54-82 in Grosse Geographen: Pioniere--Aussenseiter--Gelehrte by Beck (Berlin: Dietrich Reimer Verlag, 1982).

Biography of the German geographer, traveler, and political figure Georg Forster (1754-1794), son of J.R. Forster (below).

1243. Fiedler, Horst. Georg-Forster-Biblographie, 1767 bis 1970. (Deutsche Akademie der Wissenschaften zu Berlin, Zentralinstitut für Literaturgeschichte.) Berlin: Akademie-Verlag, 1971. 208 p.

A bibliography of writings relating to the German geographer and naturalist Georg Forster (1754-1794).

1244. Kersten, Kurt. Der Weltumsegler: Johann Georg Adam Forster, 1754-1794. Bern: Francke Verlag, 1957. 400 p.

Biography of the German traveler and scientist Georg Forster (1754-1794).

1245. Ramakers, Günter. "Georg Forster (1754-1794)," Geographisches Taschenbuch 1977-78 (Wiesbaden: Franz Steiner Verlag GMBH, 1977), pp. 149-164.

Part Three--Biographical Works 259

Biographical sketch of the German geographer and traveler Georg Forster (1754-1794), who accompanied Cook's second expedition (1772-1775).

1246. Schottlaender, Felix. "Georg Forster und die anfänge der Geographie in Deutschland," Neue Jahrbücher für Wissenschaft und Jugendbildung, vol. 4, no. 5 (1928), 560-565.

The German scientist and traveler Georg Forster (1754-1794) and the beginnings of modern geography in Germany.

FORSTER, Johann Reinhold

1247. Hoare, Michael E. "Johann Reinhold Forster: The Neglected 'Philosopher' of Cook's Second Voyage (1772-1775)," Journal of Pacific History, vol. 2 (1967), 215-224.

The German naturalist Johann Reinhold Forster (1729-1798), who served as the "principal scientist" on James Cook's second Pacific voyage. Author quotes Ernst Plewe (1932) to the effect that Forster's Observations (1778) "is the most important systematic contribution to modern geography before Humboldt's Kosmos."

1248. Hoare, Michael E. The Tactless Philosopher: Johann Reinhold Forster (1729-1798). Melbourne, Australia: The Hawthorn Press, 1976. x + 419 p.

Biography of the German naturalist Johann Reinhold Forster (1729-1798), who accompanied Cook's second voyage (1772-1775). Forster was "the first government-paid scientist in the British service." Father of Georg Forster (above).

FOURIER, Joseph

1249. Guiral, P. "Un aspect peu connu de Fourier: Fourier géographe et climatologue," pp. 373-378 in Livre jubilaire offert à Maurice Zimmermann (Lyon: Universite de Lyon et M. Audin, 1949).

The geographical and climatological work of the French mathematician Joseph Fourier (1768-1830).

FRANZ, Johann Michael

1250. Jakel, Reinhard. "Johann Michael Franz, 1700-1761," Geographers: Biobibliographical Studies, vol. 5 (1981), 41-48.

J.M. Franz (1700-1761) was a German geographer who taught at the University of Göttingen.

FRÖBEL, Julius

1251. Müller, Georg. Die Untersuchungen Julius Fröbels über die Methoden und die Systematik der Erdkunde und ihre Stellung im Entwicklungsgange der Geographie als Wissenschaft: Ein Beitrag zur Geschichte der geographischen Methodik. (Inaugural-Dissertation, Halle-Wittenberg.) Halle: Hofbuchdruckerei von C.A. Kaemmerer & Co., 1908. 62 p.

The methodology of geography, according to the German geologist, writer, and political figure Julius Fröbel (1805-1893).

GALLOIS, Lucien

1252. Meynier, André. "Lucien Gallois (1857-1941)," pp. 25-33 in "Les géographes français," Secrétariat d'état aux universités, Comité des travaux historiques et scientifiques, Bulletin de la section de géographie, vol. 81 (For the years 1968-1974) (Paris: Bibliothèque Nationale, 1975).

The life of the French geographer Lucien Gallois (1857-1941), who taught at the Sorbonne from 1893 to 1927.

GAUTIER Emile-Félix

1253. Larnaude, Marcel. "Emile-Félix Gautier (1864-1940) et Augustin Bernard (1865-1947)," pp. 107-118 in "Les géographes français," Secrétariat d'état aux universités, Comité des travaux historiques et scientifiques, Bulletin de la section de géographie, vol. 81 (For the years 1968-1974) (Paris: Bibliothèque Nationale, 1975).

A biographical sketch of the French geographer Emile-Félix Gautier (1864-1940), professor of geography at the University of Algiers from 1902.

Part Three--Biographical Works

GEDDES, Arthur

1254. Learmonth, Andrew T.A. "Arthur Geddes, 1895-1968," Geographers: Biobibliographical Studies, vol. 2 (1978), 45-51.

The Scottish geographer Arthur Geddes (1895-1968) taught at the University of Edinburgh, 1927-1965. Son of Patrick Geddes (below).

GEDDES, Patrick

1255. Fleure, H.J. "Patrick Geddes (1854-1932)," Sociological Review, n.s. vol. 1, no. 2 (December 1953), 5-13.

An appreciation of the life and work of the Scottish polymath (biologist, geographer, sociologist, planner, etc.) Patrick Geddes (1854-1932), by one of his geographer friends.

1256. Mairet, Philip. Pioneer of Sociology: The Life and Letters of Patrick Geddes. London: Lund Humphries, 1957. xx + 226 p.

Biography of the Scottish biologist, geographer, and sociologist Patrick Geddes (1854-1932).

1257. Robson, B.T. "Geography and Social Science: The Role of Patrick Geddes," pp. 186-207 in Geography, Ideology and Social Concern, ed. D.R. Stoddart (Oxford: Basil Blackwell, 1981).

The pioneering views of the Scottish geographer, biologist, sociologist, and planner Patrick Geddes (1854-1932). Despite earlier neglect, "many of his ideas now play a central part both in geography and in the wider social sciences" (p. 187).

1258. Stevenson, W. Iain. "Patrick Geddes and Geography: A Biobibliographical Study," University College, London, Department of Geography, Occasional Papers, no. 27 (March 1975). Unpaged.

The Scottish polymath Patrick Geddes (1854-1932) and the Geddesian tradition in British geography.

1259. Stevenson, W. Iain. "Patrick Geddes, 1854-1932," Geographers: Biobibliographical Studies, vol. 2 (1978), 53-65.

Sketch of the Scottish geographer, biologist, sociologist, and planner Patrick Geddes (1854-1932).

GEIKIE, Archibald

1260. Marsden, W.E. "Archibald Geikie, 1835-1924," Geographers: Biobibliographical Studies, vol. 3 (1979), 39-52.

Archibald Geikie (1835-1924) was a Scottish geologist and physical geographer; Professor of Geology at the University of Edinburgh, 1871-1882; and Director General of the Geological Survey of Great Britain, 1882-1901. Brother of James Geikie (below).

GEIKIE, James

1261. Marsden, W.E. "James Geikie, 1839-1915," Geographers: Biobibliographical Studies, vol. 3 (1979), 53-62.

James Geikie (1839-1915) was a Scottish geologist and physical geographer. Succeeded his brother Archibald (above) as Professor of Geology at the University of Edinburgh, 1882-1914.

GERLAND, Georg

1262. Mühlmann, W.E. "Georg Gerland," pp. 633-634 in Vol. 6 of Encyclopaedia of the Social Sciences (London: Macmillan and Co., Ltd., 1932).

The German geographer Georg Gerland (1833-1919) was professor of geography at the University of Strassburg from 1875.

GILBERT, Edmund William

1263. Freeman, T.W. "Edmund William Gilbert, 1900-1973," Geographers: Biobibliographical Studies, vol. 3 (1979), 63-71.

E.W. Gilbert (1900-1973) was an English geographer who taught at the universities of London (Bedford College), Reading, and Oxford (1936-1967 at Oxford, Professor since 1953).

1264. Robinson, Guy, and Patten, John. "Edmund W. Gilbert and the Development of Historical Geography," Journal of Historical Geography, vol. 6, no. 4 (October 1980), 409-419.

Part Three--Biographical Works 263

A biographical essay on Edmund William Gilbert (1900-1973), who taught at the University of Oxford from 1936 to 1967 (Professor from 1953), with emphasis on his contributions to the development of historical geography in England.

GILLMAN, Clement

1265. Hoyle, Brian S. "Clement Gillman, 1882-1946," Geographers: Biobibliographical Studies, vol. 1 (1977), 35-41.

Clement Gillman (1882-1946) was an Anglo-German geographer who worked in German East Africa (later Tanganyika) from 1905 onward as a railroad engineer.

1266. Hoyle, Brian S. "Clement Gillman, 1882-1946: Bibliographical Notes on a Pioneer East African Geographer," East African Geographical Review, no. 3 (April 1965), 1-16.

Clement Gillman (1882-1946) was a German geographer who worked in Tanganyika from 1905 onward and is perhaps best known for his maps of population and vegetation in that country.

1267. Weigt, E. "Clemens Gillman und die neuere geographische Erforschung Ostafrikas," Erdkunde, vol. 3, no. 4 (December 1949), 193-199.

The German railway engineer and geographer Clement Gillman (1882-1946) worked in German East Africa (later Tanganyika).

GOETHE, Johann Wolfgang von

1268. Berger, Dorothea. Goethe als Vertreter der Länderkunde im 18. Jahrhundert: Ein Beitrag zu Goethes Schaffen, ein Beitrag zur Geschichte der Länderkunde. (Inaugural-Dissertation, Greifswald.) Greifswald: Druck von Emil Hartmann, 1916. 111 p.

The German poet Johann Wolfgang von Goethe (1749-1832) as a regional geographer.

1269. Cameron, Dorothy. "Early Discoverers, XXII: Goethe--Discoverer of the Ice Age," Journal of Glaciology, vol. 5, no. 41 (June 1965), 751-754.

The German scientist and man of letters Johann Wolfgang von Goethe (1749-1832) "was one of the first to attribute the transport of erratic blocks to glaciers, and to believe that an ice sheet covered northern Germany; furthermore, he was the very first to believe in an ice age."

1270. Hederich, Reinhard. "Goethe und die physikalische Geographie," Münchener Geographische Studien, no. 5 (1898). 66 p.

The German poet Johann Wolfgang von Goethe (1749-1832) and his knowledge of physical geography.

1271. Reuther, Martin. "Goethes Beziehungen zur Geographie--inbesondere zur Landschaft: Ein Beitrag zur Geschichte der Geographie," pp. 367-384 in Studia z dziejów geografii i kartografii / Etudes d'histoire de la géographie et de la cartographie, ed. Józef Babicz (Warsaw: Polish Academy of Science, 1973).

Johann Wolfgang von Goethe (1749-1832), the German man of letters, and his connections with geography.

1272. Schmitthenner, Heinrich. "Carl Ritter und Goethe," Geographische Zeitschrift, vol. 43, no. 5 (May 1937), 161-175.

Johann Wolfgang von Goethe (1749-1832) and his influence on Carl Ritter (1779-1859) before the latter's appointment to the University of Berlin in 1820.

GOGOL, Nikolai

1273. Camena d'Almeida, Pierre. "Gogol géographe," Revue de géographie, vol. 37, no. 2 (August 1895), 81-84.

Geographical concepts in the Arabesques (1835) of the Russian writer Nikolai Gogol (1809-1852).

GOUROU, Pierre

1274. Gallais, Jean. "L'évolution de la pensée géographique de Pierre Gourou sur les pays tropicaux (1935-1970)," Annales de géographie, vol. 90, no. 498 (March-April 1981), 129-150.

The work of the French geographer Pierre Gourou (b. 1900) in tropical geography from 1935 to 1970.

Part Three—Biographical Works

GRADMANN, Robert

1275. Gradmann, Robert. Lebenserinnerungen. Edited by Karl Heinz Schröder. Stuttgart: W. Kohlhammer Verlag, 1965. ix + 164 p.

Memoirs of the German geographer Robert Gradmann (1865-1950), professor of geography at the University of Erlangen.

1276. Schröder, Karl Heinz. "Robert Gradmann, 1865-1950," Geographers: Biobibliographical Studies, vol. 6 (1982), 47-54.

The German geographer Robert Gradmann (1865-1950) was professor of geography at the universities of Tubingen and Erlangen.

GRANÖ, Johannes Gabriel

1277. Granö, Olavi. "Johannes Gabriel Granö, 1882-1956," Geographers: Biobibliographical Studies, vol. 3 (1979), 73-84.

J.G. Granö (1882-1956) was a Finnish geographer who taught at the Universities of Helsinki, Tartu, and Turku and wrote Reine Geographie (1929).

GRANT, James

1278. Casada, James A. "James A. Grant and the Royal Geographical Society," Geographical Journal, vol. 140, part 2 (June 1974), 245-253.

James Grant (1827-1892) was a British explorer and promoter of African exploration. This paper is concerned especially with his connections with the Royal Geographical Society.

GREENWOOD, George

1279. Stoddart, David R. "Colonel George Greenwood, the Father of Modern Subaerialism," Scottish Geographical Magazine, vol. 76, no. 2 (September 1960), 108-110.

The Englishman George Greenwood (1799-1875) wrote a work titled Rain and Rivers ... (1857), the main thesis of which was the supremacy of rainwash as an agent of land sculpture and the relative unimportance of streams and rivers.

GRIGORYEV, Andrei Alexandrovich

1280. Zabelin, I.M. "Andrei Alexandrovich Grigoryev, 1883-1968," Geographers: Biobibliographical Studies, vol. 5 (1981), 55-61.

Sketch of the Soviet physical geographer A.A. Grigoryev (1883-1968), professor of geography at the Leningrad and Moscow universities and the USSR Academy of Sciences.

GUTHRIE, William

1281. East, W. Gordon. "An Eighteenth-Century Geographer: William Guthrie of Brechin," Scottish Geographical Magazine, vol. 72, no. 1 (April 1956), 32-37.

The Scottish writer William Guthrie (1708-1770) published his famous Geography in 1770 (last ed. 1843).

HAECKEL, Ernst

1282. Helbing, Helmut. "Ernst Haeckels Beziehungen zur Geographie," Wissenschaftliche Zeitschrift der Friedrich-Schiller-Universität, Jena, Mathematisch-Naturwissenschaftliche Reihe, vol. 14, no. 4 (1965), 153-162.

The German biologist Ernst Haeckel (1834-1919) and his geographical interests.

HAENKE, Thaddaeus

1283. Ballivian, M.V., and Kramer, Pedro. Tadeo Haenke: Escritos precedidos de algunos apuntes para su biografía y acompañados de varios documentos illustrativos. La Paz, Bolivia: Imprenta y Litografia de "El Nacional" de Isaac V. Vila, 1898. xxxvii + 113 + iv p.

Thaddaeus (or Tadeusz, Tadeo, etc.) Haenke (1761-1816) was a Bohemian naturalist who went to South America in 1789 with the Malaspina expedition. He was interested in all aspects of physical geography. Humboldt admired his work.

1284. Beck, Hanno. "Thaddaeus Haenke und Alexander von Humboldt," Forschungen und Fortschritte, vol. 35, no. 3 (March 1961), 65-71.

The Bohemian naturalist Thaddaeus Haenke (1761-1816) and his influence on Alexander von Humboldt (1769-1859).

1285. Gicklhorn, Josef, and Gicklhorn, Renée. "Th. Haenkes Bedeutung für die Erforschung Südamerikas vor Alexander v. Humboldt," <u>Mitteilungen</u> <u>der</u> <u>Geographischen</u> <u>Gesellschaft</u> <u>in</u> <u>Hamburg</u>, vol. 47 (1941), 263-364.

Thaddaeus Haenke (1761-1816) was a Bohemian doctor, naturalist, and traveler who took part in the Malaspina expedition to the Americas. Lived in South America (Bolivia and Peru) from 1794 onward.

1286. Gicklhorn-Wien, Renée. <u>Thaddäus</u> <u>Haenkes</u> <u>Reisen</u> <u>und</u> <u>Arbeiten</u> <u>in</u> <u>Südamerika</u>. (Acta Humboldtiana, Series Historica, no. 1.) Wiesbaden: Franz Steiner Verlag GmbH, 1966. xv + 231 p.

The Bohemian naturalist and traveler Thaddaeus Haenke (1761-1816) took part in the Malaspina expedition to the Americas (1789-1794) and remained in South America. "... [Humboldt's] journey through Peru lasted only two months, whereas Haenke lived and worked in the country for 20 years."

1287. Kühnel, Josef. <u>Thaddaeus</u> <u>Haenke</u>: <u>Leben</u> <u>und</u> <u>Wirken</u> <u>eines</u> <u>Forschers</u>. (Veröffentlichen des Collegium Carolinum, vol. 9.) Munich: Verlag Robert Lerche, 1960. 276 p.

The life of the Bohemian naturalist Thaddaeus Haenke (1761-1816), who traveled to the Americas with the Malaspina Expedition and stayed in South America.

1288. Kühnel, Josef. "Thaddaeus Haenke (1761-1817)," <u>Geographisches</u> <u>Taschenbuch</u> <u>1962/63</u> (Wiesbaden: Franz Steiner Verlag GMBH, 1962), pp. 259-264.

Life of the Bohemian naturalist Thaddaeus Haenke (1761-1817) (or 1816; see Renée Gicklhorn-Wien, above), who went to the Americas with the Malaspina Expedition and stayed in South America (Peru and Bolivia) for the rest of his life.

HAHN, Eduard

1289. Mühlmann, W.E. "Eduard Hahn," pp. 244-245 in Vol. 7 of Encyclopaedia of the Social Sciences (London: Macmillan and Co., Ltd., 1932).

Sketch of the German geographer Eduard Hahn (1856-1928), who taught at the University of Berlin.

1290. Plewe, Ernst. "Eduard Hahn (1856-1928)," Geographisches Taschenbuch 1975/76 (Wiesbaden: Franz Steiner Verlag GMBH, 1975), pp. 239-246.

Sketch of the Berlin geographer Eduard Hahn (1856-1928), who is known for his work on agricultural systems and in plant and animal domestication.

HAUSHOFER, Karl

1291. Jacobsen, Hans-Adolf. "'Kampf um Lebensraum': Zur Rolle des Geopolitikers Karl Haushofer im Dritten Reich," German Studies Review, vol. 4, no. 1 (February 1981), 79-104.

The German geographer Karl Haushofer (1869-1945) and the influence of his geopolitical views in Nazi Germany.

1292. Jacobsen, Hans-Adolf. Karl Haushofer, Leben und Werk. 2 vols. ("Schriften des Bundesarchivs," 24/I-II.) Boppard am Rhein: Harald Boldt Verlag, 1979. Vol. 1, x + 660 p.; Vol. 2, xviii + 629 p.

Life of the German geographer Karl Haushofer (1869-1945), professor at the University of Munich (1921-1939), plus selected texts in geopolitics (Vol. 1, pp. 483-646) and selected correspondence, 1917-1946 (whole of Vol. 2).

1293. Kloster Ullman, Elba E. "El General Haushofer y la geopolítica," Anuario de Geografía (Mexico City), vol. 7 (1967), 121-141.

Concerns the German geographer Karl Haushofer (1869-1946), his work in geopolitics, and the work of his predecessors, such as Friedrich Ratzel, Alfred Mahan, Halford Mackinder, and Rudolf Kjellén.

1294. Schöller, Peter. "Die rolle Karl Haushofers für entwicklung und Ideologie nationalsozialistischer

Geopolitik," Erdkunde, vol. 36, no. 3 (September 1982), 160-167.

The German geographer Karl Haushofer (1869-1945) and the development of geopolitics in the Nazi period.

HAVILAND, Alfred

1295. Freeman, T.W. "19th-Century Medical Geographer," Geographical Magazine, vol. 51, no. 2 (November 1978), 90.

Concerns the English doctor Alfred Haviland (1825-1902), who wrote Climate, Weather and Diseases (1855) and Geographical Distribution of Disease in Great Britain (1875).

HEDIN, Sven

1296. Beck, Hanno. "Sven Hedin (1865-1952)," Geographisches Taschenbuch 1964/65 (Wiesbaden: Franz Steiner Verlag GMBH, 1964), pp. 290-302.

Sketch of the Swedish explorer in Asia, Sven Hedin (1865-1952).

1297. Pedreschi, L. "I viaggi di Sven Hedin e la loro importanza geografica," Rome, Università, Facoltà di Magistero, Istituto di scienze geografiche e cartografiche, Memorie geografiche, vol. 1 (1954), 143-196.

Bibliographical guide to about 100 items on the Asian expeditions of the Swedish explorer Sven Hedin (1865-1952). [LG]

1298. Tiessen, Ernst, ed. Meister und Schüler: Ferdinand von Richthofen an Sven Hedin. Berlin: Verlag von Dietrich Reimer, 1933. 148 p.

Connections between the German geographer Ferdinand von Richthofen (1833-1905) and the Swedish explorer Sven Hedin (1865-1952). Pp. 71-148 contain letters from Richthofen to Hedin (1889-1905).

HERBERTSON, Andrew John

1299. Fleure, H.J. "The Later Developments in Herbertson's Thought: A Study in the Application of Darwin's Ideas," Geography, vol. 37, part 2 (April 1952), 97-103.

After an introductory section about the British geographer A.J. Herbertson (1865-1915), Fleure expands on his own notions about environmental influences, especially climatic.

1300. Gilbert, E.W., et al. "A.J. Herbertson Centenary Issue," Geography, vol. 50, part 4 (November 1965), 313-372.

Several articles on the British geographer Andrew John Herbertson (1865-1915), who taught at the University of Oxford, 1899-1915. Articles by E.W. Gilbert, H.J. Fleure, J.F. Unstead, and L.J. Jay.

1301. Jay, L.J. "Andrew John Herbertson, 1865-1915," Geographers: Biobibliographical Studies, vol. 3 (1979), 85-92.

The British geographer A.J. Herbertson (1865-1915) taught at the University of Oxford, 1899-1915 (Professor since 1910).

HERDER, Johann Gottfried von

1302. Grundmann, Johannes. Die geographischen und völkerkundlichen Quellen und Anschauungen in Herders 'Ideen zur Geschichte der Menschheit.' (Inaugural-Dissertation, Leipzig.) Berlin: Weidmannsche Buchhandlung, 1900. vi + 139 p.

An examination of the geographical and ethnological sources and views in Ideen zur Geschichte der Menschheit by the German historian, philosopher, and theologian Johann Gottfried von Herder (1744-1803).

1303. Hauck, Paul. "Johann Gottfried Herder als Geograph," Petermanns Geographische Mitteilungen, vol. 123, no. 1 (1979), 49-58.

Concerning the German historian J.G. Herder (1744-1803). (From English abstract) "Herder's geographical ideas, which are primarily rooted in his 'Ideas on the Philosophy Concerning the History of Humanity,' were influenced by Spinoza's pantheism, by Kant and the 18th-century geographers."

1304. Hauck, Paul. "Johann Gottfried Herders Stellung zur Schulgeographie," Zeitschrift für den Erdkundeunterricht, vol. 13, no. 12 (1961), 450-463.

Geographical instruction in the 18th century and the influence of the work of the German philosopher and historian J.G. Herder (1744-1803).

1305. Schwarz, Gabriele. "Johann Gottfried Herder: Seine Stellung zur Landschaft und seine Bedeutung für die Geographie," pp. 169-187 in <u>Landschaft</u> und <u>Land, der Forschungsgegenstand der Geographie (Festschrift Erich Obst zum 65. Geburtstag)</u>, ed. Kurt Kayser (Remagen: Verlag des Amtes fur Landeskunde, 1951).

The German historian and theologian Johann Gottfried von Herder (1744-1803) and the significance of his works to geography.

HETTNER, Alfred

1306. Beck, Hanno. "Alfred Hettner--der einflussreiche Methodiker (1859-1941)," pp. 213-228 in <u>Grosse Geographen: Pioniere--Aussenseiter--Gelehrte</u> by Beck (Berlin: Dietrich Reimer Verlag, 1982).

Biographical essay on the German geographer Alfred Hettner (1859-1941), professor of geography at the University of Heidelberg, 1899-1928.

1307. Hartshorne, Richard. "Alfred Hettner," pp. 354-356 in Vol. 6 of <u>International Encyclopedia of the Social Sciences</u> (New York: The Macmillan Company and The Free Press, 1968).

Sketch of the German geographer Alfred Hettner (1859-1941), who founded <u>Geographische Zeitschrift</u> (1895) and taught at Heidelberg University (1899-1928).

1308. Pfeifer, Gottfried, ed. <u>Alfred Hettner *6.8.1859: Gedenkschrift zum 100. Geburtstag</u>. (Heidelberger Geographische Arbeiten, 6.) Heidelberg and Munich: Keysersche Verlagsbuchhandlung, 1960. 88 p.

Papers on the German geographer Alfred Hettner (1859-1941) by Erich Maschke, Ernst Plewe, and Friedrich Metz. Also includes 3 autobiographical pieces by Hettner.

1309. Pfeifer, Gottfried. "Alfred Hettner zum 100. Geburtstage," <u>Kosmos</u>, vol. 55, no. 8 (August 1959), 351-353.

Biographical sketch of the German geographer Alfred Hettner (1859-1941), professor at the University of Heidelberg.

1310. Plewe, Ernst. "Alfred Hettner, 1859-1941," Geographers: Biobibliographical Studies, vol. 6 (1982), 55-63.

The German geographer Alfred Hettner (1859-1941) taught at the University of Heidelberg from 1899 to 1928.

1311. Schmitthenner, Heinrich. "Alfred Hettner," pp. xi-xliv in Allgemeine Geographie des Menschen by Alfred Hettner (ed. Schmitthenner), Vol. 1 (Stuttgart: W. Kohlhammer Verlag, 1947).

Biographical sketch of the German geographer Alfred Hettner (1859-1941).

HIMLY, Louis-Auguste

1312. Berdoulay, Vincent. "Louis-Auguste Himly, 1823-1906," Geographers: Biobibliographical Studies, vol. 1 (1977), 43-47.

Louis-Auguste Himly (1823-1906) was a French historian and geographer who taught geography at the Sorbonne for 40 years (Dean of the Faculty of Arts, 1881-1898).

HINKS, Arthur Robert

1313. Steers, J.A. "A.R. Hinks and the Royal Geographical Society," Geographical Journal, vol. 148, part 1 (March 1982), 1-7.

An appreciation of the English geographer A.R. Hinks (1873-1945), with emphasis on his work with the Royal Geographical Society, of which he was the long-time Secretary and Editor.

HÖHNEL, Ludwig von

1314. Hermann, Annemarie. "Ludwig von Höhnel, 1857-1942," Geographers: Biobibliographical Studies, vol. 4 (1980), 43-47.

Ludwig von Höhnel (1857-1942) was an Austrian naval officer and East African explorer.

Part Three--Biographical Works

HOFF, Karl Ernst Adolf von

1315. Reich, Otto. Karl Ernst Adolf von Hoff, der Bahnbrecher moderner Geologie: Eine wissenschaftliche Biographie. Leipzig: Verlag von Veit & Comp., 1905. vi + 144 p.

K.E.A. von Hoff (1771-1837) was a German geologist who held views similar to Lyell's and clearly recognized the role of erosion in the formation of valleys.

HÜTTNER, Johann Christian

1316. Gedan, Paul. "Johann Christian Hüttner: Ein Beitrag zur Geschichte der Geographie," Mitteilungen des Vereins für Erdkunde zu Leipzig, 1897, pp. 1-37.

Johann Christian Hüttner (1766-1847) was a German who went to England at age 25 and remained there for the rest of his life, working as a translator in the Foreign Office. He accompanied the British ambassador to China in the 1790s and published a book about China in German in 1797.

HUMBOLDT, Alexander von

1317. Almagià, Roberto. "Alessandro von Humboldt," Vie del Mundo, vol. 21 (1959), 493-502.

A sketch of the work of the German geographer Alexander von Humboldt (1769-1859), with an introduction on the advances in geographical knowledge in the second half of the 18th century. Includes maps of Humboldt's itineraries in the Americas and Russia. [LG]

1318. Banse, Ewald. "Alexander von Humboldt," p. 549 of Vol. 7 of Encyclopaedia of the Social Sciences (New York: The Macmillan Company, 1932).

Biographical sketch of the German geographer and naturalist Alexander von Humboldt (1769-1859).

1319. Banse, Ewald. Alexander von Humboldt: Erschliesser einer neuen Welt. (Grosse Naturforscher, Vol. 14.) Stuttgart: Wissenschaftliche Verlagsgesellschaft M.B.H., 1953. viii + 146 p.

Biography of the German geographer Alexander von Humboldt (1769-1859). Emphasis on American travels

(1799-1804), including preparations and subsequent publications.

1320. Beck, Hanno. "Alexander von Humboldt," pp. 545-546 in Vol. 6 of International Encyclopedia of the Social Sciences (New York: The Macmillan Company and The Free Press, 1968).

   Biographical sketch of the German geographer Alexander von Humboldt (1769-1859).

1321. Beck, Hanno. Alexander von Humboldt. 2 vols. Wiesbaden: Franz Steiner Verlag GmbH, 1959-1961. Vol. 1, Von der Bildungsreise zur Forschungsreise, 1769-1804, xvi + 303 p.; Vol. 2, Von Reisewerk zum 'Kosmos,' 1804-1859, xii + 439 p.

   Definitive biography of the German geographer Alexander von Humboldt (1769-1859).

1322. Beck, Hanno. "Alexander v. Humboldt--der grösste Geograph der neueren Geschichte (1769-1859)," pp. 83-102 in Grosse Geographen: Pioniere--Aussenseiter--Gelehrte by Beck (Berlin: Dietrich Reimer Verlag, 1982).

   An essay on the life and contributions of the German geographer Alexander von Humboldt (1769-1859).

1323. Beck, Hanno. "Zur Geschichte der Alexander-von-Humboldt Forschung," pp. 483-496 in Alexander von Humboldt: Werk und Weltgeltung, ed. Heinrich Pfeiffer (Munich: R. Piper & Co. Verlag, 1969).

   Humboldtian scholarship over the years.

1324. Beck, Hanno, ed. Gespräche Alexander von Humboldts. Berlin: Akademie-Verlag, 1959. xxxii + 492 p.

   Record of conversations or encounters with Alexander von Humboldt, 1785-1859.

1325. Beck, Hanno. "Hinweise auf Gespräche Alexander von Humboldt," pp. 249-276 in Alexander von Humboldt: Werk und Weltgeltung, ed. Heinrich Pfeiffer (Munich: R. Piper & Co. Verlag, 1969).

   Additional mentions of Humboldt. Sequel to Beck's book Gespräche (above).

1326. Beck, Hanno. "Das Ziel der grossen Reise Alexander von Humboldts," Erdkunde, vol. 12, no. 1 (February 1958), 42-50.

   The preparation of the German geographer Alexander von Humboldt (1769-1859) for his American travels (1799-1804).

1327. Biermann, Kurt-R. "Alexander von Humboldt als Initiator und Organisator internationaler Zusammenarbeit auf geophysikalischen Gebiet," Proceedings of the 15th International Congress of Historical Sciences (Edinburgh), 1978, pp. 126-138.

   The German geographer Alexander von Humboldt (1769-1859) as an initiator and organizer of international cooperation in establishing networks of observation in geomagnetism and climatology. [HB]

1328. Biermann, Kurt-R. "Alexander von Humboldts Forschungsprogramm von 1812 und dessen Stellung in Humboldts indischen und sibirischen Reiseplänen," pp. 471-483 in Studia z dziejów geografii i kartografii /Etudes d'histoire de la géographie et de la cartographie, ed. Józef Babicz (Warsaw: Polish Academy of Science, 1973).

   Alexander von Humboldt's research plan of 1812 and its connection with his later travels to Siberia (1829).

1329. Bitterling, Richard. Alexander von Humboldt. ("Lebenswege in Bildern," ed. Ernst Hermann.) Munich and Berlin: Deutscher Kunstverlag, 1959. 116 p.

   Pictorial biography of the German geographer Alexander von Humboldt (1769-1859).

1330. Bitterling, Richard. "Alexander von Humboldts Amerikareise in zeitgenössischer Darstellung: Zur 150. Wiederkehr der Reiseabschlusses aus dem Französischen übersetzt und eingeleitet," Petermanns Geographische Mitteilungen, vol. 98 (1954), 161-171.

   The American travels (1799-1804) of the German geographer Alexander von Humboldt (1769-1859). From contemporary account by Jean-Claude de La Métherie.

1331. Botting, Douglas. <u>Humboldt</u> <u>and</u> <u>the</u> <u>Cosmos</u>. New York, etc.: Harper & Row, Publishers, 1973. 295 p.

A popular biography of the German geographer Alexander von Humboldt (1769-1859).

1332. Bowen, Margarita J. "Mind and Nature: The Physical Geography of Alexander von Humboldt," <u>Scottish Geographical Magazine</u>, vol. 86, no. 3 (December 1970), 222-233.

The role of the German geographer Alexander von Humboldt (1769-1859) in the systematization of "the unformed science of physical geography." "Science is mind applied to nature."

1333. Brand, Donald D. "Humboldts Essai Politique sur le Royaume de la Nouvelle-Espagne," pp. 123-141 in <u>Alexander</u> <u>von</u> <u>Humboldt: Studien</u> <u>zu</u> <u>seiner</u> <u>universalen</u> <u>Geisteshaltung</u>, ed. Joachim Schultze (Berlin: Verlag Walter De Gruyter & Co., 1959).

The <u>Essai</u> <u>politique</u> <u>sur</u> <u>le</u> <u>royaume</u> <u>de</u> <u>la</u> <u>Nouvelle-Espagne</u> (1811) of the German geographer Alexander von Humboldt (1769-1859) is called "the first modern regional economic geography."

1334. Browne, C.A. "Alexander von Humboldt as Historian of Science in Latin America," <u>Isis</u>, vol. 35, part 2 (Spring 1944), 134-139.

Concerns the German geographer Alexander von Humboldt (1769-1859). (Page 138) "It has been the purpose of the present paper not to sketch Humboldt's own contributions to science but to point out a few of the numerous ways in which his writings [on Latin America] can be helpful to the student of the history of science."

1335. Bunksé, Edmunds V. "Humboldt and an Aesthetic Tradition in Geography," <u>Geographical</u> <u>Review</u>, vol. 71, no. 2 (April 1981), 127-146.

(From Abstract, p. 250) "Alexander von Humboldt ... directed his attention to the aesthetic qualities of the tropical and cordilleran landscapes of South America and concurrently probed the human aesthetic experiences of nature. He influenced the development of a distinct trend in American landscape painting ...

Humboldt provided the basis for an aesthetic tradition in geography."

1336. De Terra, Helmut. Humboldt: The Life and Times of Alexander von Humboldt. New York: Alfred A. Knopf, 1955. xii + 386 + ix (index) p.

Life of the German geographer and naturalist Alexander von Humboldt (1769-1859). German translation published in 1956, Alexander von Humboldt und seine Zeit (Wiesbaden: F.A. Brockhaus).

1337. Döring, Lothar. Wesen und Aufgaben der Geographie bei Alexander von Humboldt. Frankfurter Geographische Hefte, 5th year (1931), no. 1. 173 p.

The life and work of the German geographer Alexander von Humboldt (1769-1859), with emphasis on the nature of his geographical work. Three chapters (pp. 64-153) treat geography as the study of distributions, geography as the study of landscape, and geography as a bridging science.

1338. Friis, Herman R. "Alexander von Humboldts Besuch in den Vereinigten Staaten von Amerika vom 20. Mai bis zum 30. Juni 1804," pp. 142-195 in Alexander von Humboldt: Studien zu seiner universalen Geisteshaltung, ed. Joachim H. Schultze (Berlin: Verlag Walter De Gruyter & Co., 1959).

Humboldt's visit to the United States in 1804 and his significant meetings with President Thomas Jefferson.

1339. Friis, Herman R. "Baron Alexander von Humboldt's Visit to Washington, D.C., June 1 through June 13, 1804," Records of the Columbia Historical Society (Washington, D.C.), 1963, pp. 1-35.

Humboldt's visit to Washington in 1804. Part of a longer paper (above) published in German in 1959.

1340. Gerassimow, I.P. "Alexander von Humboldt und Wassilij Wassiljewitsch Dokutschajew," pp. 43-48 in Alexander von Humboldt: Vorträge und Aufsätze Anlässlich der 100. Wiederkehr seines Todestages am 6. Mai 1959, ed. Johannes F. Gellert (Geographische Gesellschaft der Deutschen Demokratischen Republik, Wissenschaftliche Abhandlungen, Vol. 2) (Berlin: VEB Deutscher Verlag der Wissenschaften, 1960).

A description of the works of the German geographer Alexander von Humboldt (1769-1859) and those of V.V. Dokuchaev (1846-1903), a Russian geographer of a later era.

1341. Hard, Gerhard. "'Kosmos' und 'Landschaft': Kosmologische und landschaftsphysiognomische Denkmotive bei Alexander von Humboldt und in der geographischen Humboldt-Auslegung des 20. Jahrhunderts," pp. 133-177 in Alexander von Humboldt: Werk und Weltgeltung, ed. Heinrich Pfeiffer (Munich: R. Piper & Co. Verlag, 1969).

The concepts of "cosmos" and "landscape" in the works of the German geographer Alexander von Humboldt (1769-1859) and in 20th-century writings about Humboldt.

1342. Herneck, Friedrich. "Alexander von Humboldt und seine Verdienste um die Wissenschaft von der Erde," Geologie, vol. 19, no. 6 (July 1970), 625-636.

The German geographer, naturalist, and explorer Alexander von Humboldt (1769-1859), from his student days at Frankfurt an der Oder (1787) to the publication of Kosmos (1845-), with emphasis on his service to the earth sciences.

1343. Honigmann, Peter. "Alexander von Humboldts Journale seiner russisch-sibirischen Reise 1829," Petermanns Geographische Mitteilungen, vol. 127, no. 2 (1983), 103-108.

Concerns the diaries kept by the German geographer Alexander von Humboldt (1769-1859) on his journey to western Siberia in 1829.

1344. Kehr, Kurt. "Alexander von Humboldt: Forscherpersönlichkeit und Wissenschaftskonzeption," Naturwissenschaftliche Rundschau, vol. 34, no. 8 (August 1981), 327-333.

A brief overview of Alexander von Humboldt's major contributions, with emphasis on his American travels (1799-1804), and some remarks on the continuing relevance of his scientific conceptions.

1345. Kellner, L. Alexander von Humboldt. London: Oxford University Press, 1963. 246 p.

Part Three--Biographical Works

A biography of the German geographer and natural scientist Alexander von Humboldt (1769-1859).

1346. Kick, Wilheim. "Alexander von Humboldts Wirken für die Hochgebirgsforschung in Asien, besonders über die Brüder Schlagintweit," <u>Petermanns Geographische Mitteilungen</u>, vol. 113, no. 2 (May 1969), 89-99.

The interest of the German geographer Alexander von Humboldt (1769-1859) in mountain research in Asia and his influence on the Schlagintweit brothers (Hermann, Adolf, and Robert).

1347. Krammer, Mario. <u>Alexander von Humboldt: Mensch, Zeit, Werk</u>. Berlin and Munich: Gebrüder Weiss Verlag, 1954. 439 p.

The life and works of the German geographer Alexander von Humboldt (1769-1859). Part 1, pp. 9-105, treats the life of Humboldt; Part 2, pp. 109-173, deals with his correspondence, 1789-1859; and Part 3, pp. 177-425, consists of extracts from Humboldt's <u>Ansichten der Natur</u> and <u>Kosmos</u>.

1348. Krause, Fritz, ed. <u>Kosmos und Humanität: Alexander von Humboldts Werk in Auswahl</u>. Bremen: Carl Schünemann Verlag, 1960. xl + 452 p.

Editor's introduction, pp. ix-xl, is an appreciation of the German geographer Alexander von Humboldt (1769-1859) and his works. The rest of the book is an anthology of selections from Humboldt's <u>Ansichten der Natur</u> and <u>Kosmos</u>.

1349. Kühn, Arthur. "Alexander von Humboldt und das geographische Weltbild seiner Zeit," <u>Geographische Rundschau</u>, vol. 11, no. 5 (May 1959), 172-179.

The German geographer Alexander von Humboldt (1769-1859), his geographical contributions, and the state of geographical knowledge in his time.

1350. Lentz, Eduard. "Alexander von Humboldt's Aufbruch zur Reise nach Süd-Amerika: Nach ungedruckten Briefen A. v. Humboldt's an Baron v. Forell," <u>Wissenschaftliche Beiträge zum Gedächtnis der hundertjährigen Wiederkehr des Antritts von Alexander von Humboldt's Reise nach Amerika am 5. Juni 1799</u> (Issued by the

Berlin Geographical Society on the occasion of the 7th International Geographical Congress) (Berlin: W.H. Kühl, 1899). 54 p.

Concerns Humboldt's departure on his journey to South America in 1799. Also published in Zeitschrift der Gesellschaft für Erdkunde zu Berlin, vol. 34 (1899), 311-362.

1351. Macpherson, Anne. "The Human Geography of Alexander von Humboldt," Unpublished Ph.D. thesis in Geography, University of California, Berkeley, 1971. 595 p.

The German geographer Alexander von Humboldt (1769-1859) was strongly influenced by the works of Immanuel Kant. In his writings about Latin America, Humboldt was consciously developing Kant's idea that the order of nature could be a model for the social order of mankind. [AM]

1352. Maull, Otto. "Alexander von Humboldt," Geographische Rundschau, vol. 11, no. 5 (May 1959), 169-170.

Brief sketch of the life of the German geographer Alexander von Humboldt (1769-1859).

1353. Meurers, Joseph, ed. "Alexander-von-Humboldt-Colloquium der Academia Cosmologica Nova, Salzburg 1978," Philosophia Naturalis, vol. 17, no. 4 (1979), 403-512.

7 papers by German philosophers on various philosophical issues connected with the Kosmos of Alexander von Humboldt. Papers in German with Spanish translations.

1354. Meyer-Abich, Adolf. Alexander von Humboldt in Selbstzeugnissen und Bilddokumenten. Reinbek bei Hamburg: Rowohlt Taschenbuch Verlag GmbH, 1967. 187 p.

The life of the German geographer Alexander von Humboldt (1769-1859).

1355. Minguet, Charles. Alexandre de Humboldt, historien et géographe de l'Amérique espagnole, 1799-1804. Paris: François Maspero, 1969. 693 p.

The Latin American travels (1799-1804) of the German geographer Alexander von Humboldt (1769-1859), his previous experiences, and his later views and

Part Three--Biographical Works

publications on American topics (Indians, Blacks, physical geography, Christopher Columbus, economic matters, and towns).

1356. Muthmann, Friedrich. *Alexander von Humboldt und sein Naturbild im Spiegel der Goethezeit.* Zürich and Stuttgart: Artemis-Verlag, 1955. 154 p.

The German geographer Alexander von Humboldt (1769-1859), with emphasis on his views of nature.

1357. Oberhummer, Eugen. "L'oeuvre géographique d'Alexandre de Humboldt au Mexique," *Annales de géographie,* vol. 20, no. 109 (15 January 1911), 65-69.

Concerns the visit of the German geographer Alexander von Humboldt (1769-1859) to Mexico (1803-1804) and his subsequent writings about that country.

1358. Otero Pedrayo, R., et al. [Special issue devoted to Alexander von Humboldt], *Estudios Geográficos* (Madrid), vol. 20, no. 76 (August 1959), 317-425.

4 articles on the German geographer Alexander von Humboldt (1769-1859) by R. Otero Pedrayo, Enrique Alvarez López, German Bleiberg, and Amando Melón.

1359. Otremba, Erich. "Die Llanos des Orinoco und des Apure in der Landschafts- und Reisebeschreibung Alexander von Humboldts," pp. 69-89 in *Alexander von Humboldt: Studien zu seiner universalen Geisteshaltung* ed. by Joachim Schultze (Berlin: Verlag Walter De Gruyter & Co., 1959).

(From English summary, p. 69) "A comparison of Humboldt's impression of the Llanos in Venezuela with the aspect of this area today, as observed by the present author on an expedition which followed Humboldt's footsteps, affords very good insight into Humboldt's view of landscape and his methods of investigating it."

1360. Pfeifer, Gottfried. "Die Neue Welt in der Perspektive Alexander von Humboldts," *Erdkunde,* vol. 13, no. 4 (December 1959), 395-411.

Much has been made of Alexander von Humboldt (1769-1859) as a natural scientist, but this paper shows his importance as a human geographer, with reference to his American travels, 1799-1804.

1361. Pfeifer, Gottfried. "Ritter, Humboldt und die moderne Geographie," <u>Verhandlungen des Deutschen Geographentages</u> (Berlin, 1959), vol. 32 (Wiesbaden: Franz Steiner Verlag GMBH, 1960), 69-83.

An appreciation of the German geographers Carl Ritter (1779-1859) and Alexander von Humboldt (1769-1859) on the 100th anniversary of their deaths. Emphasis on their influence on the work of Alfred Hettner and later geographers.

1362. Pfeiffer, Heinrich, ed. <u>Alexander von Humboldt: Werk und Weltgeltung</u>. Munich: R. Piper & Co. Verlag, 1969. 505 p.

Essays by various scholars on the German geographer Alexander von Humboldt (1769-1859), arranged in 3 sections: "Humboldts Einfluss auf die Geisteswissenschaften," "Humboldt und die Moderne Naturwissenschaft," and "Humboldt in Gesprächen und Begegnungen." Includes chapters by the geographers Gerhard Hard, Carl Troll, and Hanno Beck (listed separately in this bibliography).

1363. Plewe, Ernst. "Alexander von Humboldt (1768[sic--1769]-1859)," <u>Geographisches Taschenbuch 1958/59</u> (Wiesbaden: Franz Steiner Verlag GMBH, 1958), pp. 494-500.

Sketch of the German geographer Alexander von Humboldt (1769-1859).

1364. Sáenz de la Calzada, Carlos, et al. "Homenaje a Alejandro de Humboldt a los 200 años de su nacimiento," Universidad Nacional Autónoma de México, Facultad de Filosofía y Letras, <u>Anuario de Geografía</u>, vol. 9 (1969), 9-116.

7 articles (pp. 11-116) on the life and work of the German geographer Alexander von Humboldt on the 200th anniversary of his birth. Articles by Carlos Sáenz de la Calzada, Leopoldo Zea, Manuel Maldonado-Koerdell (2 articles), Angel Bassols Batalla, Rayfred L. Stevens, and Jorge A. Vivó Escoto.

Part Three--Biographical Works 283

1365. Sanke, Heinz, et al. "Alexander von Humboldt, 1769-1859," Zeitschrift für den Erdkundeunterricht, vol. 11, no. 4 (1959), 90-144.

6 articles on the German geographer Alexander von Humboldt (1769-1859) by Heinz Sanke, Johannes F. Gellert, Inge Paulukat, Georg von Vietinghoff, Walter Bieler, and Gerhard Engelmann.

1366. Sanke, Heinz, et al. "Zur zweihundertsten Wiederkehr des Geburtstages von Alexander von Humboldt am 14. September 1969," Petermanns Geographische Mitteilungen, vol. 113, no. 2 (May 1969), 81-111.

4 articles on the German geographer Alexander von Humboldt (1769-1859) by Heinz Sanke, Wilhelm Kick, Gerhard Engelmann, and Paul Holz.

1367. Schmidt, Erwin. "Carl Ritter als Zeitgenosse Alexander von Humboldts," pp. 89-99 in Alexander von Humboldt: Vorträge und Aufsätze Anlässlich der 100. Wiederkehr seines Todestages am 6. Mai 1959, ed. Johannes F. Gellert (Geographische Gesellschaft der Deutschen Demokratischen Republik, Wissenschaftliche Abhandlungen, vol. 2) (Berlin: VEB Deutscher Verlag der Wissenschaften, 1960).

Connections between the German geographers Carl Ritter (1779-1859) and Alexander von Humboldt (1769-1859). Article deals mostly with Ritter.

1368. Schmieder, Oskar. "Alexander von Humboldt: Persönlichkeit, wissenschaftliches Werk und Auswirkung auf die moderne Länderkunde," Geographische Zeitschrift, vol. 52, no. 2 (1964), 81-95.

The life and work of the German geographer Alexander von Humboldt (1769-1859), with some comments about his influence on later German geographers, such as Alfred Hettner.

1369. Schultze, Joachim H., ed. Alexander von Humboldt: Studien zu seiner universalen Geisteshaltung. Berlin: Verlag Walter De Gruyter & Co., 1959. xxiv + 277 p.

Essays by geographers and others commemorating the 100th anniversary of the death of the German geographer Alexander von Humboldt (1769-1859). See especially the articles by Rayfred Stevens, Erich Otremba, Georg Wüst, Donald Brand, Herman Friis, and Carl Troll (listed separately in this bibliography). Articles in German with English abstracts.

1370. Scurla, Herbert. <u>Alexander von Humboldt: Sein Leben und Wirken</u>. 6th ed. Berlin: Verlag der Nation, 1968. 419 p.

Biography of the German geographer and traveler Alexander von Humboldt (1769-1859).

1371. Sinnhuber, Karl. "Alexander von Humboldt, 1769-1859," <u>Scottish Geographical Magazine</u>, vol. 75, no. 2 (September 1959), 89-101.

Biographical essay on the German geographer, naturalist, and traveler Alexander von Humboldt (1769-1859).

1372. Stams, Werner. "Alexander von Humboldts Reise zur Mitte Asiens: Zur Erinnerung an seine Forschungsreise durch Russland vor 150 Jahren," <u>Geographische Berichte</u>, vol. 24, no. 4 (June 1979), 241-253.

The journey of the German geographer Alexander von Humboldt (1769-1859) to Central Asia in 1829.

1373. Stevens, Rayfred L. "Humboldt als wissenschaftlicher Reisender und als Naturbeobachter," pp. 1-35 in <u>Alexander von Humboldt: Studien zu seiner universalen Geisteshaltung</u>, ed. Joachim H. Schultze (Berlin: Verlag Walter De Gruyter & Co., 1959).

The German geographer Alexander von Humboldt (1769-1859) as a scientific traveler in the Americas (1799-1804).

1374. Stevens-Middleton, Rayfred L. <u>La obra de Alexander von Humboldt en Mexico: Fundamento de la geografía moderna</u>. Instituto Panamericano de Geografía e Historia, Publicación no. 202 (Mexico City, 1956). xxi + 269 p.

The observations and publications of the German geographer Alexander von Humboldt (1769-1859) concerning Mexico. Chapter 8 (pp. 199-245), "Epistemological Aspects of Humboldt's Work," deals with geography generally and Humboldt's influence. It also treats Humboldt's *Essai politique sur le royaume de la Nouvelle-Espagne* (1811) as a prototype of modern regional geography.

1375. Stoetzer, Carlos. "Humboldt: A Hundred Years After," *Américas*, vol. 11, no. 5 (May 1959), 2-8.

   A sketch of the life of the German geographer Alexander von Humboldt (1769-1859), with emphasis on his American travels (1799-1804).

1376. Suchowa, N.G. *Alexander von Humboldt in der russischen Literatur: Bibliographie*. Leipzig: VEB Verlag für Buch- und Bibliothekswesen, 1960. 96 p.

   Bibliography of Russian editions of the works of Alexander von Humboldt and of writings about Humboldt. Useful 12-page index of personal names with brief biographical information.

1377. Theodorides, Jean. "Humboldt and England," *British Journal for the History of Science*, vol. 3, part 1 (June 1966), 39-55.

   The English connections of the German geographer Alexander von Humboldt (1769-1859).

1378. Troll, Carl. "Alexander von Humboldt," *Die Grossen Deutschen*, vol. 3 (1956), 175-188.

   Biographical sketch of the German geographer Alexander von Humboldt (1769-1859).

1379. Troll, Carl. "Alexander von Humboldts wissenschaftliche Sendung," pp. 258-277 in *Alexander von Humboldt: Studien zu seiner universalen Geisteshaltung*, ed. Joachim H. Schultze (Berlin: Verlag Walter De Gruyter & Co., 1959).

   The scientific mission of the German geographer Alexander von Humboldt (1769-1859). (From English abstract, p. 258) "Humboldt considered his mission to be the elucidation of nature's secrets by means of free empirical research throughout the world, the preservation of progressive, liberal humanity and the defence of human rights, civil liberty and social justice."

1380. Troll, Carl. "Die Lebensformen der Pflanzen: Alexander von Humboldts Ideen in der ökologischen Sicht von heute," pp. 197-246 in Alexander von Humboldt: Werk und Weltgeltung, ed. Heinrich Pfeiffer (Munich: R. Piper & Co. Verlag, 1969).

Plant geography and the enduring significance of the ideas of the German geographer Alexander von Humboldt (1769-1859), with examples from the Tropics.

1381. Troll, Carl. "The Work of Alexander von Humboldt and Carl Ritter: A Centenary Address," The Advancement of Science, vol. 16, no. 64 (March 1960), 441-452.

Sketches of the German geographers Alexander von Humboldt (1769-1859) and Carl Ritter (1779-1859) on the 100th anniversary of their deaths.

1382. Tulard, Jean, ed. L'Amérique espagnole en 1800, vue par un savant allemand, Humboldt. Paris: Calman-Lévy, 1965. 298 p.

Extracts from the correspondence and publications of the German geographer Alexander von Humboldt (1769-1859) concerning Latin America at the beginning of the 19th century. Editor's introduction on pp. 11-26.

1383. Whitaker, Arthur P. "Alexander von Humboldt and Spanish America," Proceedings of the American Philosophical Society, vol. 104, no. 3 (15 June 1960), 317-322.

The enduring significance of Humboldt's travels in Latin America, 1799-1804.

1384. Wilhelmy, Herbert, et al. Alexander von Humboldt: eigene und neue Wertungen der Reisen, Arbeit und Gedankenwelt. (Geographische Zeitschrift, Beihefte: Erdkundliches Wissen, Schriftenfolge für Forschung und Praxis, 23.) Wiesbaden: Franz Steiner Verlag, 1970. 73 p.

Contains 3 papers on Humboldt: by Herbert Wilhelmy on Humboldt's American travels, 1799-1804 (pp. 1-22); by Gerhard Engelmann on Humboldt's work on Kosmos (5 vols., 1845-1862), beginning in the 1820s (pp. 23-48); and by Gerhard Hard on the concept of landscape in Humboldt's work and in other authors' writings on art (c. 1800) (pp. 49-73).

1385. Wüst, Georg. "Alexander von Humboldts Stellung in der Geschichte der Ozeanographie," pp. 90-104 in Alexander von Humboldt: Studien zu seiner universalen Geisteshaltung, ed. Joachim H. Schultze (Berlin: Verlag Walter De Gruyter & Co., 1959).

The German geographer Alexander von Humboldt (1769-1859) as a pioneer in oceanography.

HURAULT, Louis

1386. Clos-Arceduc, Albert. "Le Général Louis Hurault (1886-1973)," pp. 199-202 in "Les géographes français," Secrétariat d'état aux universités, Comité des travaux historiques et scientifiques, Bulletin de la section de géographie, vol. 81 (For the years 1968-1974) (Paris: Bibliothèque Nationale, 1975).

Essay on the geographical activities of the French general Louis Hurault (1886-1973), chief of the Geographical Service of the Army, where he supervised work in surveying and mapping.

HUTCHINGS, Geoffrey Edward

1387. Wheeler, Keith. "Geoffrey Edward Hutchings, 1900-1964," Geographers: Biobibliographical Studies, vol. 2 (1978), 67-71.

G.E. Hutchings (1900-1964) was an English geographer and naturalist, known for his works on regional survey and landscape drawing.

HUXLEY, Thomas Henry

1388. Stoddart, David R. "'That Victorian Science': Huxley's Physiography and Its Impact on Geography," Transactions of the Institute of British Geographers, no. 66 (November 1975), 17-40.

(Page 17) "No work was more influential in the transformation of geography at all levels than Thomas Henry Huxley's [1825-1895] Physiography, first published in 1877."

JACKSON, Julian

1389. Goudie, Andrew S. "Colonel Julian Jackson and His Contribution to Geography," Geographical Journal, vol. 144, part 2 (July 1978), 264-270.

Julian Jackson (1790-1853) was an English military officer in India and Russia. Between 1841 and 1847 he served the Royal Geographical Society in various offices: secretary, editor, and librarian. (Page 267) "... his real contribution to geography was his work in physical geography, and in particular his studies of rivers and lakes."

JOHNSTON, Harry H.

1390. Casada, James A. "Sir Harry H. Johnston as a Geographer," Geographical Journal, vol. 143, part 3 (November 1977), 393-406.

The geographical activities of the English Africanist Harry Johnston (1858-1927).

JONES, Llewellyn Rodwell

1391. Wise, Michael J. "Llewellyn Rodwell Jones, 1881-1947," Geographers: Biobibliographical Studies, vol. 4 (1980), 49-53.

L. Rodwell Jones (1881-1947) was an English economic geographer who taught at Leeds University and the London School of Economics (Professor, 1925-1945).

KANITZ, Felix

1392. Weiss, J. "Felix Kanitz, ein Pionier der Balkanforschung," Mitteilungen der Geographischen Gesellschaft in Wien, vol. 73, nos. 1-3 (1930), 5-21.

The Austrian geographer, archeologist, and ethnographer Felix Kanitz (1829-1904) as a pioneer in Balkan studies.

KANT, Immanuel

1393. Adickes, Erich. Kant als Naturforscher. 2 vols. Berlin: Verlag W. De Gruyter & Co., 1924-1925. Vol. 1, 1924, xx + 378 p.; Vol. 2, 1925, viii + 494 p.

Views of natural science, including physical geography, of the German philosopher Immanuel Kant (1724-1804).

1394. Adickes, Erich. Kants Ansichten über Geschichte und Bau der Erde. Tübingen: Verlag von J.C.B. Mohr (Paul Siebeck), 1911. viii + 207 p.

The evolution of Immanuel Kant's (1724-1804) views on cosmology, geology, and physical geography from the 1750s to the 1790s.

1395. Adickes, Erich. Untersuchungen zu Kants physischer Geographie. Tübingen: Verlag von J.C.B. Mohr (Paul Siebeck), 1911. vii + 344 p.

An examination of physical geography as taught by the German philosopher Immanuel Kant (1724-1804) at the University of Königsberg in the second half of the 18th century. From various manuscript and published versions of Kant's lectures.

1396. Büttner, Manfred. "Kant and the Physico-Theological Consideration of the Geographical Facts," Organon (Warsaw), no. 11 (1975), 231-249.

The German philosopher Immanuel Kant (1724-1804) emancipated geography from the physico-theological philosophy. Geography is theologically neutral. Pp. 234-239 treat the development of geography from the Reformation to the early 18th century, and pp. 240-246 deal with Kant and geography since the middle of the 18th century.

1397. Büttner, Manfred. "Kant und die Uberwindung der physikotheologischen Betrachtung der geographisch-kosmologischen Fakten: Ein Beitrag zur Geschichte der Geographie in ihren Beziehungen zur Theologie und Philosophie," Erdkunde, vol. 29, no. 3 (October 1975), 162-166.

Covers the same topic as Büttner's paper in English (above).

1398. Büttner, Manfred, and Hoheisel, Karl. "Immanuel Kant, 1724-1804," Geographers: Biobibliographical Studies, vol. 4 (1980), 55-67.

Immanuel Kant (1724-1804) was a Prussian philosopher and geographer who taught physical geography at the University of Königsberg from 1756/57.

1399. Eckardt, Wilhelm R. "Uber Immanuel Kants Bedeutung für die moderne Naturwissenschaft," Naturwissenschaftliche Wochenschrift, n.s. vol. 6, no. 43 (27 October 1907), 679-681.

The German philosopher Immanuel Kant (1724-1804) and his significance for scientific works of the 19th century (Humboldt, Dove, etc.).

1400. Gerland, Georg. "Immanuel Kant, seine geographischen und anthropologischen Arbeiten," Kant-Studien, vol. 10, nos. 1-2 (1 March 1905), 1-43; nos. 4-5 (1 November 1905), 417-547.

Geographical and anthropological works of the German philosopher Immanuel Kant (1724-1804).

1401. Hauck, Paul. "Immanuel Kant als Geograph," Petermanns Geographische Mitteilungen, vol. 124, no. 4 (1980), 263-274.

The German philosopher Immanuel Kant (1724-1804) and his view of geography.

1402. Hoheisel, Karl. "Immanuel Kant und die Konzeption der Geographie am Ende des 18. Jahrhunderts," pp. 263-276 in Wandlungen im geographischen Denken von Aristoteles bis Kant, ed. Manfred Büttner ("Abhandlungen und Quellen zur Geschichte der Geographie und Kosmologie," vol. 1) (Paderborn, etc.: Ferdinand Schöningh, 1979).

Sources, evolution, and immediate influence (on Humboldt and Ritter) of Immanuel Kant's conception of geography.

1403. Lehmann, F.W. Paul. "Kants Bedeutung als akademischer Lehrer der Erdkunde," Verhandlungen des Sechsten Deutschen Geographentages (Berlin: Verlag von Dietrich Reimer, 1886), pp. 119-157.

The German philosopher Immanuel Kant (1724-1804) as a professor of geography at the University of Königsberg.

1404. Livingstone, D.N., and Harrison, R.T. "Immanuel Kant, Subjectivism, and Human Geography: A Preliminary Investigation," Transactions of the Institute of British Geographers, vol. 6, no. 3 (1981), 359-374.

The German philosopher Immanuel Kant (1724-1804), his place in the history of geographical thought, and his continuing indluence.

Part Three--Biographical Works

1405. May, Joseph A. Kant's Concept of Geography and Its Relation to Recent Geographical Thought. (University of Toronto, Department of Geography, Research Publications, no. 4.) Toronto: University of Toronto Press, 1970. xi + 280 p.

Immanuel Kant (1724-1804) lectured on geography for 40 years at the University of Königsberg. Chapter 2, pp. 25-50, "Philosophy and Geography, an Historical Sketch," ranges from the Presocratic philosophers to the 20th century. Chapter 3, pp. 51-83, concerns "The Origins, Development, and Influence of Kant's Concept of Geography."

1406. Richards, Paul. "Kant's Geography and Mental Maps," Transactions of the Institute of British Geographers, no. 61 (March 1974), 1-16.

The geographical writing and teaching of the German philosopher Immanuel Kant (1724-1804), with special concern for his anticipation of the cognitive or "mental" map.

1407. Schöne, Gustav Hermann. "Die Stellung Immanuel Kants innerhalb der geographischen Wissenschaft," Altpreussische Monatsschrift, vol. 33, nos. 3-4 (April-June 1896), 217-296.

The German philosopher Immanuel Kant (1724-1804) and his geographical works.

KAPP, Ernst

1408. Geiser, Samuel W. "Dr. Ernst Kapp, Early Geographer in Texas," Field and Laboratory, vol. 14, no. 1 (January 1946), 16-31.

The German geographer Ernst Kapp (1808-1896) spent the years 1849-1865 in Texas.

1409. Valk, Jan Gerard van der. Ernst Kapp, 1808-1896: De Beteeknis van zijn Denkbeelden voor de Sociale Geographie van de 20ste Eeuw. Utrecht: Kemink en Zoon N.V., n.d. [1939?]. 112 p.

Concerns the German geographer Ernst Kapp (1808-1896) and the significance of his ideas for human geography in the late 19th and early 20th centuries. Begins with brief treatment of 18th-century geography, Immanuel Kant, and Carl Ritter. Besides Kapp, the book treats such geographers as Peschel, Ratzel, Hettner, Schlüter, Brunhes, and Vidal de la Blache. [JAvG]

KIRCHHOFF, Alfred

1410. Meynen, Emil. "Alfred Kirchhoff, 1838-1907," Geographers: Biobibliographical Studies, vol.4 (1980), 69-76.

Alfred Kirchhoff (1838-1907) was a professor of geography at the University of Halle, 1873-1904.

1411. Schmidt, Max Georg. "Alfred Kirchhoff als akademischer Lehrer," Geographischer Anzeiger, vol. 39, no. 10 (20 May 1938), 217-224.

The German geographer Alfred Kirchhoff (1838-1907) taught at the University of Halle, 1873-1904. Appreciation of his teaching on the 100th anniversary of his birth.

KNAPP, Charles

1412. Spinner, Henri. "Hommage à Charles Knapp," Bulletin de la Société neuchâteloise de géographie, vol. 51, no. 5 (n.s., no. 10) (1954-1955), 1-12.

The Swiss geographer Charles Knapp (1855-1921) founded the Neuchâtel Geographical Society in 1885.

KÖPPEN, Wladimir

1413. Wegener-Köppen, Else. Wladimir Köppen: Ein Gelehrtenleben für die Meteorologie. ("Grosse Naturforscher," 18.) Stuttgart: Wissenschaftliche Verlagsgesellschaft M.B.H., 1955. 194 p.

The life of the German meteorologist and climatologist Wladimir Peter Köppen (1846-1940).

KOHL, Johann Georg

1414. Alexander, Anneli. "J.G. Kohl und seine Bedeutung für die deutsche Landes- und Volksforschung," Deutsche Geographische Blätter, vol. 43, nos. 1-2 (1940), 7-126.

Johann Georg Kohl (1808-1878), German historian of exploration and cartography, and his contributions to geography.

Part Three--Biographical Works

1415. Peucker, Thomas K. "Johann Georg Kohl, A Theoretical Geographer of the 19th Century," Professional Geographer, vol. 20, no. 4 (July 1968), 247-250.

The German geographer and librarian J.G. Kohl (1808-1878) and his studies of location and networks--studies that anticipate the work of theoretical geographers in the 20th century.

1416. Pfeifer, Gottfried. "... man sollte J.G. Kohl nicht vergessen!," pp. 221-236 in Mensch und Erde: Festschrift für Wilhelm Müller-Wille zum 20 Okt. 1976, ed. Karl-Friedrich Schreiber and Peter Weber (Westfälische Geographische Studien, 33 1976).

The contributions of the German geographer and librarian J.G. Kohl (1808-1878) to theoretical geography.

KOMAROV, Vladimir Leontyevitch

1417. Ilyina, T.D. "Vladimir Leontyevitch Komarov, 1869-1945," Geographers: Biobibliographical Studies, vol. 1 (1977), 55-58.

V.L. Komarov (1869-1945) was a Russian botanist and plant geographer at St. Petersburg (later Leningrad) University who published studies of the vegetation of Central and East Asia.

KRASNOV, Andrey Nikolaevich

1418. Alexandrovskaya, Olga Andreyevna. "Andrey Nikolaevich Krasnov, 1862-1914," Geographers: Biobibliographical Studies, vol. 4 (1980), 77-86.

A.N. Krasnov (1862-1914) was a Russian botanist and geographer who taught both subjects in Kharkov.

KRAUSE, Carl Christian Friedrich

1419. Köhler, Arthur. Der Philosoph Carl Christian Friedrich Krause als Geograph. (Inaugural-Dissertation, Leipzig.) Leipzig: Druck von Bar & Hermann, n.d. [1904]. 94 p.

The geographical works of the German polymath C.C.F. Krause (1781-1832).

KROPOTKIN, Peter

1420. Alexandrovskaya, Olga. "Pyotr Alexeivich Kropotkin, 1842-1921," Geographers: Biobibliographical Studies, vol. 7 (1983), 57-62.

　　The Russian geographer and anarchist Peter Kropotkin (1842-1921), who lived in exile in Western Europe from 1876 to 1917. Emphasis on Russian sources (cf. S.R. Potter, below).

1421. Breitbart, Myrna Margulies. "Peter Kropotkin, the Anarchist Geographer," pp. 134-153 in Geography, Ideology and Social Concern, ed. D.R. Stoddart (Oxford: Basil Blackwell, 1981).

　　Essay on the Russian geographer and anarchist Peter Kropotkin (1842-1921).

1422. Mondolfo, Rodolfo. "Prince Petr Alexeyevich Kropotkin," pp. 602-604 in Vol. 8 of Encyclopaedia of the Social Sciences (London: Macmillan and Co., Ltd., 1932).

　　The Russian geographer and anarchist Peter Kropotkin (1842-1921), who lived outside Russia from 1876 to 1917.

1423. Potter, S.R. "Peter Alexeivich Kropotkin, 1842-1921," Geographers: Biobibliographical Studies, vol. 7 (1983), 63-69.

　　A biographical sketch of the Russian geographer Peter Kropotkin (1842-1921), with emphasis on English-language sources (cf. O. Alexandrovskaya, above).

1424. Slatter, John. "The Kropotkin Papers," Geographical Magazine, vol. 53, no. 14 (November 1981), 917-921.

　　Biographical sketch of the Russian geographer Peter Kropotkin (1842-1921), with emphasis on material drawn from over 100 letters in the correspondence between Kropotkin and John Scott Keltie of the Royal Geographical Society.

KUBARY, Jan Stanislaw

1425. Slabczynski, Waclaw. "Jan Stanislaw Kubary, 1846-1896," Geographers: Biobibliographical Studies, vol. 4 (1980), 87-89.

Jan Stanislaw Kubary (1846-1896) was a Polish naturalist, ethnographer, and explorer of Oceania.

KÜTTNER, Carl Gottlob

1426. Friedrich, Felix. Carl Gottlob Küttner: Ein Beitrag zur Geschichte der Geographie und des deutschen Geisteslebens am Ausgange des 18. Jahrhunderts. (Inaugural-Dissertation, Leipzig.) Crimmitschau: Druck von Robert Raab, 1903. 76 p.

The life of the German traveler and travel writer Carl Gottlob Küttner (1755-1805), who wrote about northern and western Europe.

LA MÉTHERIE, Jean-Claude de

1427. Günther, Siegmund. "Die Bedeutung De la Métherie's für die Entwicklung der physikalischen Erdkunde," Mittheilungen der Kaiserlich Königlichen Geographischen Gesellschaft in Wien, vol. 43 (1900), 3-14.

The significance of Jean-Claude de La Métherie's Théorie de la terre (3 vols., Paris, 1795) for the development of physical geography.

LAPPARENT, Albert de

1428. Broc, Numa. "De la géologie à la géographie: Albert de Lapparent (1839-1908)," Revue de géographie de Lyon, vol. 52, no. 3 (1977), 273-279.

The French geologist Albert de Lapparent (1839-1908) and his contributions to geography. Famous for his textbook Traité de géologie (1882).

LARCOM, Thomas Aiskew

1429. Andrews, J.H. "Thomas Aiskew Larcom, 1801-1879," Geographers: Biobibliographical Studies, vol. 7 (1983), 71-74.

The English military officer and surveyor Thomas Larcom (1801-1879) worked in the Ordnance Survey of Ireland from 1826 to 1846 and later served on the Irish Board of Works.

LAUTENSACH, Hermann

1430. Beck, Hanno. "Hermann Lautensach--Meister der Länderkunde (1886-1971)," pp. 241-272 in Grosse

Geographen: Pioniere--Aussenseiter--Gelehrte by Beck (Berlin: Dietrich Reimer Verlag, 1982).

Biographical essay on the German geographer Hermann Lautensach (1886-1971).

1431. Tilley, Philip D. "Hermann Lautensach, 1886-1971," Geographers: Biobibliographical Studies, vol. 4 (1980), 91-101.

Hermann Lautensach (1886-1971) was a German geographer who taught at the universities of Giessen, Braunschweig, Greifswald, and Stuttgart.

LEAKE, William Martin

1432. Marshall-Cornwall, James. "Three Soldier-Geographers," Geographical Journal, vol. 131, part 3 (September 1965), 357-365.

The geographical activities of 3 British army officers--William Martin Leake (1777-1860), Edward Sabine (1788-1883), and Francis Rawdon Chesney (1789-1872).

LELEWEL, Joachim

1433. Rzepa, Zbigniew. "Joachim Lelewel, 1786-1861," Geographers: Biobibliographical Studies, vol. 4 (1980), 103-112.

Joachim Lelewel (1786-1861) was a Polish historian of geography and cartography who lived in Belgium for the last 28 years of his life.

LENCEWICZ, Stanislaw

1434. Kondracki, Jerzy. "Stanislaw Lencewicz, 1899-1944," Geographers: Biobibliographical Studies, vol. 5 (1981), 77-81.

The Polish geographer Stanislaw Lencewicz (1899-1944) "was the first professor of geography at Warsaw University and the founder of the Warsaw school of physical geography."

LENIN, Vladimir

1435. Rado, Sándor. "Lenin und die Geographie," Petermanns Geographische Mitteilungen, vol. 114, no. 1 (12 January 1970), 1-13.

The Russian revolutionary and political leader Vladimir Lenin (1870-1924) and his views on geography, particularly economic geography.

LE PLAY, Frédéric

1436. Herbertson, Dorothy. "The Life of Frédéric Le Play" (ed. Victor Branford and Alexander Farquharson), Sociological Review, vol. 38, section 2 (1946--pub. 1950), [89]-204.

Life of the French sociologist Frédéric Le Play (1806-1882), author of Les ouvriers européens (1855) and other works that are still of considerable interest to geographers. Manuscript originally prepared by Dorothy Herbertson, wife of the geographer A.J. Herbertson, between 1897 and 1899.

LEVASSEUR, Emile

1437. Nardy, Jean-Pierre. "Emile Levasseur, 1828-1911," Geographers: Biobibliographical Studies, vol. 2 (1978), 81-88.

Emile Levasseur (1828-1911) was a French economist and geographer who held a chair in geography, history, and political science in the Collège de France from 1872.

1438. Nardy, Jean-Pierre. "Levasseur, géographe," pp. 35-89 in Pour le cinquantenaire de la mort de Paul Vidal de la Blache: études d'histoire de la géographie (Cahiers de géographie de Besançon, 16; Annales littéraires de l'Université de Besançon, 93) (Paris: Les Belles Lettres, 1968).

The life and work of the French geographer and economist Emile Levasseur (1828-1911).

LEWIS, William Vaughan

1439. King, Cuchlaine A.M. "William Vaughan Lewis, 1907-1961," Geographers: Biobibliographical Studies, vol. 4 (1980), 113-120.

W.V. Lewis (1907-1961) was a British geomorphologist who taught at the University of Cambridge, 1929-1961.

LICHTENBERG, Georg Christoph

1440. Günther, Siegmund. "G.C. Lichtenberg und die Geophysik," Abhandlungen der Kaiserlich Königlichen Geographischen Gesellschaft in Wien, vol. 1 (1899), 119-135.

The German scientist Georg Christoph Lichtenberg (1744-1799) taught a course in physical geography at the University of Göttingen.

LINTON, David Leslie

1441. Gold, John R.; Haigh, Martin J.; and Warwick, G.T. "David Leslie Linton, 1906-1971," Geographers: Biobibliographical Studies, vol. 7 (1983), 75-83.

The British geomorphologist David Linton (1906-1971), professor of geography at Sheffield University (1945-1958) and Birmingham University (1958-1971).

LIVINGSTONE, David

1442. Siddle, D.J. "David Livingstone: A Mid-Victorian Field Scientist," Geographical Journal, vol. 140, part 1 (February 1974), 72-79.

The Scottish missionary-explorer David Livingstone (1813-1873). (Page 72) "The purpose of this paper is to attempt an assessment of David Livingstone, not as 'the greatest geographer that Africa has ever seen' (Debenham ...), but as a mid-Victorian field scientist whose work clearly reflected the uneasy relationship between science and religion which was a characteristic of that age."

LÖFFLER, Ernst

1443. Garboe, Axel. "Ernst Löffler (1835-1911): Den første professor i geografi ved Københavns Universitet," Geografisk Tidsskrift, vol. 63, no. 1 (June 1964), 1-27.

The Danish physical geographer Ernst Löffler (1835-1911) was the first professor of geography (1898) at the University of Copenhagen.

LOMONOSOV, Mikhail Vasilyevich

1444. Alexandrovskaya, O.A. "Mikhail Vasilyevich Lomonosov, 1711-1765," Geographers: Biobibliographical Studies, vol. 6 (1982), 65-70.

The Russian scientist Mikhail Lomonosov (1711-1765) was in charge of the Geography Department of the Academy of Sciences in St. Petersburg from 1758.

LÓPEZ, Tómas

1445. Marcel, Gabriel. Le géographe Thomas López et son oeuvre: essai de biographie et de cartographie. Madrid: Establecimiento Tipográfico de Fortanet, 1908. 123 p.

Tómas López (1731-1802) was a Spanish cartographer who had been trained in France. Cartobibliography of López' works, pp. 67-121.

LULOFS, Johann

1446. Hermans, W.F. "Johann Lulofs en zijn tijdgenoten," Tijdschrift van het Koninklijk Nederlandsch Aardrijkskundig Genootschap, vol. 71, no. 2 (April 1954), 89-104.

Concerning the Dutch scientist Johann Lulofs (1711-1768) and his contemporaries. (From English abstract, p. 89) "Manuals of physical geography have not been written in the Netherlands for two centuries. The last one, published in 1750, is the Inleiding ... by Johann Lulofs (1711-1768), Professor of Mathematics, Astronomy and Philosophy at Leyden University. The German translation of this book by Kästner (1755) made it possible that Lulofs' work influenced geography until the beginning of the 19th century, especially through the teaching of Immanuel Kant whose lectures on physical geography were largely based on Lulofs' publication."

MACKINDER, Halford John

1447. Beck, Hanno. "Sir Halford Mackinder--Schöpfer des einflussreichsten Weltbildes der Neuzeit (1861-1947)," pp. 229-240 in Grosse Geographen: Pioniere--Aussenseiter--Gelehrte by Beck (Berlin: Dietrich Reimer Verlag, 1982).

A biographical essay on the British geographer and statesman H.J. Mackinder (1861-1947).

1448. Blouet, Brian W. "Sir Halford Mackinder as British High Commissioner to South Russia, 1919-1920,"

Geographical Journal, vol. 142, part 2 (July 1976), 228-236.

The British geographer Halford Mackinder (1861-1947) was appointed British High Commissioner to South Russia in 1919 and was thus in a position to test some of the notions in his recently published book, Democratic Ideals and Reality.

1449. Blouet, Brian W. Sir Halford Mackinder, 1861-1947: Some New Perspectives. University of Oxford, School of Geography, Research Paper, no. 13 (1975). 49 p.

An essay on the British geographer and statesman H.J. Mackinder (1861-1947), who began teaching at the University of Oxford in 1887 and founded the School of Geography in 1899.

1450. Gilbert, Edmund W. "Halford Mackinder," pp. 515-517 in Vol. 9 of International Encyclopedia of the Social Sciences (New York: The Macmillan Company and The Free Press, 1968).

Biographical sketch of the English geographer Halford Mackinder (1861-1947), who founded the School of Geography at Oxford University in 1899 ("the first British university department of geography"). "He created modern British geography as a university subject."

1451. Gilbert, Edmund W. "Seven Lamps of Geography: An Appreciation of the Teaching of Sir Halford J. Mackinder," Geography, vol. 36, part 1 (March 1951), 21-43.

Concerns the English geographer Halford Mackinder (1861-1947).

1452. Gilbert, Edmund W. Sir Halford Mackinder, 1861-1947: An Appreciation of His Life and Work. London: G. Bell and Sons, Ltd., 1961. 32 p.

The English geographer Halford John Mackinder (1861-1947), Reader in geography at the University of Oxford from 1887 to 1905; London School of Economics, 1895-1925 (Director, 1903-1908); Member of Parliament, 1910-1922; etc.

Part Three--Biographical Works 301

1453. Hall, Arthur R. "Mackinder and the Course of Events," <u>Annals of the Association of American Geographers</u>, vol. 45, no. 2 (June 1955), 109-126.

   The evolution of the geopolitical views of the English geographer Halford Mackinder (1861-1947) from 1904 to 1943.

1454. Lange, Gottfried. "Sir Halford Mackinder (1861-1947)," <u>Geographisches Taschenbuch 1964/65</u> (Wiesbaden: Franz Steiner Verlag GMBH, 1964), pp. 309-316.

   The life of the English geographer and statesman Halford Mackinder (1861-1947).

1455. Parker, William H. <u>Mackinder: Geography as an Aid to Statecraft</u>. Oxford: Clarendon Press, 1982. 295 p.

   Biography of the English geographer Halford Mackinder (1861-1947), who is remembered especially for his geopolitical ideas.

1456. Saey, Pieter. "Halford J. Mackinder of het vooruitkijken naar achter," <u>Bulletin de la Société belge d'études géographiques</u>, vol. 50, no. 1 (1981), 41-61.

   A critique of the geopolitical views of the English geographer and statesman Halford Mackinder (1861-1947), especially as expressed in his book <u>Democratic Ideals and Reality</u> (1919).

1457. Semmel, Bernard. "Sir Halford Mackinder: Theorist of Imperialism," <u>Canadian Journal of Economics and Political Science</u>, vol. 24, no. 4 (November 1958), 554-561.

   The British geographer Halford Mackinder (1861-1947) and his imperialist views.

1458. Unstead, J.F. "H.J. Mackinder and the New Geography," <u>Geographical Journal</u>, vol. 113 (January-June 1949), 47-57.

   The role of the English geographer Halford Mackinder (1861-1947) in promoting geography in Great Britain from 1884 onward.

MACONOCHIE, Alexander

1459. Ward, R. Gerard. "Captain Alexander Maconochie, R.N., K.H., 1787-1860," Geographical Journal, vol. 126, part 4 (December 1960), 459-468.

The Scottish naval officer Alexander Maconochie (1787-1860) was the first holder of a professorship of geography in a British university (University College, London, 1833-1836).

MALTE-BRUN, Conrad

1460. Broc, Numa. "Un bicentenaire: Malte-Brun (1775-1975)," Annales de géographie, vol. 84, no. 466 (November-December 1975), 714-720.

The Danish-born French geographer Conrad Malte-Brun (1775-1826) wrote Précis de la géographie universelle (1810-1829) and was influential in the establishment of the Paris Geographical Society in 1821.

1461. Frenzel, Reinhard. Malthe Conrad Bruun (Malte-Brun) Frankreichs bedeutendster Geograph im ersten Viertel des 19. Jahrhunderts: Ein Beitrag zur Geschichte der geographischen Wissenschaft. (Inaugural-Dissertation, Leipzig.) Crimmitschau: Druck von Robert Raab, 1908. 127 p.

Life and work of the Danish-born French geographer Conrad Malte-Brun (1775-1826), who had been known in Denmark as Malthe Conrad Bruun. Part 1, pp. 3-19, gives an overview of geography in France around the turn of the century (i.e., 18th into 19th century).

MARINELLI, Olinto

1462. Riccardi, Riccardo. "Olinto Marinelli nel centenario della nascità," Bollettino della Società Geografica Italiana, series 10, vol. 3, nos. 1-6 (January-June 1974), 31-43.

A centennial tribute to the Italian geographer Olinto Marinelli (1874-1926).

1463. Sestini, Aldo. "La figura e l'opera di Olinto Marinelli," Rivista Geografica Italiana, vol. 81, no. 4 (December 1974), 523-544.

The life of the Italian geographer Olinto Marinelli (1874-1926), professor of geography at the University

Part Three--Biographical Works 303

of Florence, 1902-1926. Followed by "Bibliografia degli scritti di Olinto Marinelli" by Aldo Sestini (pp. 617-683). Bibliography of 474 items, 1892-1930.

MARKHAM, Clements R.

1464. Markham, Albert H. The Life of Sir Clements R. Markham. London: John Murray, 1917. xi + 384 p.

Biography of the English geographer Clements R. Markham (1830-1916) by his cousin. Clements Markham was president of the Royal Geographical Society, 1893-1905. (Page 362) "Peru was his first love, Polar exploration his second."

1465. Williams, Donovan. "Clements Robert Markham and the Geographical Department of the India Office 1867-77," Geographical Journal, vol. 134, part 3 (September 1968), 343-352.

The English geographer Clements R. Markham (1830-1916) and the reconstitution of the Geographical Department of the India Office, 1867-1877.

MARTHE, Friedrich

1466. Spethmann, Hans. Friedrich Marthe. Ein vergessener deutscher Geograph. Berlin: Reimar Hobbing, 1935. 31 p.

Biography of the German geographer Friedrich Marthe (1832-1893), showing that his methodology was the basis for the work of Ferdinand von Richthofen. [HB]

1467. Vooys, A.C. de. "Is Marthe een Voorloper van Hettner of is Hettner een Volgeling van Marthe?," Geografisch Tijdschrift, vol. 5, no. 4 (September 1971), 312-318.

An appreciation of the work of the German geographer Friedrich Marthe (1832-1893). Shows Marthe's influence on Richthofen and Hettner, and Marthe's methodological significance.

MARTONNE, Emmanuel de

1468. Dresch, Jean. "Emmanuel de Martonne," Geographisches Taschenbuch 1970/72 (Wiesbaden: Franz Steiner Verlag GMBH, 1970), pp. 280-291.

A biographical sketch of the French geographer Emmanuel de Martonne (1873-1955), professor of geography at the University of Paris from 1909.

1469. Dresch, Jean. "Emmanuel de Martonne (1873-1955)," pp. 35-48 in "Les géographes français," Secrétariat d'état aux universités, Comité des travaux historiques et scientifiques, Bulletin de la section de géographie, vol. 81 (For the years 1968-1974) (Paris: Bibliothèque Nationale, 1975).

The life and work of the French geographer Emmanuel de Martonne (1873-1955), professor of geography at the Sorbonne from 1909 to 1944.

1470. Ribiero, Orlando. "Um mestre da geografia do nosso século--Emmanuel de Martonne (1873-1955)," Finisterra (Lisbon), vol. 8, no. 16 (1973), 163-264.

The French geographer Emmanuel de Martonne (1873-1955), professor at the Sorbonne from 1909. Followed by 2 other papers on de Martonne: "Pages choisies d'Emmanuel de Martonne" by Suzanne Daveau (pp. 265-284) and "La géographie climatique dans l'oeuvre d'Emm. de Martonne" by Pierre Birot (pp. 285-291).

MAUCH, Karl

1471. Plott, Adalbert. "Karl Mauch (1837-1875)," Geographisches Taschenbuch 1966/69 (Wiesbaden: Franz Steiner Verlag GMBH, 1968), pp. 217-226.

Sketch of the German explorer Karl Mauch (1837-1875), who was in South Africa from 1865 onward.

1472. Sommerlatte, Herbert W.A. "Karl Mauch (1837-1875), ein nahezu vergessener Afrika-Reisender," Die Erde, vol. 111, no. 3 (1980), 199-211.

Karl Mauch (1837-1875) was a German explorer in southern and southeastern Africa.

MAXWELL, James Clerk

1473. Martens, Robert. "J. Cl. Maxwell's Gleichungen des denudativen Feldes: Uber Anfänge der theoretischen Geographie um 1870," Sudhoffs Archiv, vol. 50, no. 3 (September 1966), 286-296.

The geomorphological work of the British physicist James Clerk Maxwell (1831-1879).

MEHEDINŢI, Simion

1474. Mihailescu, Vintila. "Simion Mehedinţi, 1868-1962," Geographers: Biobibliographical Studies, vol. 1 (1977), 65-72.

Simion Mehedinţi (1868-1962) was a Romanian geographer who was Professor of Geography at the University of Bucharest from 1901.

MENDELSSOHN, Georg Benjamin

1475. Hohmann, Joseph. "Georg Benjamin Mendelssohn 1794-1874," Erdkunde, vol. 23, no. 3 (September 1969), 161-165.

The life and work of the German scholar Georg Benjamin Mendelssohn (1794-1874), professor of geography and statistics at the University of Bonn.

1476. Hohmann, Joseph. "Georg Benjamin Mendelssohn, 1794-1874," pp. 185-190 in Bonner Gelehrte: Beiträge zur Geschichte der Wissenschaften in Bonn, vol. 4, Mathematik und Naturwissenschaften (Bonn: H. Bouvier u. Co. Verlag/Ludwig Röhrscheid Verlag, 1970).

A biographical sketch of the German scholar Georg Benjamin Mendelssohn (1794-1874), who taught geography at the University of Bonn from 1829 onward (personal chair in 1847).

MERZ, Alfred

1477. Nöthlich, Kurt. "Alfred Merz zum hundertjährigen Geburtstag," Geographisches Taschenbuch 1981/1982 (Wiesbaden: Franz Steiner Verlag GmbH, 1981), pp. 71-94.

Life of the Austrian-born German hydrographer and oceanographer Alfred Merz (1880-1925), who died in Argentina while taking part in the Meteor Expedition.

MICHELET, Jules

1478. Corcelle, J. "Michelet géographe, à propos de son centenaire (juin 1898)," Revue de géographie, vol. 42, no. 12 (June 1898), 451-455.

Geography in the works of the French historian and writer Jules Michelet (1798-1874).

MILL, Hugh Robert

1479. Freeman, T.W. "Hugh Robert Mill, 1861-1950," Geographers: Biobibliographical Studies, vol. 1 (1977), 73-78.

H.R. Mill (1861-1950) was a British geographer who served as librarian of the Royal Geographical Society (1892-1900) and Director of the British Rainfall Organization (1901-1919).

1480. Mill, Hugh Robert. An Autobigraphy. London: Longmans, Green and Company, 1951. xii + 224 p.

The autobiography of the British geographer Hugh Robert Mill (1861-1950), librarian of the Royal Geographical Society and later Director of the British Rainfall Organization.

MILNE, Geoffrey

1481. Milne, Kathleen. "Geoffrey Milne, 1898-1942," Geographers: Biobibliographical Studies, vol. 2 (1978), 89-92.

Geoffrey Milne (1898-1942) was an English geographer, soil scientist, and agricultural chemist, known for his publications on East African soils and vegetation.

MÖLLER, Johann Heinrich

1482. Roob, Helmut. "Johann Heinrich Möller--Orientalist, Geograph und Historiker," pp. 167-173 in Leipziger Geographische Beiträge: Prof. Dr. phil. habil. Dr. Ing. E. h. Edgar Lehmann zum 60. Geburtstag (Leipzig: Deutsche Institut für Länderkunde und Geographische Institut du Karl-Marx-Universität, 1965).

The German orientalist, geographer, and historian Johann Heinrich Möller (1792-1867).

MORTON, John

1483. Bunting, B.T. "John Morton (1781-1864): A Neglected Pioneer of Soil Science," Geographical Journal, vol. 130, part 1 (March 1964), 116-119.

Observations of the British farmer and land agent John Morton (1781-1864), as recorded in his book On the Nature and Property of Soils (1838). Morton is described as "one of the great pioneers of modern geographical and applied soil science."

MURCHISON, Roderick

1484. Eyles, V.A. "Roderick Murchison, Geologist and Promoter of Science," Nature, vol. 234, no. 5329 (17 December 1971), 387-389.

Biographical sketch of the British geologist and geographer Roderick Impey Murchison (1792-1871).

1485. Gilbert, Edmund W., and Goudie, Andrew S. "Sir Roderick Impey Murchison, Bart, KCB, 1792-1871," Geographical Journal, vol. 137, part 4 (December 1971), 505-511.

Roderick Murchison (1792-1871), a British geologist and geographer, "played a major part in founding the Royal Geographical Society, Section E of the British Association, and the Hakluyt Society."

MUSHKETOV, Ivan Vasylievitch

1486. Yugai, R.L. "Ivan Vasylievitch Mushketov, 1850-1902," Geographers: Biobibliographical Studies, vol. 7 (1983), 89-91.

The Russian geographer and geologist Ivan Mushketov (1850-1902) taught at the St. Petersburg Mining Institute from 1877 onward.

NANSEN, Fridtjof

1487. Bitterling, Richard. "Fridtjof Nansen (1861-1930)," Geographisches Taschenbuch 1960/61 (Wiesbaden: Franz Steiner Verlag GMBH, 1960), pp. 438-446.

Biographical sketch of the Norwegian Arctic explorer Fridtjof Nansen (1861-1930).

1488. Nockher, Ludwig. Fridtjof Nansen, Polarforscher und Helfer der Menschheit. ("Grosse Naturforscher," vol. 19.) Stuttgart: Wissenschaftliche Verlagsgesellschaft M.B.H., 1955. 236 p.

Biography of the Norwegian explorer Fridtjof Nansen (1861-1930).

1489. Vogt, Per, et al. Fridtjof Nansen: Explorer--Scientist--Humanitarian. Oslo: Dreyers Forlag, 1961. 197 p.

Essays on the Norwegian Arctic explorer Fridtjof Nansen (1861-1930).

NEWBIGIN, Marion

1490. Freeman, T.W. "Two Ladies," Geographical Magazine, vol. 49, no. 3 (December 1976), 208.

Concerns the British geographers Marion Newbigin (1869-1934), editor of the Scottish Geographical Magazine, and Eva G.R. Taylor (1879-1966), professor at Birkbeck College, University of London.

NORDENSKIÖLD, Adolf Erik

1491. Kish, George. "Adolf Erik Nordenskiöld (1832-1901): Historian of Science and Bibliophile," Biblis, 1968, pp. 171-182.

A.E. Nordenskiöld (1832-1901), Swedish explorer and historian of cartography, was also an historian of science and collector of rare books and maps.

1492. Kish, George. "Adolf Erik Nordenskiöld (1832-1901): Polar Explorer and Historian of Cartography," Geographical Journal, vol. 134, no. 4 (December 1968), 487-505.

A.E. Nordenskiöld (1832-1901) was a Swedish Arctic Explorer and historian of cartography.

1493. Kish, George. North-east Passage: Adolf Erik Nordenskiöld, His Life and Times. Amsterdam: Nico Israel, 1973. 283 p.

Biography of the Swedish Arctic explorer and historian of cartography A.E. Nordenskiöld (1832-1901). He was the leader of the "Vega" Expedition, the first to make the sea journey around northern Asia (Northeast Passage) (1878-1879).

Part Three--Biographical Works

OBERHUMMER, Eugen

1494. Zimmermann, Susanne, and Dörflinger, Johannes. "Eugen Oberhummer, 1859-1944," Geographers: Biobibliographical Studies, vol. 7 (1983), 93-100.

The German-born geographer Eugen Oberhummer (1859-1944) was professor of geography at the University of Vienna from 1903 to 1931.

OBRUCHEV, Vladimir

1495. French, R.A. "V.A. Obruchev: The Centenary of a Great Geographer," Geographical Journal, vol. 129, part 4 (December 1963), 494-497.

Biographical essay on the Russian geologist and physical geographer Vladimir Afanas'yevich Obruchev (1863-1956).

OGILVIE, Alan Grant

1496. Baker, J.N.L. "A.G. Ogilvie and His Place in British Geography," pp. 1-6 in Geographical Essays in Memory of Alan G. Ogilvie, ed. Ronald Miller and J. Wreford Watson (London, etc.: Thomas Nelson and Sons Ltd., 1959).

Essay on the Scottish geographer A.G. Ogilvie (1887-1954), who taught at the University of Edinburgh, 1924-1953 (Professor from 1931).

ORGHIDAN, Nicolai

1497. Nedelcu, Eugen. "Nicolai Orghidan, 1881-1967," Geographers: Biobibliographical Studies, vol. 6 (1982), 77-79.

The Romanian physical geographer Nicolai Orghidan (1881-1967) was a schoolman and school inspector who taught geography in military schools.

ORMSBY, Hilda

1498. Harrison Church, R.J. "Hilda Ormsby, 1877-1973," Geographers: Biobibliographical Studies, vol. 5 (1981), 95-97.

Hilda Ormsby (1877-1973) was an English geographer who taught at the London School of Economics (with her brother, L. Rodwell Jones) and published books on the geography of London and France.

PARTSCH, Joseph

1499. Waldbaur, H. "Zur Erinnerung an Joseph Partsch (4. Juli 1851-22. Juni 1925)," Die Erde, vol. 3, no. 1 (September 1951), 60-64.

Concerning the German geographer Joseph Partsch (1851-1925), professor of geography at the University of Leipzig, 1905-1922.

PASSARGE, Siegfried

1500. Kanter, Helmut. "Siegfried Passarges Gedanken zur Geographie," Die Erde, vol. 91, no. 1 (February 1960), 41-51.

The life and thought of the German geographer Siegfried Passarge (1866-1958).

PAVLOV, Aleksei Petrovich

1501. Romanova, Maria Mikhailovna. "Aleksei Petrovich Pavlov, 1854-1929," Geographers: Biobibliographical Studies, vol. 6 (1982), 81-85.

The Russian geologist and geomorphologist Aleksei Petrovich Pavlov (1854-1929) taught at the University of Moscow from 1880.

PENCK, Albrecht

1502. Beck, Hanno. "Albrecht Penck--Geograph, bahnbrechender Eiszeitforscher und Geomorphologe (1858-1945)," pp. 191-212 in Grosse Geographen: Pioniere--Aussenseiter--Gelehrte by Beck (Berlin: Dietrich Reimer Verlag, 1982).

A biographical essay on the German geographer Albrecht Penck (1858-1945), who was a professor of geography at the universities of Vienna and Berlin.

1503. Engeln, O.D. von. "Memorial to Albrecht Penck (1858-1945)," Proceedings Volume of the Geological Society of America, Annual Report for 1958 (September 1959), 169-172.

A biographical sketch of the German geographer Albrecht Penck (1858-1945).

1504. Louis, Herbert. "Albrecht Penck und sein Einfluss auf Geographie und Eiszeitforschung," Die Erde, vol. 89, nos. 3-4 (November 1958), 161-182.

The German geographer Albrecht Penck (1858-1945) and his influence on geography and Pleistocene research.

1505. Marcinek, Joachim. "Die Bedeutung von Albrecht Penck für die Eiszeitforschung," Geographische Berichte, vol. 28, no. 3 (1983), 153-164.

An appreciation of the work of the German geographer Albrecht Penck (1858-1945), on the 125th anniversary of his birth. Emphasis on his glacial and alpine studies.

1506. Meynen, Emil. "Albrecht Penck, 1858-1945," Geographers: Biobibliographical Studies, vol. 7 (1983), 101-108.

The German physical geographer Albrecht Penck (1858-1945) was professor of geography at the universities of Vienna (1885-1906) and Berlin (1906-1926).

1507. Sölch, Johann. Albrecht Penck. (Sonderheft der Geographischen Gesellschaft Wien.) Wiener Geographische Studien, no. 17 (1948). 37 p.

Life and work of the German geographer Albrecht Penck (1858-1945).

1508. Spreitzer, Hans. "Albrecht Pencks Wirken in Wien," Mitteilungen der Osterreichischen Geographischen Gesellschaft, vol. 101, no. 3 (1959), 375-380.

The German geographer Albrecht Penck (1858-1945) was professor of geography at the University of Vienna (1885-1906) before moving to the University of Berlin.

PERTHES, Johann Georg Justus

1509. Carlberg, Berthold. "J.G. Justus Perthes 200 Jahre," Petermanns Geographische Mitteilungen, vol. 93, no. 4 (Fourth Quarter 1949), 193-196.

Biographical sketch of Johann Georg Justus Perthes (1749-1816), German founder of the geographical and cartographic establishment in Gotha that long bore his name.

PESCHEL, Oscar

1510. Sauer, Carl. "Oskar Peschel," p. 92 in Vol. 12 of Encyclopaedia of the Social Sciences (London: Macmillan and Co. Ltd., 1934).

Sketch of the German geographer Oscar Peschel (1826-1875).

1511. Schmitthenner, Heinrich. "Oscar Peschel und die Geomorphologie," Petermanns Geographische Mitteilungen, Erganzungsheft, no. 262 (1957), 1-8.

Oscar Peschel (1826-1875) and his work in geomorphology, with introductory material on his predecessors, beginning with Nicolas Desmarest (1725-1815).

1512. Valk, Jan Gerard van der. Peschels Kritick op Ritter: Een Bijdrage tot de Geschiedenis van de Probleemstelling van de Sociale Geografie. Geografisch Instituut der Rijksuniversiteit te Utrecht, Sociaal Geografische Mededeelingen, no. 1 (1944). 79 p.

Concerns criticism of the work of the German geographer Carl Ritter (1779-1859) by his countryman Oscar Peschel (1826-1875). [JAvG]

PETERMANN, August

1513. Weller, Hugo Ewald. August Petermann: Ein Beitrag zur Geschichte der geographischen Entdeckungen und der Kartographie im 19. Jahrhundert. ("Quellen und Forschungen zur Erd- und Kulturkunde," 4.) Leipzig: Otto Wigand m.b.H., 1911. x + 284 p.

The German cartographer August Petermann (1822-1878) and his contributions to cartography and exploration.

1514. Weller, Hugo Ewald. Leben und Wirken August Petermanns. Leipzig: Verlag von Julius Klinkhardt, 1914. 64 p.

Life and work of the German geographer and cartographer August Petermann (1822-1878).

PHILIPPSON, Alfred

1515. Lehmann, Herbert. "Alfred Philippson zum Gedächtnis anlässlich der 100. Wiederkehr seines Geburtstages am

Part Three--Biographical Works

1. Januar 1964," Geographische Zeitschrift, vol. 52, no. 1 (1964), 1-6.

An appreciation of the work of the German geographer Alfred Philippson (1864-1953).

1516. Lehmann, Herbert. "Alfred Philippsons Lebenswerk," Colloquium Geographicum, no. 5 (1956), 9-14.

The life of the German geographer Alfred Philippson (1864-1953). Followed by E. Kirsten's "Bibliographie A. Philippson," pp. 15-25.

1517. Lehmann, Herbert, and Troll, Carl. "Alfred Philippson, 1864-1953," pp. 205-214 in Bonner Gelehrte: Beiträge zur Geschichte der Wissenschaften in Bonn, Vol. 4, Mathematik und Naturwissenschaften (Bonn: H. Bouvier u. Co. Verlag/Ludwig Röhrscheid Verlag, 1970).

The life and work of the German geographer Alfred Philippson (1864-1953), professor of geography at the University of Bonn (1911-1929).

PINKERTON, John

1518. Sitwell, O.F.G. "John Pinkerton: An Armchair Geographer of the Early Nineteenth Century," Geographical Journal, vol. 138, part 4 (December 1972), 470-479.

The geographical views of the Scottish compiler John Pinkerton (1758-1826), especially as expressed in his Modern Geography (1802).

1519. Wilcock, Arthur A. "'The English Strabo': The Geographical Publications of John Pinkerton," Transactions of the Institute of British Geographers, no. 61 (March 1974), 35-45.

Concerns the Scottish compiler John Pinkerton (1758-1826), with emphasis on his Modern Geography (1802) and its various editions and translations. Also treats the Précis de la géographie universelle of the Danish-French geographer Conrad Malte-Brun (1775-1826).

PLAYFAIR, John

1520. Seylaz, Louis. "Early Discoverers: XV--A Forgotten Pioneer of the Glacial Theory: John Playfair (1748-

1819)," Journal of Glaciology, vol. 4, no. 31 (March 1962), 124-126.

The Scottish scientist John Playfair (1748-1819) is best known for his Illustrations of the Huttonian Theory of the Earth (1802).

POL, Wincenty

1521. Babicz, Józef. "Wincenty Pol, 1807-1872," Geographers: Biobibliographical Studies, vol. 2 (1978), 93-97.

Wincenty Pol (1807-1872) was a Polish geographer who held the chair of geography at the University of Cracow, 1849-1853.

PORTHAN, Henrik Gabriel

1522. Mead, William R. "H.G. Porthan as a Geographer," Terra, vol. 83, no. 3 (1971), 143-148.

The geographical interests of Henrik Gabriel Porthan (1759-1906), rector of Turku University in Finland.

QUELLE, Otto

1523. Brauer, Adalbert. "Otto Quelle, 1879-1959," pp. 215-222 in Bonner Gelehrte: Beiträge zur Geschichte der Wissenschaften in Bonn, Vol. 4, Mathematik und Naturwissenschaften (Bonn: H. Bouvier u. Co. Verlag/Ludwig Röhrscheid Verlag, 1970).

Sketch of the German geographer Otto Quelle (1879-1959), who taught at the University of Bonn (1912-1930), ord. prof. in 1920) and the Technical University of Berlin (from 1930).

RADDE, Gustav

1524. Breuste, Jürgen. "Gustav Radde und sein Wirken als Geograph," Petermanns Geographische Mitteilungen, vol. 127, no. 2 (1983), 109-113.

The German geographer and naturalist Gustav Radde (1831-1903), director of the Tiflis (Tbilisi) Caucasian Museum, conducted scientific exploration in the Caucasus region and Siberia.

Part Three--Biographical Works

RANKE, Leopold von

1525. Hashagen, Justus. "Ranke als Geograph," *Geographische Zeitschrift*, vol. 48, nos. 4-5 (May 1942), 173-177.

The German historian Leopold von Ranke (1795-1886) as a geographer.

RATZEL, Friedrich

1526. Beck, Hanno. "Friedrich Ratzel--der grosse Anreger der Anthropogeographie (= Geographie des Menschen) (1844-1904)," pp. 164-179 in *Grosse Geographen: Pioniere--Aussenseiter--Gelehrte* by Beck (Berlin: Dietrich Reimer Verlag, 1982).

Biographical sketch of the German geographer Friedrich Ratzel (1844-1904), professor of geography at the University of Leipzig, 1886-1904.

1527. Buttmann, Günther. *Friedrich Ratzel: Leben und Werk eines deutschen Geographen, 1844-1904*. ("Grosse Naturforscher," 40.) Stuttgart: Wissenschaftliche Verlagsgesellschaft MBH, 1977. 151 p.

Biography of the German geographer Friedrich Ratzel (1844-1904).

1528. Helmolt, Hans, ed. *Kleine Schriften von Friedrich Ratzel*. 2 vols. Munich and Berlin: Druck und Verlag von R. Oldenbourg, 1906. Vol. 1, xxxv + 530 p.; Vol. 2, ix + 542 + lxii p.

A selection of articles by the German geographer Friedrich Ratzel (1844-1904). Vol. 1, pp. xxi-xxxi, autobiographical notes by Ratzel. Pages 373-530 contain Ratzel's biographical writings on Ernst Haeckel, Carl Ritter, Oscar Peschel, Eduard Pöppig, Moritz Wagner, Eduard Vogel, Gerhard Rohlfs, Heinrich Noé, Bruno Hassenstein, Emin Pasha, and Heinrich Schurtz. At end of Vol. 2 is 58-page (Roman numerals) "Ratzel-Bibliographie 1867-1905," compiled by Viktor Hansch, followed by 4-page list of addenda by the editor, Hans Helmolt.

1529. Mikesell, Marvin W. "Friedrich Ratzel," pp. 327-329 in Vol. 13 of *International Encyclopedia of the Social Sciences* (New York: The Macmillan Company and The Free Press, 1968).

A biographical sketch of the German geographer Friedrich Ratzel (1844-1904), professor of geography at the University of Leipzig from 1886.

1530. Overbeck, Hermann. "Das politischgeographische Lehrgebäude von Friedrich Ratzel in der Sicht unserer Zeit," Die Erde, vol. 88, nos. 3-4 (December 1957), 169-192.

Evaluation of the German geographer Friedrich Ratzel's (1844-1904) view of political geography, as expressed in his writing and teaching.

1531. Saey, Pieter. "Theorie en ideologie bij Friedrich Ratzel," Bulletin de la Société belge d'études géographiques, vol. 48, no. 2 (1979), 201-232.

Theory and ideology in the works of the German geographer Friedrich Ratzel (1844-1904).

1532. Sauer, Carl O. "The Formative Years of Ratzel in the United States," Annals of the Association of American Geographers, vol. 61, no. 2 (June 1971), 245-254.

The American experiences (1873-1874) of the German geographer Friedrich Ratzel (1844-1904).

1533. Sauer, Carl O. "Friedrich Ratzel," pp. 120-121 in Vol. 13 of Encyclopaedia of the Social Sciences (London: Macmillan and Co., Ltd., 1934).

Sketch of the German geographer Friedrich Ratzel (1844-1904).

1534. Smith, Woodruff D. "Friedrich Ratzel and the Origins of Lebensraum," German Studies Review, vol. 3, no. 1 (February 1980), 51-68.

The German geographer Friedrich Ratzel (1844-1904) first used the concept of Lebensraum in its classic sense in the 1890s and "published its most cogent statement in an essay in 1901." (Page 53) "Ratzel defined Lebensraum as the geographical surface area required to support a living species at its current population size and mode of existence." The concept was used in the Nazi period to support expansionist policies.

Part Three--Biographical Works

1535. Steinmetzler, Johannes. <u>Die Anthropogeographie Friedrich Ratzels und ihre ideengeschichtlichen Wurzeln</u>. Bonner geographische Abhandlungen, no. 19 (1956). 151 p.

Anthropogeography of the German geographer Friedrich Ratzel (1844-1904) and its intellectual roots (evolutionary thought, positivism, Herder, Ritter, etc.).

1536. Wanklyn, Harriet. <u>Friedrich Ratzel: A Biographical Memoir and Bibliography</u>. Cambridge: Cambridge University Press, 1961. x + 96 p.

Biography of the German geographer Friedrich Ratzel (1844-1904). Part 2, pp. 55-94, is "A Bibliography of Ratzel's Work."

RAVENSTEIN, Ernst Georg

1537. Grigg, David B. "Ernst Georg Ravenstein, 1834-1913," <u>Geographers: Biobibliographical Studies</u>, vol. 1 (1977), 79-82.

E.G. Ravenstein (1834-1913) was a German-born English geographer and cartographer, who "deserves to rank as one of the founders of theoretical geography" for his studies of population and the laws of migration.

1538. Grigg, David B. "The First English New Geographer," <u>Geographical Magazine</u>, vol. 46, no. 6 (March 1974), 246-247.

Concerns the English geographer Ernst Georg Ravenstein (1834-1913), with emphasis on his work on migration. "Instead of emphasizing differences between places, the heart of traditional geography, the New Geography looks for similarities in form, function and process, and seeks to establish generalizations and laws."

RECLUS, Élisée

1539. Beck, Hanno. "Elisée Reclus--Revolutionär, Anarchist und grösster Geograph Frankreichs (1830-1905)," pp. 121-148 in <u>Grosse Geographen: Pioniere--Aussenseiter--Gelehrte</u> by Beck (Berlin: Dietrich Reimer Verlag, 1982).

Essay on the French geographer and anarchist Elisée Reclus (1830-1905).

1540. Dunbar, Gary S. "Elisée Reclus, An Anarchist in Geography," pp. 154-164 in Geography, Ideology and Social Concern, ed. David Stoddart (Oxford: Basil Blackwell, 1981).

On the French geographer and anarchist Elisée Reclus (1830-1905).

1541. Dunbar, Gary S. Elisée Reclus, Historian of Nature. Hamden, Connecticut: Archon Books, 1978. 193 p.

Biography of the French geographer Elisée Reclus (1830-1905), best known for his Nouvelle géographie universelle (19 vols., 1876-1894).

1542. Dunbar, Gary S. "Elisée Reclus in Louisiana," Louisiana History, vol. 23, no. 4 (Fall 1982), 341-352.

The activities of the French geographer Elisée Reclus (1830-1905) in Louisiana in the period 1853-1855.

1543. Giblin, Béatrice. "Elisée Reclus, 1830-1905," Geographers: Biobibliographical Studies, vol. 3 (1979), 125-132.

A sketch of the life of the French geographer Elisée Reclus (1830-1905).

1544. Giblin, Béatrice. "Elisée Reclus: geographie, anarchisme," Hérodote, no. 2 (April-June 1976), 30-49.

An essay on the life of the French geographer and anarchist Elisée Reclus (1830-1905).

1545. Giblin, Béatrice, et al. "Elisée Reclus, un géographe libertaire," Hérodote, no. 22 (Third Quarter 1981), 3-128.

A special issue of the radical French geographical journal Hérodote devoted mostly to articles about Elisée Reclus (1830-1905), including 3 articles by Béatrice Giblin. Other articles on Reclus by Yves Lacoste, Pierre Gentelle, Pierre-Yves Péchoux, Martin Zemliak, and Marie-Larence Netter.

1546. Mikesell, Marvin W. "Observations on the Writings of Elisée Reclus," Geography, vol. 44, no. 4 (November 1959), 221-226.

The life and work of the French geographer Elisée Reclus (1839-1905).

1547. Nettlau, Max. Elisée Reclus, Anarchist und Gelehrter (1830-1905). (Beiträge zur Geschichte des Sozialismus, Syndikalismus, Anarchismus, Vol. 4.) Berlin: Verlag "Der Syndikalist," Fritz Kater, 1928. 345 p.

Biography of the French geographer and anarchist Elisée Reclus (1830-1905), with emphasis on his political side. Also published in an enlarged Spanish edition, Eliseo Reclus, la vida de un sabio justo y rebelde, translated by V. Orobón Fernández, 2 vols. ("Biblioteca de La Revista Blanca") (Barcelona: Publicaciones de "La Revista Blanca," n. d. [1929]).

1548. Nettlau, Max. "Jacques Elisée Reclus," pp. 164-165 in Vol. 13 of Encyclopaedia of the Social Sciences (London: Macmillan and Co., Ltd., 1934).

A sketch of the French geographer and anarchist Elisée Reclus (1830-1905).

REEVES, Edward A.

1549. Reeves, Edward A. The Recollections of a Geographer. London: Seeley, Service & Co., Ltd., n.d. 1935. 223 p.

Autobiography of E.A. Reeves (1862-1945), map curator and instructor in surveying at the Royal Geographical Society. Associated with the RGS from 1878 to 1933.

REIN, Johannes Justus

1550. Hohmann, Joseph. "Johannes Justus Rein, 1835-1918," pp. 199-204 in Bonner Gelehrte: Beiträge zur Geschichte der Wissenschaften in Bonn, Vol. 4, Mathematik und Naturwissenschaften (Bonn: H. Bouvier u. Co. Verlag/Ludwig Röhrscheid Verlag, 1970).

A sketch of the German geographer Johannes Justus Rein (1835-1918), professor of geography at the University of Bonn (1883-1910).

RENNELL, James

1551. Frenzel, Curt Arthur. <u>Major James Rennell der Schöpfer der neueren englischen Geographie: Ein Beitrag zur Geschichte der Erdkunde.</u> (Inaugural-Dissertation, Leipzig.) Pulsnitz: Druck von E.L. Försters Erben, 1904. viii + 194 p.

Biography of the English geographer James Rennell (1742-1830). Chapter 1, pp. 1-17, treats the development of European geography before Rennell (13th-18th centuries), and Chapter 8, pp. 184-190, deals with Rennell's significance as a geographer, his place in the history of geography, and the further development of geography in England after his death.

1552. Heaney, G.F. "Rennell and the Surveyors of India," <u>Geographical Journal</u>, vol. 134, part 3 (September 1968), 318-327.

The English geographer James Rennell (1742-1830) was the first Surveyor General of Bengal (1767-1777). Also discusses other Surveyor Generals--William Lambton, George Everest, Colin Mackenzie, etc.--and the Indian Himalayan explorers ("Pundits"). Describes the work of the Survey of India down to the end of the British period (1947).

1553. Markham, Clements R. <u>Major James Rennell and the Rise of Modern English Geography.</u> (The Century Science Series.) New York: Macmillan & Co., 1895. 232 p.

Life of James Rennell (1742-1830), "the greatest geographer that Great Britain has yet produced" (p. v).

REVERT, Eugene

1554. Papy, Louis. "Cavaillès, Arqué and Revert: Three Geographers of Bordeaux," <u>Geographers: Biobibliographical Studies</u>, vol. 7 (1983), 5-9. (Revert on pp. 8-9)

French geographer Eugene Revert (1895-1957), who taught in 9 lycées from 1924 to 1947 and served as professor of geography at the University of Bordeaux from 1947 to 1952.

RICHTHOFEN, Ferdinand von

1555. Beck, Hanno. "Ferdinand Freiherr von Richthofen, 1833-1905," pp. 191-198 in <u>Bonner Gelehrte: Beiträge</u>

zur Geschichte der Wissenschaften in Bonn, Vol. 4, Mathematik und Naturwissenschaften (Bonn: H. Bouvier u. Co. Verlag/Ludwig Röhrscheid Verlag, 1970).

Life and work of the German geographer Ferdinand von Richthofen (1833-1905), professor of geography at the universities of Bonn (1878-1883), Leipzig (1883-1886), and Berlin (1886-1905).

1556. Beck, Hanno. "Ferdinand Freiherr v. Richthofen--vorbildlicher China-Forscher und anerkanntester Geograph seiner Zeit (1833-1905)," pp. 149-163 in Grosse Geographen: Pioniere--Aussenseiter--Gelehrte by Beck (Berlin: Dietrich Reimer Verlag, 1982).

Essay on the German geographer Ferdinand von Richthofen (1833-1905), who conducted research in California and China in the 1860s and early 1870s.

1557. Drygalski, Erich von. "Ferdinand von Richthofen und die Deutsche Geographie," Zeitschrift der Gesellschaft für Erdkunde zu Berlin, 1933, nos. 3-4 (June), 88-97.

The German geographer Ferdinand von Richthofen (1833-1905) on the centennial of his birth. Richthofen is here called "the greatest president" that the Berlin Geographical Society has had.

1558. Freitag, Ulrich. "Ferdinand von Richthofens 'Atlas von China' (Idee--Durchführung--Ergebnis)," Die Erde vol. 114, nos. 2-3 (1983), 119-134.

(From English abstract, p. 119) "The development of v. Richthofen's concepts [and] methods of mapping, his travels in China 1868-1872, [and] his compilation of the maps of an atlas to accompany his books on China are discussed as part of the development of geographical knowledge of Central and East Asia in Central Europe during the 19th century."

1559. Kolb, Albert. "Ferdinand Freiherr von Richthofen, 1833-1905," Geographers: Biobibliographical Studies, vol. 7 (1983), 109-115.

The German geographer and geologist Ferdinand von Richthofen (1833-1905) taught geography in the universities of Bonn, Leipzig, and Berlin and became rector of the University of Berlin in 1903.

1560. Solger, Friedrich. "Erinnerung an Ferdinand von Richthofen," Wissenschaftliche Zeitschrift der Humboldt-Universität zu Berlin, Mathematisch-Naturwissenschaftliche Reihe, vol. 5, no. 2 (1955/1956), 107-111.

An appreciation of the German geographer Ferdinand von Richthofen (1833-1905).

1561. Stäblein, Gerhard. "Der Lebensweg des Geographen, Geomorphologen und China-Forschers Ferdinand von Richthofen," Die Erde, vol. 114, no. 2-3 (1983), 90-102.

Life of the German geographer, geomorphologist, and China scholar Ferdinand von Richthofen (1833-1905).

1562. Tiessen, Ernst, ed. Meister und Schüler: Ferdinand von Richthofen an Sven Hedin. Berlin: Verlag von Dietrich Reimer, 1933. 148 p.

Connections between the German geographer Ferdinand von Richthofen (1833-1905) and the Swedish explorer Sven Hedin (1865-1952). Pp. 71-148 contain letters from Richthofen to Hedin, 1889-1905.

1563. Tiessen, Ernst, comp. Die Schriften Ferdinand Freiherr von Richthofens. (Offprint from Männer der Wissenschaft, no. 4.) Leipzig: Verlag von Wilhelm Weicher, 1906. 18 p.

Bibliography of the writings of the German geographer Ferdinand von Richthofen (1833-1905), 1856-1905.

1564. Wegener, Georg, and Wissmann, Hermann von. "Ferdinand Freiherr von Richthofen 1833-1905," Die Grossen Deutschen, Vol. 5 (1957), 390-398.

Biographical sketch of the German geographer Ferdinand von Richthofen (1833-1905).

1565. Yamato, Kasai. "Von Richthofen and Modern Geography: System and Unity of the Geographic Sciences," Tohoku University (Japan), Science Reports, 7th Series (Geography), Vol. 25 (1975), 95-103.

The German geographer Ferdinand von Richthofen (1833-1905) and his views on geographical methodology.

1566. Zuckermann, Brigitta. "Die Bedeutung Ferdinand von Richthofens als Geologe und Geograph," Geographische Berichte, vol. 28, no. 1 (1983), 1-15.

An appreciation of the geographical and geological work of the German geographer Ferdinand von Richthofen (1833-1905), published on the occasion of his 150th birthday.

RITTER, Carl

1567. Bader, Frido J. Walter. "Carl Ritter und die Gesellschaft für Erdkunde zu Berlin," Die Erde, vol. 110, no. 4 (1979), 309-314.

The German geographer Carl Ritter (1779-1859) and his relations with the Berlin Geographical Society, of which he was one of the founders in 1828.

1568. Beck, Hanno. "Carl Ritter--Christ und Geograph," Gesnerus, vol. 38, nos. 1-2 (1981), 259-262.

A brief sketch of the German geographer Carl Ritter (1779-1859), with consideration of his teleological views.

1569. Beck, Hanno. Carl Ritter, Genius der Geographie: Zu seinem Leben und Werk. Berlin: Dietrich Reimer Verlag, 1979. 132 p.

Biography of the German geographer Carl Ritter (1779-1859), who taught at the University of Berlin from 1820 onward.

1570. Beck, Hanno. "Carl Ritter--Genius der Geographie (1779-1859)," pp. 103-120 in Grosse Geographen: Pioniere--Aussenseiter--Gelehrte by Beck (Berlin: Dietrich Reimer Verlag, 1982).

Essay on the German geographer Carl Ritter (1779-1859).

1571. Bitterling, Richard. "Carl Ritter zum Gedächtnis an seinem 150. Geburtstage: 7. August 1929," Geographischer Anzeiger, vol. 30, no. 8 (1929), 233-264.

Concerning the German geographer Carl Ritter (1779-1859). Includes 2 previously unpublished Ritter letters (1815 and 1838) and 2 travel accounts by Ritter (Chamonix, 1811, and Constantinople, 1837).

1572. Döring, Lothar. "Carl Ritter und die Schulgeographie," Geographische Rundschau, vol. 11, no. 5 (May 1959), 197-204.

The influence of the German geographer Carl Ritter (1779-1859) on geographical textbooks and teaching in his own and succeeding generations.

1573. Dörries, Hans. "Carl Ritter und die Entwicklung der Geographie in heutige Beurteilung," Die Naturwissenschaften, vol. 17, no. 32 (9 August 1929), 627-631.

The German geographer Carl Ritter (1779-1859) and his role in the development of geography in Germany.

1574. Gage, William L. Geographical Studies by the Late Professor Carl Ritter of Berlin. Boston: Gould and Lincoln, 1863. 356 p.

Concerning the German geographer Carl Ritter (1779-1859), professor of geography at the University of Berlin from 1820. Contains "A Sketch of the Life of Carl Ritter" by Gage (pp. 13-32), "An Account of Prof. Ritter's Geographical Labors" by Heinrich Bögekamp (pp. 33-51), and Gage's translations of some of Ritter's geographical writings (pp. 55-356).

1575. Hartshorne, Richard, and Gurgel, Klaus. "Zu Carl Ritters Einfluss auf die Entwicklung der Geographie in den Vereinigten Staaten von Amerika," pp. 201-219 in Carl Ritter: Zur europäisch-amerikanischen Geographie an der Wende vom 18. zum 19. Jahrhundert, ed. by Manfred Büttner (Abhandlungen und Quellen zur Geschichte der Geographie und Kosmologie, Vol. 2) (Paderborn: Ferdinand Schöningh, 1980).

The influence of the German geographer Carl Ritter (1779-1859) on American geography, particularly through the works of Arnold Guyot, Matthew Fontaine Maury, Francis Parker, and William Morris Davis.

1576. Hözel, Emil. "Das geographische Individuum bei Karl Ritter und seine Bedeutung für den Begriff des Naturgebietes und der Naturgrenze," Geographische Zeitschrift, vol. 2, no. 7 (1896), 378-396; no. 8, 433-444.

The idea of geographical individuality in the work of the German geographer Carl Ritter (1779-1859) and its significance for the concept of the natural region. Preliminary remarks on the attempts to determine natural regions and their boundaries before the time of Ritter, with emphasis on German geographers of the 18th and early 19th centuries.

1577. Korinman, Michel. "Carl Ritter (1779-1859): un des premiers grands géographes universitaires," Hérodote, no. 22 (Third Quarter 1981), 129-148.

An essay on the life and work of the German geographer Carl Ritter (1779-1859).

1578. Kramer, Fritz L. "A Note on Carl Ritter, 1779-1859," Geographical Review, vol. 49, no. 3 (July 1959), 406-409.

With excerpts from the published lectures and letters of the German geographer Carl Ritter (1779-1859), the author shows that Ritter was not a very good field observer or scientist. Compares Ritter's description of the Verona-Vicenza area unfavorably with the earlier observations of Goethe.

1579. Kramer, Gustav. Carl Ritter: Ein Lebensbild nach seinem handschriftlichen Nachlass dargestellt. 2 vols. Halle: Verlag der Buchhandlung des Waisenhauses, 1864-1870. Vol. 1, 1864, x + 482 p.; Vol. 2, 1870, x + 454 p.

Life of the German geographer Carl Ritter (1779-1859) by his brother-in-law Gustav Kramer.

1580. Lehmann, Edgar. "Carl Ritters kartographische Leistung," Die Erde, vol. 90, no. 2 (April 1959), 184-222.

The cartographic interests and activities of the German geographer Carl Ritter (1779-1859). Issue consists of 9 articles on Ritter.

1581. Linke, Max. "Carl Ritter, 1779-1859," Geographers: Biobibliographical Studies, vol. 5 (1981), 99-108.

The life of the German geographer Carl Ritter (1779-1859), who taught at the University of Berlin from 1820 onward.

1582. Maull, Otto. "Carl Ritter," Geographische Rundschau, vol. 11, no. 5 (May 1959), 171.

Brief sketch of the life of the German geographer Carl Ritter (1779-1859).

1583. Mehedinți, Simion. "La géographie comparée d'après Ritter et Peschel," Annales de géographie, vol. 10, no. 49 (15 January 1901), 1-9.

An essay on the views of the German geographers Alexander von Humboldt, Carl Ritter, and Oscar Peschel. Essentially a critique of Ritter's views. Cites Humboldt and Peschel approvingly.

1584. Müller, Alice. Carl Ritter: Eine Auswahl aus Reisetagebuchern und Briefen. Museumsbucherei Quedlinburg, vol. 5 (1959). 48 p.

Anthology of selected letters and excerpts from travel diaries of the German geographer Carl Ritter (1779-1859). [HB]

1585. Nicolas-Obadia, Georges. "Carl Ritter et la formation de l'axiomatique géographique," pp. 3-32 in Introduction à la géographie générale comparée by Carl Ritter (translated by Danielle Nicolas-Obadia, introduction and notes by Georges Nicolas-Obadia) (Cahiers de géographie de Besançon, 22) (Paris: Les Belles Lettres, 1974).

Concerning the German geographer Carl Ritter (1779-1859). Followed by a Ritter bibliography (pp. 243-247) and G. Nicolas-Obadia's biographical sketch of Ritter (pp. 249-253). Most of the work is taken up by D. Nicolas-Obadia's translation of Ritter's Introduction à la géographie générale comparée (1852) and G. Nicolas-Obadia's notes to Ritter's work (pp. 33-215). Also included is Elisée Reclus' translation of Ritter's "De la configuration des continents ..." (pp. 217-241).

1586. Ostuni, Josefina. "Carlos Ritter," Boletín de Estudios Geográficos (Mendoza, Argentina), vol. 9, no. 35 (April-June 1962), 56-86.

The life and work of the German geographer Carl Ritter (1779-1859).

1587. Ostuni, Josefina. "Carlos Ritter," Boletim Geográfico (Rio de Janeiro), vol. 26, no. 196 (January-February 1967), 30-47.

The German geographer Carl Ritter (1779-1859), his publications and influence.

1588. Paassen, Chr. van. "Carl Ritter anno 1959," Tijdschrift van het Koninklijk Nederlandsch Aardrijkskundig Genootschap, vol. 76, no. 4 (October 1959), 327-351.

An essay on the German geographer Carl Ritter (1779-1859), on the occasion of the 100th anniversary of his death.

1589. Pfeifer, Gottfried. "Ritter, Humboldt und die moderne Geographie," Verhandlungen des Deutschen Geographentages (Berlin, 1959), vol. 32 (Wiesbaden: Franz Steiner Verlag GMBH, 1960), 69-83.

An appreciation of the German geographers Carl Ritter (1779-1859) and Alexander von Humboldt (1769-1859) on the 100th anniversary of their deaths. Emphasis on their influence on the work of Alfred Hettner and later geographers.

1590. Plewe, Ernst. "Carl Ritter," pp. 517-520 in Vol. 13 of International Encyclopedia of the Social Sciences (New York: The Macmillan Company and The Free Press, 1968).

Sketch of the German geographer Carl Ritter (1779-1859).

1591. Plewe, Ernst. "Carl Ritter (1779-1859)," Geographisches Taschenbuch 1958/59 (Wiesbaden: Franz Steiner Verlag GMBH, 1958), pp. 501-503.

Concerns the German geographer Carl Ritter (1779-1859).

1592. Plewe, Ernst, ed. Die Carl-Ritter-Bibliothek. (Geographische Zeitschrift: Beihefte.) (Erdkundliches Wissen, no. 50.) Wiesbaden: Franz Steiner Verlag GMBH, 1978. xxiv + 565 p.

Reprint of Verzeichniss der Bibliothek und Kartensammlung ... Dr. Carl Ritter in Berlin ... (Leipzig: T.O. Weigel, 1861), with a new preface (pp.

vii-xvi) on Ritter's legacy by Ernst Plewe. The book is an auction catalogue listing the books and maps in the library left by the German geographer Carl Ritter (1779-1859).

1593. Plewe, Ernst. "Carl Ritters 'produktenkundliche' Monographien im Rahmen seiner wissenschaftlichen Entwicklung," Geographische Zeitschrift, vol. 67, no. 1 (First Quarter 1979), 12-28.

Concerning the German geographer Carl Ritter (1779-1859) and the "product monographs" (about 55 in number) embedded in his monumental Erdkunde (19 vols. in 21 parts, 1822-1859). Shows the evolution of Ritter's geography from a descriptive "Staatenkunde" to a problem-oriented cultural geography. [HB]

1594. Plewe, Ernst. "Carl Ritters Stellung in der Geographie," Verhandlungen des Deutschen Geographentages (Berlin, 1959), vol. 32 (Wiesbaden: Franz Steiner Verlag GMBH, 1960), 59-68.

An appreciation of the work of the German geographer Carl Ritter (1779-1859) on the centennial of his death.

1595. Plewe, Ernst. "Heinrich Barth und Carl Ritter: Briefe und Urkunden," Die Erde, vol. 96, no. 4 (1965), 245-278.

Letters and documents illustrating the relations between the German geographers Carl Ritter (1779-1859) and Heinrich Barth (1821-1865) in the years 1848-1858.

1596. Plott, Adalbert. "Bibliographie der Schriften Carl Ritters," Die Erde, 94th year, no. 1 (1963), 13-36.

Bibliography of the writings of the German geographer Carl Ritter (1779-1859). Works published 1802-1895.

1597. Richter, Otto. Der teleologische Zug im Denken Carl Ritters. (Inaugural-Dissertation, Leipzig.) Borna-Leipzig: Buchdruckerei Robert Noske, 1905. xviii + 105 p.

The teleological thought of the German geographer Carl Ritter (1779-1859). Introduction (pp. 1-11) discusses writings about Ritter since 1867. Bibliography of Ritter's works (pp. ix-xiv) and writings about Ritter (pp. xiv-xvii).

Part Three--Biographical Works 329

1598. Sauer, Carl. "Karl Ritter," p. 395 of Vol. 13 of Encyclopaedia of the Social Sciences (London: Macmillan and Co., Ltd., 1934).

Sketch of the German geographer Carl Ritter (1779-1859).

1599. Schmidt, Erwin. "Carl Ritter als Zeitgenosse Alexander von Humboldts," pp. 89-99 in Alexander von Humboldt: Vorträge und Aufsätze Anlässlich der 100. Wiederkehr seines Todestages am 6. Mai 1959, ed. Johannes F. Gellert (Geographische Gesellschaft der Deutschen Demokratischen Republik, Wissenschaftliche Abhandlungen, Vol. 2) (Berlin: VEB Deutscher Verlag der Wissenschaften, 1960).

Connections between the German geographers Carl Ritter (1779-1859) and Alexander von Humboldt (1769-1859). Article deals mostly with Ritter.

1600. Schmidt, Erwin. "Carl Ritters Beitrag zur Herausbildung der Geographie als Wissenschaft," Geographische Bericht, vol. 24, no. 2 (January 1979), 97-104.

The German geographer Carl Ritter (1779-1859) and his views on geography as a scientific discipline.

1601. Schmitthenner, Heinrich. "Carl Ritter, 1779-1859," Der Grossen Deutschen, vol. 3 (1956), 189-200.

Biographical sketch of the German geographer Carl Ritter (1779-1859), professor of geography at the University of Berlin from 1820.

1602. Schmitthenner, Heinrich. "Carl Ritter und Goethe," Geographische Zeitschrift, vol. 43, no. 5 (May 1937), 161-175.

Johann Wolfgang Goethe (1749-1832) and his influence on Carl Ritter (1779-1859) before the latter's appointment to the University of Berlin in 1820.

1603. Schmitthenner, Heinrich. Studien über Carl Ritter. Frankfurter Geographische Hefte, no. 29 (25th year, no. 4, 1951). 100 p.

The life and work of the German geographer Carl Ritter (1779-1859).

1604. Schultze, Joachim H., ed. "Carl Ritter zum Gedächtnis," Die Erde, vol. 90, no. 2 (April 1959), 98-254.

Special issue of Die Erde honoring the German geographer Carl Ritter (1779-1859) on the 100th anniversary of his death. Articles by Ernst Plewe, Ernst Kirsten, Edgar Lehmann, Wolfgang Tichy, Helmut Preuss, and Hanno Beck.

1605. Schulz, Heinz. "Carl Ritter (1779-1859)-- Weltanschauung, Weltbild und geographische Ideen," Petermanns Geographische Mitteilungen, vol. 124, no. 3 (1980), 201-206.

The German geographer Carl Ritter (1779-1859) and his philosophical and methodological positions.

1606. Schulz, Heinz. "Carl Ritter (1779-1859)-- Wirksamkeiten in Berlin und regional-geographischer Aspekt," Wissenschaftliche Zeitschrift der Humboldt-Universität zu Berlin, Mathematisch-Naturwissenschaftliche Reihe, vol. 29, no. 2 (1980), 175-179.

The work of the German geographer Carl Ritter (1779-1859) and the influence of Schelling and Herder on his thought.

1607. Schulz, Heinz. "Der Geograph Carl Ritter (1779 bis 1859)--Aspekte seines Verhältnisses zu Philosophie und Weltanschauung," Zeitschrift für den Erdkundeunterricht, vol. 34, no. 8/9 (1982), 305-311.

An interpretation of the philosophical positions of the German geographer Carl Ritter (1779-1859).

1608. Schulze, Bruno. Charakter und Entwicklung der Länderkunde Karl Ritters. (Inaugural-Dissertation, Halle-Wittenberg.) Halle: Druck von Wischan & Wettengel, 1902. 61 p.

Methodology of Carl Ritter's regional geography from Europa (1804) onward.

1609. Sinnhuber, Karl. "Carl Ritter, 1779-1859," Scottish Geographical Magazine, vol. 75, no. 3 (December 1959), 153-163.

The German geographer Carl Ritter (1779-1859), professor of geography at the University of Berlin from 1820.

1610. Troll, Carl. "The Work of Alexander von Humboldt and Carl Ritter: A Centenary Address," The Advancement of Science, vol. 16, no. 64 (March 1960), 441-452.

Sketches of the German geographers Alexander von Humboldt (1769-1859) and Carl Ritter (1779-1859) on the 100th anniversary of their deaths.

1611. Valk, Jan Gerard van der. Peschels Kritick op Ritter: Een Bijdrage tot de Geschiedenis van de Probleemstelling van de Sociale Geografie. Geografisch Instituut der Rijksuniversiteit te Utrecht, Sociaal Geografische Mededeelingen, no. 1 (1944). 79 p.

Concerns criticism of the work of the German geographer Carl Ritter (1779-1859) by his countryman Oscar Peschel (1826-1875). [JAvG]

1612. Wünsche, Ernst Alwin. Die geschichtliche Bewegung und ihre geographische Bedingtheit bei Carl Ritter und bei seinen hervorragendsten Vorgängern in der Anthropo-Geographie. (Inaugural-Dissertation, Leipzig.) Dresden: Druck von Otto Francke, 1899. ix + 167 p.

Concerns the human geography of the German geographer Carl Ritter (1779-1859) and the influence of his predecessors, especially Strabo, Montesquieu, and Herder.

1613. Zögner, Lothar. Carl Ritter in seiner Zeit, 1779-1859. (Staatsbibliothek Preussischer Kulturbesitz, Ausstellungskataloge 11.) Berlin: Staatsbibliothek Preussischer Kulturbesitz, 1979. 128 p.

Catalogue of the exhibit honoring the German geographer Carl Ritter (1779-1859) at the Prussian State Library (1 November 1979 to 12 January 1980).

ROBEQUAIN, Charles

1614. Delvert, Jean. "Charles Robequain (1897-1963)," pp. 145-151 in "Les géographes français," Secrétariat d'état aux universités, Comité des travaux historiques et scientifiques, Bulletin de la section de géographie, vol. 81 (For the years 1968-1974) (Paris: Bibliothèque Nationale, 1975).

Essay on the French geographer Charles Robequain (1897-1963), professor of colonial geography at the University of Paris from 1938.

ROMER, Eugeniusz

1615. Babicz, Józef. "L'école géographique polonaise d'Eugeniusz Romer," pp. 45-48 in Les écoles géographiques, ed. Babicz (Warsaw: PWN--Polish Scientific Publishers, 1980).

The life of the Polish geographer Eugeniusz Romer (1871-1954).

1616. Babicz, Józef. "Eugeniusz Romer, 1871-1954," Geographers: Biobibliographical Studies, vol. 1 (1977), 89-96.

A biographical sketch of the Polish geographer Eugeniusz Romer (1871-1954), who taught at the University of Lwow, 1899-1931.

ROXBY, Percy Maude

1617. Freeman, T.W. "Percy Maude Roxby, 1880-1947," Geographers: Biobibliographical Studies, vol. 5 (1981), 109-116.

P.M. Roxby (1880-1947) was an English geographer who taught at the University of Liverpool, 1904-1944.

RÜHL, Alfred

1618. Harke, Hellmut. "Alfred Rühls unvergessene Verdienste um die Geographie," Geographie Berichte, vol. 27, no. 3 (1982), 193-199.

An appreciation of the contributions of the German geographer Alfred Rühl (1882-1935).

SABINE, Edward

1619. Georgi, Johannes. "Edward Sabine, ein grosser Geophysiker der 19. Jahrhunderts," Deutsche Hydrographische Zeitschrift, vol. 11, no. 6 (December 1958), 225-239.

The English explorer and scientist Edward Sabine (1788-1883), "the first all-round geophysicist."

1620. Marshall-Cornwall, James. "Three Soldier-Geographers," Geographical Journal, vol. 131, part 3 (September 1965), 357-365.

The geographical activities of 3 British army officers--William Martin Leake (1777-1860), Edward Sabine (1788-1883), and Francis Rawdon Chesney (1789-1872).

SAPPER, Karl

1621. Termer, Franz. "Carlos Sapper.--Explorador de Centro América (1866-1945)," Anales de la Sociedad de Geografía e Historia de Guatemala, vol. 29, nos. 1-4 (January-December 1956), 55-101.

The Central American researches of the German geographer Karl Sapper (1866-1945). Followed by Termer's bibliography of Sapper's writings, 1888-1948 (pp. 102-130).

1622. Termer, Franz. Karl Theodor Sapper, 1866-1945: Leben und Wirken eines deutschen Geographen und Geologen. (Deutsche Akademie der Naturforscher Leopoldina, Lebensdarstellungen deutscher Naturforscher, 12.) Leipzig: Johann Ambrosius Barth, 1966. 89 p.

Life of the German geographer Karl Sapper (1866-1945).

SCHLÜTER, Otto

1623. Käubler, Rudolf. "Otto Schlüters Bedeutung für die geographische Wissenschaft," Die Erde, vol. 95, no. 1 (1964), 5-15.

The German geographer Otto Schlüter (1872-1959) was professor of geography at the University of Halle.

1624. Schick, Manfred. "Otto Schlüter, 1872-1959," Geographers: Biobibliographical Studies, vol. 6 (1982), 115-122.

The German geographer Otto Schlüter (1872-1959) was professor of geography at the University of Halle, 1911-1951.

SCHMIEDER, Oskar

1625. Schmieder, Oskar. Lebenserinnerungen und Tagebuchblätter eines Geographen. Schriften des Geographischen Instituts der Universität Kiel, vol. 40 (1972). 186 p.

Autobiographical work by the German geographer Oskar Schmieder (1891-1980). Covers the period 1910-1936.

SCHRADER, Franz

1626. Broc, Numa. "Franz Schrader, 1844-1924," Geographers: Biobibliographical Studies, vol. 1 (1977), 97-103.

Franz Schrader (1844-1924) was a French geographer and cartographer who worked for the Hachette publishing house in Paris from 1877 onward.

1627. Broc, Numa. "Pour le cinquantenaire de la mort de Franz Schrader (1844-1924)," Revue géographique des Pyrénées et du Sud-Ouest, vol. 45, no. 1 (January 1974), 5-16.

Life of the French geographer, cartographer, and alpinist Franz Schrader (1844-1924).

1628. Laget, Gustave de. "La vie et l'oeuvre de Franz Schrader," Bulletin de la Société de géographie et d'études coloniales de Marseille, vol. 63 (1944-1947), 91-102.

Life and work of the French geographer and cartographer Franz Schrader (1844-1924).

SCHULTZ, Woldemar

1629. Kohlhepp, Gerd. "Woldemar Schultz, ein Pionier geographischer Forschung in Brasilien," Geographische Zeitschrift, vol. 56, no. 3 (October 1968), 225-228.

Concerns the German geographer Woldemar Schultz (1833-1866) and his field work in Rio Grande do Sul and Santa Catarina states, Brazil, 1858-1860.

SCHWEINFURTH, Georg

1630. Linke, Max. "Georg Schweinfurth und sein Wirken als Geograph," Petermanns Geographische Mitteilungen, vol. 121, no. 4 (1977), 247-251.

The geographical work of Georg Schweinfurth (1836-1925), who was a German explorer in Africa, especially in the Nile region. In 1875 he founded the Egyptian Geographical Society.

Part Three--Biographical Works 335

SCORESBY, William

1631. Waites, Bryan. "William Scoresby, 1789-1857," Geographers: Biobibliographical Studies, vol. 4 (1980), 139-147.

William Scoresby (1789-1857) was an English whaler, Arctic explorer, and divine.

SEMENOV, Petr Petrovich

1632. Fischer, Dora. "Peter Petrowitsch Semjonow-Tian-Schanskij," Die Erde, 1953, no. 1 (March), 67-70.

Sketch of the Russian geographer and traveler Petr Petrovich Semenov (or Semenov-Tian-Shanskii) (1827-1914).

1633. Lincoln, W. Bruce. Petr Petrovich Semenov-Tian-Shanskii: The Life of a Russian Geographer. (Russian Biography Series, no. 8.) Newtonville, Massachusetts: Oriental Research Partners, 1980. x + 118 p.

Biography of the Russian geographer Petr Petrovich Semenov (1827-1914).

SILVA TELLES, Francisco Xavier da

1634. Ribeiro, Orlando. "Silva Telles, introdutor do ensino da geografia em Portugal," Finisterra, vol. 11, no. 21 (1976), 12-36.

Francisco Xavier da Silva Telles (1860-1930), originally a medical doctor, was the first holder of the chair of geography at the University of Lisbon (1904).

SIMONY, Friedrich

1635. Penck, Albrecht. Friedrich Simony: Leben und Wirken eines Alpenforschers. University of Vienna, Geographische Institut, Geographische Abhandlungen, vol. 6, no. 3 (1898). 72 p.

Biography of Friedrich Simony (1813-1896), professor of geography at the University of Vienna and master of the art of making relief models. [HB]

SMITH, George Adam

1636. Middleton, Dorothy. "George Adam Smith, 1856-1942," Geographers: Biobibliographical Studies, vol. 1 (1977), 105-106.

George Adam Smith (1856-1942) was a British theologian, Principal of Aberdeen University (1909-1935), and author of Historical Geography of the Holy Land (1894).

SMOLENSKI, Jerzy

1637. Leszczycki, Stanislaw. "Jerzy Smolenski, 1881-1940," Geographers: Biobibliographical Studies, vol. 6 (1982), 123-127.

The Polish geomorphologist Jerzy Smolenski (1881-1940) taught at the University of Cracow.

SÖLCH, Johann

1638. Matznetter, Josef. "Johann Sölch, 1883-1951," Geographers: Biobibliographical Studies, vol. 7 (1983), 117-124.

The Austrian geographer Johann Sölch (1883-1951) was professor of geography at the University of Vienna from 1935.

SOMERVILLE, Mary

1639. Baker, J.N.L. "Mary Somerville and Geography in England," Geographical Journal, vol. 111, nos. 4-6 (September 1948), 207-222.

The British scientist Mary Somerville (1780-1872), who wrote Physical Geography (1848). Long description of the state of geography in British universities at the time of the publication of her book and for about a quarter of a century afterward.

1640. Oughton, Marguerita. "Mary Somerville, 1780-1872," Geographers: Biobibliographical Studies, vol. 2 (1978), 109-111.

Mary Somerville (1780-1872) was a British geographer and scientific writer who wrote Physical Geography (1848).

1641. Patterson, Elizabeth C. "The Case of Mary Somerville: An Aspect of Nineteenth-Century Science," Proceedings of the American Philosophical Society, vol. 118, no. 3 (June 1974), 269-275.

Concerns the British scientist Mary Somerville (1780-1872), who wrote Physical Geography (1848).

1642. Patterson, Elizabeth C. "Mary Somerville," British Journal for the History of Science, vol. 4, part 4 (December 1969), 311-339.

The life of the British scientist Mary Somerville (1780-1872), whose most successful book, Physical Geography, was published in 1848.

1643. Sanderson, Marie. "Mary Somerville: Her Work in Physical Geography," Geographical Review, vol. 64, no. 3 (July 1974), 410-420.

The British scientist Mary Somerville (1780-1872) "wrote the first textbook in physical geography in the English language" (Physical Geography, 1848).

SORRE, Maximilien

1644. George, Pierre. "Max. Sorre (1880-1962)," pp. 185-195 in "Les géographes français," Secrétariat d'état aux universités, Comité des travaux historiques et scientifiques, Bulletin de la section de géographie, vol. 81 (For the years 1968-1974) (Paris: Bibliothèque Nationale, 1975).

Biographical sketch of the French geographer Maximilien Sorre (1880-1962), professor of geography at the University of Paris, 1941-1948. Earlier he had been involved in university administration in Lille, Clermont-Ferrand, and Aix-Marseille. Reprinted from Annales de géographie, vol. 71 (1962), 449-459.

STEIN, Mark Aurel

1645. Iklé, Frank W. "Sir Aurel Stein: A Victorian Geographer in the Tracks of Alexander," Isis, vol. 59, part 2 (Summer 1968), 144-155.

Mark Aurel Stein (1862-1943) was a Hungarian-born British explorer, archeologist, and geographer in Southwest, South, and Central Asia.

STRZELECKI, Pawel Edmund

1646. Babicz, Józef; Slabczynski, Waclaw; and Vallance, Thomas G. "Pawel Edmund Strzelecki, 1797-1873," Geographers: Biobibliographical Studies, vol. 2 (1978), 113-118.

Strzelecki (1797-1873) was a Polish explorer of Australia (1839-1843) who resided in England after 1843 and promoted Australian and Arctic exploration and Irish famine relief.

SYDOW, Emil von

1647. Cramer, W. "Uber die Bedeutung Emil von Sydows für die Entwicklung der wissenschaftlichen Erdkunde," Verhandlungen des dritten Deutschen Geographentages, 1883, pp. 93-102.

The German geographer and cartographer Emil von Sydow (1812-1873).

TAYLOR, Eva G.R.

1648. Freeman, T.W. "Two Ladies," Geographical Magazine, vol. 49, no. 3 (December 1976), 208.

Concerns the British geographers Marion Newbigin (1869-1934), editor of the Scottish Geographical Magazine, and E.G.R. Taylor (1879-1966), professor of geography at Birkbeck College, University of London.

TEILHARD DE CHARDIN, Pierre

1649. Bywater, Vincent. "Pierre Teilhard de Chardin, 1881-1955," Geographers: Biobibliographical Studies, vol. 7 (1983), 129-133.

French Jesuit anthropologist and archeologist Pierre Teilhard de Chardin (1881-1955), with some concern for his relevance to geographers.

TELEKI, Paul

1650. Tilkovszky, L. Pál Teleki (1879-1941): A Biographical Sketch. (Studia Historica Academiae Scientiarum Hungaricae, 86.) Budapest: Akadémiai Kiadó, 1974. 70 p.

Biography of the Hungarian geographer and statesman Paul Teleki (1879-1941). Emphasis on his political life and suicide. Slightly abridged version of the original Hungarian edition that appeared in 1969.

THOMSON, John

1651. Parker, Elliott S. "John Thomson, 1837-1921: RGS Instructor in Photography," Geographical Journal, vol. 144, part 3 (November 1978), 463-471.

Concerns the Scottish photographer John Thomson (1837-1921), "the first photo-journalist in modern times," with emphasis on his work as Instructor of Photography for the Royal Geographical Society (1886-), "where he was able to tutor many of the explorers of the time while promoting the use of photography as a documentary tool."

THÜNEN, Johann Heinrich von

1652. Johnson, Hildegard Binder. "A Note on Thünen's Circles," Annals of the Association of American Geographers, vol. 52, no. 2 (June 1962), 213-220.

Concerns "the origin of the use of circular models in attempts to produce rational spatial order." Emphasis on the various editions of Der Isolierte Staat (1826) by Johann Heinrich von Thünen (1783-1850).

TILLO, Alexey Andreyevich

1653. Fedosseyev, A.I. "Alexey Andreyevich Tillo, 1839-1900," Geographers: Biobibliographical Studies, vol. 3 (1979), 155-159.

A.A. Tillo (1839-1900) was "a Russian geographer, geophysicist, geodesist, and cartographer," who was "known for his works on the hypsometry of Russia" (p. 155).

TOPELIUS, Zachris

1654. Mead, William R. "Zachris Topelius," Norsk Geografisk Tidsskrift, vol. 22, no. 2 (1968), 89-100.

The life of the Finnish historian Zachris Topelius (1818-1898), who lectured on the human geography of Finland at the University of Helsinki between 1854 and 1858.

1655. Mead, William R. "Zachris Topelius, 1818-1898," Geographers: Biobibliographical Studies, vol. 3 (1979), 161-163.

Zachris Topelius (1818-1898) was a Finnish historian and geographer, professor of history and then rector of the University of Helsinki. He was "the first Finn to identify geography as an independent discipline."

TROLL, Carl

1656. Beck, Hanno. "Carl Troll--ein Geograph im Geist Alexander v. Humboldts (1899-1975)," pp. 273-281 in Grosse Geographen: Pioniere--Aussenseiter--Gelehrte (Berlin: Dietrich Reimer Verlag, 1982).

An essay on the German geographer Carl Troll (1899-1975), professor of geography at the University of Bonn from 1938.

1657. Lautensach, Hermann. "Carl Troll--ein Forscherleben," Erdkunde, vol. 13, no. 4 (December 1959), 245-258.

The German geographer Carl Troll (1899-1975) and his geographical research since 1922. Pp. 252-258 contain Troll's bibliography, 1922-1959 (206 items).

VALLAUX, Camille

1658. Carré, François. "Camille Vallaux, 1870-1945," Geographers: Biobibliographical Studies, vol. 2 (1978), 119-126.

Camille Vallaux (1870-1945) was a French geographer who taught in Brittany, 1895-1913, and Paris, 1913-1932, and who wrote on Brittany, oceanography, political geography, and a general work, Les sciences géographiques (1925).

VÂLSAN, George

1659. Popp, Nicolae. "George Vâlsan, 1885-1935," Geographers: Biobibliographical Studies, vol. 2 (1978), 127-133.

George Vâlsan (1885-1935) was a Romanian geomorphologist who also wrote on the human aspects of geography. Professor first at Cluj University and after 1929 in Bucharest.

VERNADSKY, Vladimir Ivanovich

1660. Krout, Igor V. "Vladimir Ivanovich Vernadsky, 1863-1945," Geographers: Biobibliographical Studies, vol. 7 (1983), 135-144.

The Russian earth scientist V.I. Vernadsky (1863-1945), professor at Moscow University (1898-1911), was later affiliated with the Academy of Sciences in St. Petersburg (Leningrad) and Moscow.

Part Three--Biographical Works

VERNE, Jules

1661. Giblin, Béatrice. "Jules Verne, la géographie et 'l'Ile mystérieuse': pour le 150e anniversaire de sa naissance," Hérodote, no. 10 (1978), 76-85.

The French writer Jules Verne (1828-1905) used the works of geographers, such as Elisée Reclus, in the writing of his novels.

VIDAL DE LA BLACHE, Paul

1662. Beck, Hanno. "Paul Vidal de la Blache--der einflussreichste französische Geograph (1845-1918)," pp. 180-190 in Grosse Geographen: Pioniere--Aussenseiter--Gelehrte by Beck (Berlin: Dietrich Reimer Verlag, 1980).

The life of the French geographer Paul Vidal de la Blache (1845-1918), professor of geography at the Ecole Normale Supérieure (1877-1898) and then at the University of Paris.

1663. Claval, Paul. "Vidal de la Blache et la géographie française," pp. 91-125 in Pour le cinquantenaire de la mort de Paul Vidal de la Blache: études d'histoire de la géographie (Cahiers de géographie de Besançon, 16; Annales littéraires de l'Université de Besançon, 93) (Paris: Les Belles Lettres, 1968).

Essay on the French geographer Paul Vidal de la Blache (1845-1918) and the school that he built up in Paris.

1664. Monbeig, Pierre. "Paul Vidal de la Blache," pp. 316-318 in Vol. 16 of International Encyclopedia of the Social Sciences (New York: The Macmillan Company and The Free Press, 1968).

The French geographer Paul Vidal de la Blache (1845-1918) was a Sorbonne professor and "the leader of the French school of geography."

1665. Nicolas-O [sic--Obadia], Georges. "Paul Vidal de la Blache entre la filosofía francesa y la geografía alemana," Geo-Crítica (Barcelona), no. 35 (September 1981). 41 p.

The philosophical background and milieu of the French geographer Paul Vidal de la Blache (1845-1918).

1666. Pinchemel, Philippe. "Paul Vidal de la Blache," Geographisches Taschenbuch 1970/72 (Wiesbaden: Franz Steiner Verlag GMBH, 1970), pp. 266-279.

The French geographer Paul Vidal de la Blache (1845-1918), professor at the Sorbonne from 1898. Reprinted in "Les géographes français," Secrétariat d'état aux universités, Comité des travaux historiques et scientifiques, Bulletin de la section de géographie, vol. 81 (For the years 1968-1974) (Paris: Bibliothèque Nationale, 1975), 9-23.

1667. Ribeiro, Orlando. "En relisant Vidal de la Blache," Annales de géographie, vol. 77, no. 424 (November-December 1968), 641-662.

An appreciation of the life and work of the French geographer Paul Vidal de la Blache (1845-1918) on the 50th anniversary of his death.

1668. Robic, M.C. "La conception de la géographie humaine chez Vidal de la Blache d'après les 'Principes de géographie humaine,'" Les Cahiers de Fontenay (Ecole Normale Supérieure de Fontenay-aux-Roses) (1976), no. 4 (Géographie), pp. 1-76.

A critical analysis of Vidal's posthumously published book, Principes de géographie humaine (1922). Treats the fundamental features of Vidalian human geography and tries to explain its ambiguities by the contradiction between the "naturalism" on which (in Vidal's eyes) the worth of the discipline is based and a "humanism" closer to academic thought in the Third Republic. Includes bibliography and commentaries on Vidal. [MCR]

1669. Sion, Jules. "L'art de la description chez Vidal de la Blache," pp. 479-487 in Mélanges de philologie, d'histoire et de littérature offerts à Joseph Vianey (Paris: Les Presses Universitaires, 1934).

The art of regional description in the teaching and writings of the French geographer Paul Vidal de la Blache (1845-1918).

1670. Vallaux, Camille. "Paul Marie Joseph Vidal de la Blache," pp. 251-252 in Vol. 15 of Encyclopaedia of the Social Sciences (London: Macmillan and Co., Ltd., 1935).

Sketch of the French geographer Paul Vidal de la Blache (1845-1918), professor at the University of Paris from 1898.

VIVIEN DE SAINT-MARTIN, Louis

1671. Lagarde, Luci. "Louis Vivien de Saint-Martin, 1802-1896," Geographers: Biobibliographical Studies, vol. 6 (1982), 133-138.

The French geographical writer and editor Louis Vivien de Saint-Martin (1802-1896) was known for such works as Histoire de la géographie et des découvertes géographiques (1873) and the Atlas universel de géographie (1876-1905).

VOLNEY, Constantin François Chasse-Boeuf, Comte de

1672. Vallaux, Camille. "Deux précurseurs de la géographie humaine: Volney et Charles Darwin," Revue de synthèse, vol. 15, no. 2 (June 1938), 81-93.

If human geography is a product of the late 19th century with such geographers as Friedrich Ratzel, there were numerous earlier workers who were actually doing human geography before it was defined. Among these precursors were the French savant Constantin François Chasse-Boeuf, Comte de Volney (1757-1820) and the English naturalist Charles Darwin (1809-1882).

VOLTAIRE, François-Marie Arouet de

1673. Weinert, Hermann K. "Voltaire und die Geographie im Zeitalter der Aufklärung," pp. 239-249 in Festschrift zum 70. Geburtstag des ord. Professors der Geographie Dr. Ludwig Mecking (Published by the Geographical Institute of the University of Hamburg and the Akademie für Raumforschung und Landesplanung, Hannover, 1949).

The French philosopher Voltaire (François-Marie Arouet) (1694-1778) and his views on geography.

VOYEIKOV, Alexander Ivanovitch

1674. Fedosseyev, A.I. "Alexander Ivanovitch Voyeikov, 1842-1916," Geographers: Biobibliographical Studies, vol. 2 (1978), 135-141.

A.I. Voyeikov (various spellings) (1842-1916) was a Russian geographer and professor at St. Petersburg University, best known for his work in climatology.

VUJEVIĆ, Pavle

1675. Dukic, Dusan. "Pavle Vujević, 1881-1966," Geographers: Biobibliographical Studies, vol. 5 (1981), 129-131.

Pavle Vujević (1881-1966) was a Yugoslav meteorologist and climatologist who taught at Belgrade University from 1907 onward.

WAGNER, Hermann

1676. Behrmann, Walter. "Hermann Wagner als Akademischer Lehrer," Die Erde, vol. 6, nos. 3-4 (November 1954), 362-368.

The teaching of the German geographer Hermann Wagner (1840-1929), professor of geography at the University of Göttingen from 1880.

WAGNER, Moritz

1677. Beck, Hanno. "Moritz Wagner als Geograph," Erdkunde, vol. 7, no. 2 (June 1953), 125-128.

The geographical works of the German scholar Moritz Wagner (1813-1887). From Beck's doctoral thesis (below).

1678. Beck, Hanno. "Moritz Wagner in der Geschichte der Geographie," Doctoral thesis, University of Marburg, 1951. 369 p.

A study of the German naturalist and scientific traveler Moritz Wagner (1813-1887) and his influence on geographers, particularly Friedrich Ratzel. Concerns Wagner's theories of the migrations of organisms, his critique of Darwinian ideas, and his use of genetic thinking in geographical problems. [HB]

WAIBEL, Leo Heinrich

1679. Pfeifer, Gottfried. "Leo Heinrich Waibel, 1888-1951," Geographers: Biobibliographical Studies, vol. 6 (1982), 139-147.

Part Three--Biographical Works   345

>   The German geographer Leo Waibel (1888-1951) taught
>   at the universities of Kiel (from 1922) and Bonn (1929-
>   1937) and worked and taught in the United States and
>   Brazil from 1939 to 1951.

1680.   Troll, Carl. "Leo Waibel, 1888-1951," pp. 223-230 in
        Bonner Gelehrte: Beiträge zur Geschichte der Wissenschaften in Bonn, Vol. 4, Mathematik und Naturwissenschaften (Bonn: H. Bouvier u. Co. Verlag/Ludwig Röhrscheid Verlag, 1970).

>   The life of the German geographer Leo Waibel (1888-
>   1951), professor of geography at the University of Bonn
>   (1929-1937), who later undertook research and taught in
>   the Americas.

1681.   Valverde, Orlando. "Contribuição de Leo Waibel à geografia brasileira," Revista Brasileira de Geografia, vol. 30, no. 1 (January-March 1968), 74-83.

>   The contributions of the German geographer Leo Waibel
>   (1888-1951) to Brazilian geography.

WEBER, Alfred

1682.   Gregory, Derek. "Alfred Weber and Location Theory," pp. 165-185 in Geography, Ideology and Social Concern, ed. David Stoddart (Oxford: Basil Blackwell, 1981).

>   Location theory of the German economist Alfred Weber
>   (1868-1958), professor of political economy and social
>   science at the University of Heidelberg.

WEGENER, Alfred

1683.   Gellert, Johannes F. "Alfred Wegener und seine Bedeutung für die Geowissenschaften," Geographische Berichte, vol. 25, no. 3 (April 1980), 153-164.

>   The contributions of the German geophysicist Alfred
>   Wegener (1880-1930), including his polar research and
>   the theory of continental drift.

WEGENER, Georg

1684.   Winkler, Arno. "Georg Wegener (1863-1939)," Geographisches Taschenbuch 1964/65 (Wiesbaden: Franz Steiner Verlag GMBH, 1964), pp. 302-309.

Sketch of the German geographer Georg Wegener (1863-1939), professor of geography in the Handels-(Wirtschafts-) Hochschule, Berlin.

WELLS, Herbert George

1685. Blouet, Brian W. "H.G. Wells and the Evolution of Some Geographic Concepts," Area, vol. 9, no. 1 (1977), 49-52.

Concerning the English writer Herbert George Wells (1866-1946). (Summary, p. 49) "H.G. Wells wrote widely on social, economic, and scientific matters. Speculation on the long-term factors influencing settlement patterns led Wells into an early statement of central place theory. Wells's acquaintance with Mackinder may have had some influence upon the form of the Heartland thesis."

WEULERSSE, Jacques

1686. Gourou, Pierre. "Jacques Weulersse, 1905-1946," Geographers: Biobibliographical Studies, vol. 1 (1977), 107-112.

Jacques Weulersse (1905-1946) was a French geographer who worked in Africa and the Middle East. He was named professor of colonial geography at the University of Aix-Marseille in 1943.

YOUNGHUSBAND, Francis

1687. Seaver, George. Francis Younghusband, Explorer and Mystic. London: John Murray, 1952. xi + 391 p.

Biography of the English explorer and geographer Francis Younghusband (1863-1942), president of the Royal Geographical Society (1919-1922) and founder of the World Congress of Faiths.

ZACH, Franz Xaver von

1688. Friedrich, Klaus. "Die Bedeutung Franz Xaver von Zachs für die Entwicklung der astronomischen Geographie in Deutschland," Petermanns Geographische Mitteilungen, vol. 117, no. 2 (April 1973), 147-153.

The German astronomer Franz Xaver von Zach (1754-1832) and his work in "astronomical" geography (geodesy, navigation, and cartography). Edited Allgemeine Geographische Ephemeriden from 1798.

ZANNONI, Giovanni Antonio Rizzi

1689. Blessich, Aldo. "Un geografo italiano del secolo XVIII, Giovanni Antonio Rizzi Zannoni," Bollettino della Società Geografica Italiana, 3rd series, vol. 11, no. 1 (January 1898), 12-23; no. 2 (February 1898), 56-69; no. 4 (April 1898), 183-203; no. 9 (September 1898), 453-466; and no. 11 (November 1898), 523-537.

Zannoni (1736-1814) was an Italian geographer, cartographer, and traveler.

ZEUNE, Johann August

1690. Preuss, Helmut. "Johann August Zeune, der Hauptvertreter der 'reinen' Geographie," Erdkunde, vol. 12, no. 4 (December 1958), 277-284.

The German geographer J.A. Zeune (1778-1853), author of Gea, Versuch einer wissenschaftlichen Erdbeschreibung (1808).

XXII. OTHER COUNTRIES

BOSE, Nirmal Kumar

1691. Mookerjee, Sitanshu. "Nirmal Kumar Bose, 1901-1972," Geographers: Biobibliographical Studies, vol. 2 (1978), 9-11.

The Indian geographer and anthropologist N.K. Bose (1901-1972) was the first Reader in Geography at Calcutta University, Director of the Anthropological Survey of India (1959-1964), and a close associate of Mahatma Gandhi.

CALDAS, Francisco José de

1692. Bateman, Alfredo D. "Cronología de Caldas," Boletín de la Sociedad Geográfica de Colombia, vol. 26, no. 99 (Third Quarter 1968), 147-160.

Francisco José de Caldas (1768-1816) was a Colombian naturalist and geographer. Issue contains other articles on Caldas.

1693. Ruiz, José Ignazio. "Caldas, primer geógrafo do Colombia," Boletín de la Sociedad Geográfica de Colombia, vol. 24, nos. 91-92 (Third and Fourth Quarters 1966), 114-117.

Francisco José de Caldas (1768-1816) was a Colombian geographer and natural scientist.

CODAZZI, Agustín

1694. Acevedo Latorre, Eduardo. "Codazzi en Colombia," Revista de la Academia Colombiana de Ciencias Exactas, Físicas y Naturales, vol. 10, no. 41 (August 1959), xxv-xxxi.

The work of the Italian-born geographer Agustín Codazzi (1793-1859) in Colombia. Other papers in this issue deal with Codazzi and Alexander von Humboldt.

1695. Longhena, Mario. "Agostino Codazzi e la sua opera scientifica," Bollettino della Società Geografica Italiana, series 9, vol. 1, nos. 6-8 (June-August 1960), 289-302.

Concerns Agustín Codazzi (1793-1859), Italian-born explorer and geographer in Colombia and Venezuela.

1696. Longhena, Mario. "Agostino Codazzi viaggiatore nel Venezuela e nella Colombia (1793-1859)," Rivista Geografica Italiana, vol. 66, no. 1 (March 1959), 1-28.

Agustín Codazzi (1793-1859) was a geographer, cartographer, and explorer in northern South America.

1697. Rozo M., Darío, et al. "Centenario del geógrafo Agustín Codazzi," Boletín de la Sociedad Geográfica de Colombia, vol. 17, nos. 61-62 (First and Second Quarters 1959), 5-75.

10 papers commemorating the 100th anniversary of the death of Agustín Codazzi (1793-1859), Italian-born geographer in Venezuela and Colombia.

1698. Veggi Donati, Maria Angela. "Agostino Codazzi e la sua opera di esploratore e di cartografo," L'Universo, vol. 28, no. 2 (March-April 1948), 187-191.

Agustín Codazzi (1793-1859) was an Italian-born geographer in Venezuela and Colombia.

COTTON, Charles Andrew

1699.   Soons, Jane M., and Gage, Maxwell. "Charles Andrew Cotton, 1885-1970," Geographers: Biobibliographical Studies, vol. 2 (1978), 27-32.

Charles Cotton (1885-1970) was a New Zealand geologist and geomorphologist, known for his textbooks.

GARCÍA, Pedro Andrés

1700.   Furlong, Guillermo. "Bicentenario del primer geógrafo de la nación Argentina: Pedro Andrés García (1758-1958)," Anales de la Academia Argentina de Geografía, vol. 2 (1958), 176-177.

Colonel Pedro Andrés García (1758-1833) was a Spanish-born Argentine explorer and geographer.

1701.   Reguera Sierra, Ernesto. "El primo geógrafo de la Argentina: Pedro Andrés García," Revista Geográfica Americana (Buenos Aires), vol. 40, no. 238 (February 1956), 170-178.

Pedro Andrés García (1758-1833) was a Spanish-born officer who is called "the first military geographer" of Argentina.

GOYDER, George Woodroofe

1702.   Powell, J.M. "George Woodroofe Goyder, 1826-1898," Geographers: Biobibliographical Studies, vol. 7 (1983), 47-50.

The English-born Australian surveyor George Goyder (1826-1898) was Surveyor General of South Australia for more than 30 years and author of the famous "Goyder's Line," which marked the edge of reliable agricultural land.

HO, Robert

1703.   Cho, George C.H. "Robert Ho, 1921-1972," Geographers: Biobibliographical Studies, vol. 1 (1977), 49-54.

Robert Ho (1921-1972) was a Malayan geographer who taught at the University of Malaya (Singapore and Kuala Lumpur) and then served as Senior Research Fellow at the Australian National University (1965-1972).

HOLMES, James Macdonald

1704. Powell, J.M. "James Macdonald Holmes, 1896-1966," Geographers: Biobibliographical Studies, vol. 7 (1983), 51-55.

The Scottish-born Australian geographer James Macdonald Holmes (1896-1966) taught at the University of Sydney from 1929 to 1961, becoming "Australia's first full professor of geography" in 1945.

JOBBERNS, George

1705. Johnston, W.B. "George Jobberns, 1895-1974," Geographers: Biobibliographical Studies, vol. 5 (1981), 73-76.

A sketch of the New Zealand geographer and geomorphologist George Jobberns (1895-1974), professor of geography at Canterbury University College (University of Canterbury) and "Father of New Zealand Geography."

MITCHELL, Thomas Livingstone

1706. Powell, J.M. "Thomas Livingstone Mitchell, 1792-1855," Geographers: Biobibliographical Studies, vol. 5 (1981), 83-87.

Thomas Livingstone Mitchell (1792-1855) was a Scottish-born explorer and surveyor in Australia.

MUELLER, Ferdinand Jakob Heinrich von

1707. Powell, J.M. "Ferdinand Jakob Heinrich von Mueller, 1825-1896," Geographers: Biobibliographical Studies, vol. 5 (1981), 89-93.

F.J.H. von Mueller (1825-1896) "was a botanist and explorer prominent in the growth of the natural and social sciences in colonial Australia." Born in Schleswig-Holstein and arrived in Australia in 1847.

OGAWA, Takuji

1708. Tsujita, Usao. "Takuji Ogawa, 1870-1941," Geographers: Biobibliographical Studies, vol. 6 (1982), 71-71-76.

The Japanese geographer Takuji Ogawa (1870-1941) was professor of geography and founder of the department of geography at the University of Kyoto, 1908-1921, and professor of geology, 1921-1930.

Part Three—Biographical Works 351

PRICE, Archibald Grenfell

1709. Powell, J.M. "Archibald Grenfell Price, 1892-1977," Geographers: Biobibliographical Studies, vol. 6 (1982), 87-92.

The Australian geographer, historian, and educator A. Grenfell Price (1892-1977) was Master of St. Mark's College of the University of Adelaide from 1925 to 1957.

TAMAYO, Jorge Leonides

1710. Gutiérrez de MacGregor, Maria Teresa. "Jorge Leonides Tamayo, 1912-1978," Geographers: Biobibliographical Studies, vol. 7 (1983), 125-128.

The Mexican geographer Jorge Tamayo (1912-1978) taught geography in the National Autonomous University (UNAM), Mexico City, from 1943 onward.

TAYLOR, Thomas Griffith

1711. Hanley, W.S. "Griffith Taylor's Antarctic Achievements: A Geographical Foundation," Australian Geographical Studies, vol. 18, no. 1 (April 1980), 22-36.

The participation of the Australian geographer Griffith Taylor (1880-1963) in Robert Scott's British Antarctic Expedition (1910-1912).

1712. Marshall, Ann. "Griffith Taylor's Correlative Science," Australian Geographical Studies, vol. 18, no. 2 (October 1980), 184-193.

The Australian geographer Griffith Taylor (1880-1963), professor of geography at the universities of Sydney, Chicago, and Toronto, with emphasis on his view of geography as a correlative science. Includes extended commentary on Carl Sauer of the University of California and compares Taylor and Sauer.

1713. Powell, J.M. "Thomas Griffith Taylor, 1880-1963," Geographers: Biobibliographical Studies, vol. 3 (1979), 141-153.

Griffith Taylor (1880-1963) was an English-born Australian geographer who taught at the universities of Sydney, Chicago, and Toronto.

1714. Sanderson, Marie. "Griffith Taylor: A Geographer to Remember," Canadian Geographer, vol. 26, no. 4 (Winter 1982), 293-299.

The life of the Australian geographer Griffith Taylor (1880-1963), professor at the universities of Sydney, Chicago, and Toronto.

1715. Taylor, Thomas Griffith. Journeyman Taylor. Ed. Alasdair Alpin MacGregor. London: Robert Gale, Ltd., 1958. 352 p.

Autobiography of the Australian geographer Griffith Taylor (1880-1963), who taught at the universities of Sydney, Chicago, and Toronto.

THOMPSON, David

1716. Tyrrell, J.B. "David Thompson, A Great Geographer," Geographical Journal, vol. 37, no. 1 (January 1911), 49-58.

David Thompson (1770-1857) was an English-born fur trader and explorer in Canada ("the greatest land geographer that the British race has produced").

YAMASAKI, Naomasa

1717. Tsujita, Usao. "Naomasa Yamasaki, 1870-1929," Geographers: Biobibliographical Studies, vol. 1 (1977), 113-117.

Naomasa Yamasaki (1870-1929) was a Japanese geographer who taught at the University of Tokyo from 1908 onward.

AUTHOR INDEX
(includes editors, compilers, and translators)

Aay, Henry 41
Abbagnano, Nicola 45
Abrahams, Paul P. 739
Acevedo Latorre, Eduardo 1694
Achelis, Thomas 1155
Adickes, Erich 1393-1395
Agostini, Enrico de 506-507
Ahlmann, Hans W. 927
Ahmad, Kazi S. 42
Alexander, Anneli 1414
Alexandrovskaya, Olga
  Andreyevna 1418, 1420, 1444
Alinhac, G. 285
Allan, Douglas A. 1154
Allen, John L. 740
Altengarten, James S. 1
Almagià, Roberto 43-45, 463, 904-905, 1317
Alvarez López, Enrique 1358
Amiran, David H. K. 999
Amorim Girão, Aristides de 928
Amsler, Jean 270
Anderson, John R. L. 214
Andrews, John H. 286, 1429
Annaheim, Hans 663
Anstey, Robert L. 1055-1057
Antoniol, G. B. 46
Anz, Heinrich 902
Aragon, Augustín 595
Arden-Close, Charles 287, 622
Arena, Gabriella 906
Arnberger, Erik 288
Arnhold, Helmut 664
Arnold, Adolf 508
Asúa, Miguel de 509
Atwood, Wallace W. 665
Auerbach, Bertrand 47

Babicz, Józef 48-50, 66, 89, 132, 176, 408, 747, 768, 808, 929, 983, 1328, 1521, 1615-1616, 1646
Bacon, H. Philip 1120
Bader, Frido J. Walter 510, 642, 1567
Badey, Lucien 1186
Baetens, R. 511
Bagrow, Leo 289-290
Baker, John N. L. 51-52, 215, 798-799, 1496, 1639
Baker, Marcus 741-742
Baker, Victor R. 399
Balchin, William G. V. 666
Baldacci, Osvaldo 907
Ballivian, Manuel Vicente 1283
Banse, Ewald 53-54, 216, 1146, 1318-1319
Barbier, Joseph-Victor 217
Barnes, Harry Elmer 464
Barr, Brenton M. 472
Barra, Francisco L. de la 595
Barton, Thomas F. 512
Baschin, Otto 513-514, 667
Bassin, Mark 515
Bassols Batalla, Angel 1364
Bateman, Alfredo D. 1692
Baulig, Henri 1034, 1048
Baumgärtel, Hans 400
Bay, Helmuth 291
Beaglehole, John Cawte 218-219, 1205
Beaubois, Henry 220
Beaujeu, Jean 270
Beaujeu-Garnier, Jacqueline 844
Beauregard, Ludger 553
Beaver, Stanley H. 544, 623-624

Beck, Hanno 55-64, 105, 221-223, 401, 866-868, 930-931, 1151, 1162, 1231, 1235, 1242, 1284, 1296, 1306, 1320-1325, 1362, 1430, 1447, 1502, 1526, 1539, 1555-1556, 1568-1570, 1604, 1655, 1662, 1677-1678
Becker, Jeronimo 932
Becker, Moritz Alois 933
Beckinsale, Robert P. 407, 1035-1036, 1038
Beckman, Leif Olof 224
Beguinot, A. 402
Behm, Ernst 4-6, 9-14
Behrmann, Walter 225, 516, 1676
Ben-Arieh, Yehoshua 999
Benedict, Peter 1005
Benison, Saul 1010
Berdoulay, Vincent R. H. 65-66, 828-829, 1312
Berg, Lev Semenovich 934
Berger, Dorothea 1268
Berghaus, Heinrich 226
Beriot, Agnès 227
Berman, Mildred 1008, 1113
Bernardy, Amy A. 918
Bernhardt, Peter 643
Bernleithner, Ernst 292, 517, 668-670, 935
Berry, Brian J. L. 67, 482, 484, 497
Bertacchi, Cosimo 293, 908-909
Bestermann, Theodore 860
Bettex, Albert 228
Biddle, C. A. 294
Bieler, Walter 1365
Biermann, Kurt-R. 1327-1328
Billington, Ray Allen 1127
Bindelli, Pietro 611
Bird, James 644
Bisscn, Jean 1189
Bitterling, Richard 1329-1330, 1487, 1571
Blache, Jules 671
Bladen, Wilford A. 743
Blakemore, M. J. 295
Blakeney, T. S. 625
Bleiberg, German 1358
Blessich, Aldo 1689

Block, Robert H. 744, 1128
Blouet, Brian W. 41, 433, 474, 740, 745-746, 749, 759-760, 769, 777, 779-780, 786, 790-791, 795, 1035, 1104, 1106, 1122, 1448-1449, 1685
Blüthgen, Joachim 936-937, 939
Blume, Helmut 869
Bobek, Hans 69
Bögekamp, Heinrich 1574
Bösiger, Kurt 518
Bonacker, Wilhelm 296-299
Bonetti, Eliseo 70-71
Bonifacio, Antoine 72
Borchert, Günter 870
Botting, Douglas 1331
Bouvier, René 1174
Bowden, Martyn J. 482, 694, 747, 775
Bowditch, Nathaniel 397
Bowen, Emrys G. 544, 672
Bowen, Margarita 73, 1332
Bowman, Isaiah 74
Brack, E. V. 692
Brand, Donald D. 1333, 1369
Branford, Victor 1436
Brauer, Adalbert 1523
Braun, Gustav 871
Breitbart, Myrna Margulies 1421
Breuste, Jürgen 1524
Bridges, R. C. 519-520
Brigham, Albert Perry 521, 677
Broc, Numa 2, 75, 229, 522, 626, 830-838, 1037, 1114, 1141, 1165, 1181, 1208-1209, 1227, 1229, 1239, 1428, 1460, 1627-1628
Broek, Jan O. M. 748
Bronson, Judith Conoyer 1115
Brouillette, Benoît 839
Brown, Eric 542
Brown, Ralph Hall 403, 1092-1093
Brown, Ralph Minthorne 1081
Brown, T. Nigel N. 523, 570
Browne, Charles Albert 1334
Browne, Janet 404
Brunhes, Jean 464
Brusa, Alfio 76

Author Index

Brusa, Carlo 911
Bucher, Walter H. 1074
Buchholz, Hanns J. 1199
Buck, Peter H. 230
Bühler, Alfred 663
Büttner, Manfred 48, 77-78, 1184, 1396-1398, 1402, 1575
Buffin, Frédéric 840
Bunksé, Edmunds V. 1335
Bunting, Brian T. 1483
Burky, Charles 301
Buschick, Richard 231
Bushong, Allen D. 749-750, 1116
Buttimer, Anne 79-80, 841
Buttmann, Günther 1527
Butze, Herbert 232
Bykov, V. I. 983
Bywater, Vincent 1649

Calef, Wesley 1023
Camena d'Almeida, Pierre 1273
Cameron, Dorothy 1269
Cameron, Hector C. 1145
Cameron, Ian 524
Campbell, Eila M. J. 282
Campbell, John A. 81, 673, 1236
Campbell, Robert 41, 108
Camu, Pierre 525
Cannon, Walter F. 405
Cantor, Leonard Martin 526
Capel, Horacio 82-83
Cappon, Lester J. 751
Caraci, Giuseppe 910
Caraci, Ilaria 106
Carazzi, Maria 527
Carlberg, Berthold 1509
Carré, François 1658
Carter, George F. 1018
Carter, Harold 672
Casada, James A. 1390
Castellanos, A. 1175
Castner, Henry W. 290
Caswell, John E. 528
Cerulli, Ernesta 1168
Chabot, Georges 842-844
Chaix, Emile 677
Charliat, Pierre-Jacques 270

Chatterjee, Shiba Prasad 991
Chester, Colby Mitchell 752
Chevalier, Michel 845
Chisholm, George G. 677
Chisholm, Michael 712
Cho, George C. H. 1703
Chorley, Richard J. 211, 406-407, 739, 1036, 1038
Claparède, Arthur de 531, 546
Clark, Austin H. 1068
Clark, Rose B. 74
Claval, Paul 84-89, 105, 408, 466-467, 846-847, 1663
Clos-Arceduc, Albert 1386
Close, Charles 287, 532
Clozier, René 90, 844
Coates, Donald R. 444
Colby, Charles C. 674, 753
Corbin, Diana Fontaine Maury 1082
Corcelle, Joseph 1478
Corley, Nora T. 91
Corna-Pellegrini, Giacomo 911
Cortesão, Armando 302
Cotet, Petre 1223
Cotter, Charles H. 1083
Coughlan, Robert 1013
Court, Arnold 409
Cox, Kevin R. 754
Cramer, W. 1647
Crone, Gerald R. 92-93, 303, 533, 628, 800
Cucu, Vasile 981
Cumming, Duncan 534
Cunningham, Frank F. 1240
Curnow, Irene J. 304

Dainelli, Giotto 233
Dainville, François de 305, 848
Dalla Vedova, Giuseppe 94, 535
Daly, Charles P. 410
Daly, Reginald 413, 675, 1039
Daniel, Hermann Adalbert 185
Darby, Henry Clifford 801-802, 1021
Dardel, Eric 95
Darmstaedter, Ludwig 96
Darrah, William C. 1097

Daumas, Maurice 72
Davenport, Charles B. 1028
Davies, Gordon L. 411, 536, 803, 1204
Davies, Wayne K. D. 166
Davis, William Morris 97, 412-413, 675, 755-756, 1049
Dawson, John A. 468
Day, Alan Edwin 234
Deacon, Margaret 498
Dean, James R. 499
Debenham, Frank 235
De Blij, Harm 560
Deffontaines, Pierre 1179
Delamarre, Mariel Jean-Brunhes 1180
Delpar, Helen 236
Delvert, Jean 1614
Denucé, Jean 98
Deprez, Eugène 237
Dept, Gaston G. 757
Derruau, Max 1139
De Terra, Helmut 1336
De Vorsey, Louis 1044
Dickinson, Robert E. 41, 99-101, 804
Diderot, Denis 102
Dittrich, Mauritz 414
Dmitrevskiy, Yuriy D. 150
Dockès, Pierre 469
Dörflinger, Johannes 102, 1494
Döring, Lothar 1337, 1572
Dörries, Hans 1573
Donazzolo, Pietro 103
Downes, Alan 104
Drapeyron, Ludovic 1182
Dresch, Jean 415, 1468-1469
Dryer, Charles R. 758
Drygalski, Erich von 872, 1557
Du Bois-Reymond, René 96
Dukic, Dusan 1675
Dunbar, Gary S. 105, 306, 470, 500, 537, 629, 676-678, 759, 1009, 1029, 1069, 1540-1542
Duncan, James S. 471
Dunmore, John 238
Dunn, Antony J. 407, 1038
Dury, George H. 805
Dussart, Frans 938

Du Val, Miles 1084

East, W. Gordon 1281
Eckardt, Wilhelm R. 1399
Eckert, Max 307
Eder, Herbert M. 645
Edwards, Kenneth Charles 806
Egerton, Frank N. 416
Ehrenberg, Ralph E. 559
Ekirch, Arthur A. 1075
Ellenberger, François 849
Engelmann, Gerhard 308-315, 646, 679, 1151, 1163-1164, 1365-1366, 1384
Engeln, Oscar Diedrich von 1503
Engler, Adolf 417
Entrikin, J. Nicholas 1095, 1101
Errera, Carlo 1144
Esakov, Vasily Alexeyevich 1138, 1225
Evans, E. Estyn 673, 1237
Evers, Wilhelm 939
Eyles, Victor A. 1484
Eyre, John Douglas 1129

Fairbairn, Kenneth J. 472
Fairchild, Wilma B. 239, 647, 662
Falkenstein, Constantin Karl 240
Farmer, Bertram Hughes 807
Farquharson, Alexander 1436
Fauser, Alois 316
Faybusovich, E. L. 1076
Fead, Margaret Irene 317
Fedosseyev, Aleksandr Ivanovich 1653, 1674
Feeken, Erwin H. J. 241
Feeken, Gerda E. E. 241
Feldman, Douglas A. 473
Fenneman, Nevin M. 418
Ferrell, Edith H. 1058
Ferro, Gaetano 106
Fèvre, Joseph 107
Fevret de Fontette, Charles-Marie 2
Fick, Karl E. 538
Fiechter, Alfredo 940
Fiedler, Horst 1243

Author Index

Fischer, Dora 1632
Fischer, Eric 41, 108, 899
Fitzgerald, Walter 109
Fleiuss, Max 539
Flemal, Ronald C. 425
Fleure, Herbert John 540, 618, 680, 1255, 1299-1300
Fochler-Hauke, Gustav 110
Forbes, Eric G. 318
Forbes, Vernon S. 242
Fordham, Herbert G. 319
Foster, Alice 674
Fox, Harold S.A. 648
Fraser, J. Keith 553
Freeman, Thomas Walter 41, 111-113, 541-542, 649-650, 681, 808-809, 1234, 1263, 1295, 1479, 1490, 1648
Freitag, Ulrich 320, 1558
French, Richard Antony 1495
Frenzel, Curt Arthur 1551
Frenzel, Reinhard 1461
Friberg, Nils 927
Friederichsen, Ludwig 543, 601
Friedrich, Felix 1426
Friedrich, Klaus 1688
Friis, Herman R. 243-244, 320-321, 760, 1338-1339, 1369
Frisinger, H. Howard 419
Fryer, Douglas Henry 323
Fuchs, Gerhard 761-762
Fuchs, Roland J. 682
Fulvi, Fulvio 114
Furlong, Guillermo 1700
Fuson, Robert H. 115

Gade, Daniel W. 763, 1024
Gage, Maxwell 1699
Gage, William L. 1574
Gallais, Jean 1274
Gallois, Lucien 116, 677
Galon, Rajmund 941
Gambi, Lucio 912
Garboe, Axel 1443
Gardiner, Leslie 324
Gardiner, R. A. 325
Garnett, Alice 544
Gedan, Paul 1316

Geiger, Pedro P. 993
Geiser, Samuel W. 764, 1408
Gelfand, Lawrence 1117
Gellert, Johannes F. 308, 414, 450, 1226, 1340, 1365, 1367, 1683
Gentelle, Pierre 1545
Genthe, Martha Krug 389, 873
George, Pierre 859, 942, 1644
Georgi, Johannes 630, 874, 1619
Gerasimov, Innokentiy P. 943-944, 1160, 1226, 1340
Gerland, Georg 1400
Gerson, Ran 999
Gibbs, David 733
Giblin, Béatrice 1543-1545, 1661
Gibson, James R. 374
Gicklhorn, Josef 1285
Gicklhorn-Wien, Renée 1285-1286
Gilbert, Edmund William 245, 545, 684-685, 810, 1206, 1300, 1450-1452, 1485
Gillespie, James Edward 246
Ginkel, Hans (J.A.) van 686
Glacken, Clarence J. 118, 1187
Glaser, Hugo 247
Goegg, Edmond 531, 546
Goetzmann, William H. 248
Gold, John R. 1441
Gómez Pérez, José 1203
Goode, J. Paul 326, 547
Goodman, Edward J. 249
Gordejuela, Ruiz de 164, 970
Gottschalk, Maria Karoline Elisabeth 945
Goudie, Andrew S. 1207, 1210, 1389, 1485
Gould, Peter 1134
Gourou, Pierre 1686
Gradmann, Robert 1275
Grady, Alison Dorothy 811
Graf, Johann Heinrich 548
Granö, Olavi 119, 1277
Gras, Jacques 1197
Graul, Hans 787
Greeley, Adolphus W. 1070
Gregory, Derek 1682
Gregory, Stanley 812

Gribaudi, Dino 913
Gribaudi, Pietro 813, 914
Grigg, David B. 1537-1538
Grigor'yev, Andrei
  Alexandrovitch 947
Grivot, Françoise 39, 1212
Grob, Richard 327
Grosjean, Georges 631, 687
Gross, Walter E. 420
Grosvenor, Gilbert H. 250-251,
  549-550
Gruber, Christian 734
Grundmann, Johannes 1302
Günther, Siegmund 120-121, 421-422,
  551, 1427, 1440
Gugiuman, Ion 1218
Guichonnet, Paul 1171
Guiral, P. 1249
Gurgel, Klaus 1575
Gutiérrez de MacGregor,
  Maria Teresa 1710

Haefke, Fritz 688
Hägerstrand, Torsten 946
Haggett, Peter 211, 712
Hahn, Friedrich Gustav 122
Haigh, Martin J. 1441
Halkin, Joseph 875
Hall, Alfred Rupert 323, 375
Hall, Arthur R. 1453
Hall, D. H. 423
Hallett, Robin 552
Hamelin, Louis-Edmond 553, 994, 1171
Hammond, Harold E. 1026
Hanley, W. S. 1711
Hansch, Viktor 1528
Hansen, Robert C. 1085
Hard, Gerhard 123, 1341, 1362, 1384
Harke, Hellmut 1618
Harley, J. Brian 295
Harms, Hans 328
Harris, Chauncy D. 947, 1147
Harrison, R. T. 1404
Harrison Church, Richard J. 850, 1498
Hartke, Wolfgang 689
Hartshorne, Richard 124-126, 1040,
  1307, 1575

Harvey, Paul Dean Adshead 329
Hashagen, Justus 1525
Hassert, Kurt 252
Hassinger, Hugo 948
Hauck, Paul 1303-1304, 1401
Hawley, Arthur J. 1118
Heaney, G. F. 330, 1232, 1552
Hederich, Reinhard 1270
Helbing, Helmut 1282
Hellmann, Gustav 424, 554, 567
Helmfrid, Staffan 949
Helmolt, Hans 1528
Henze, Dietmer 253
Herbertson, Dorothy 1436
Herbst, Jurgen 765
Hermann, Albert 37-38, 696
Hermann, Annemarie 1314
Hermann, Ernst 1329
Hermans, Willem Frederik 1446
Herneck, Friedrich 1342
Herstel, Theodor 127
Hervé, Roger 39
Heslinga, M. W. 686, 1230
Hettner, Alfred 105, 128-129,
  1308, 1311
Heydenreich, Adolf 1142
Higgins, Charles G. 425
Hill, Gillian 331
Himer, Kurt 735
Hoare, Michael E. 1247-1248
Hobbs, William Herbert 254
Hodgkiss, Alan G. 332
Hodgson, H. B. 736
Hönsch, Fritz 1148
Hözel, Emil 1576
Hoffmeister, J. Edward 426
Hofmann, Robert 690
Hoheisel, Karl 78, 1398, 1402
Hohmann, Joseph 651, 1475-1476,
  1550
Holmyard, Eric John 323, 375
Holt-Jensen, Arild 130
Holz, Paul 1366
Honigmann, Peter 1343
Hooson, David J. M. 131-132, 950-
  952, 1106
Horn, Werner 333-334, 632-633

Author Index

Hottes, Karlheinz 1198-1199
Hottes, Ruth 1198, 1200-1201
Howarth, Osbert John Radcliffe 41, 101, 142, 634
Howe, G. Melvyn 953
Hoyle, Brian S. 1265-1266
Hudson, Brian 133, 1159
Hudson, John C. 766
Huender, Wilhelmina Johanna 814
Hugues, Luigi 255-256
Humboldt, Alexander von 104, 134, 427
Hume, Edgar Erskine 428
Huntington, Ellsworth 1041
Hurtig, Theodor 691

Iklé, Frank W. 1645
Ilyina, Titiana D. 1417
Inouye, Syuzi 995
Ippolito, Guglielmo 915
Irwin, B. St. G. 335
Irwin, Daniel 336
Isachenko, Anatoliy G. 954

Jackson, Stanley Percival 557
Jacobsen, Hans-Adolf 1291-1292
Jaja, Goffredo 851, 1213
Jakel, Reinhard 1184, 1250
James, Preston E. 41, 135-137, 558-561, 739, 767-772, 1019, 1025, 1050, 1098
Janssens, Emile 429
Jantzen, Günther 562
Jarcho, Saul 337
Jaudel, L. 430
Jay, Leslie J. 1190, 1220, 1300-1301
Jefferson, Mark S. W. 677
Jennings, Joseph N. 1002
Jenny, Hans 1061
Jeon, Sang-woon 996
Jervis, Walter Willson 338
Joerg, Wolfgang Louis Gottfried 138, 773
Johnson, Douglas W. 139
Johnson, Hildegard B. 1652
Johnston, Ronald J. 140, 474, 692
Johnston, William B. 1705

Jones, Clarence F. 771
Jones, Leonard C. 1059-1060
Judson, Sheldon 1042
Juillard, Etienne 1156-1157

Kádár, László 563
Käubler, Rudolf 1623
Kahn, Ely Jacques, Jr. 1132
Kalesnik, Stanislav V. 955
Kanter, Helmut 1500
Karan, Pradyumna P. 743
Kayser, Kurt 1305
Kehr, Kurt 1344
Kellner, Lotte (Charlotte) 1345
Kelly, Christine 564
Keltie, John Scott 141-142, 815-816, 819
Kennedy, J. Gerald 1027
Kersten, Kurt 1244
Kettler, Julius I. 876
Keuning, Hendrik Jakob 956
Key, Charles E. 257
Kick, Wilhelm 1346, 1366
Kielczewska-Zaleska, Maria 957
King, Cuchlaine A. M. 431, 1439
King, Philip B. 413
King, William F. 1030
Kington, J. A. 432
Kinzel, Hella 877
Kinzl, Hans 1178
Kirk, William 673
Kirsten, Ernst 1516, 1604
Kirwan, Laurence P. 258, 565
Kish, George 143, 339-340, 1491-1493
Klapp, Orrin E. 1086
Klar, Maximilian 121
Klein, Claude 1157-1158
Klimm, Lester E. 607
Klinefelter, Walter 1046
Kloster Ullman, Elba E. 1293
Klute, Fritz 146, 149
Knadler, George A. 1014
Köhler, Arthur 1419
Köhler, Franz 652, 958
Koelsch, William A. 433, 475, 694-695, 774-775, 1006, 1121, 1130

Kohlhepp, Gerd 1629
Kohn, Clyde 560
Kolb, Albert 878, 1559
Kollm, Georg 29, 32, 35
Kondracki, Jerzy 959, 1434
Koner, Wilhelm David 554, 567
Konstantinov, D. A. 960
Korinman, Michel 1577
Kosmachev, Kirill Petrovich 983
Kostrowicki, Jerzy 961
Krämer, Walter 259
Král, Jiří 962
Kramer, Fritz L. 476, 1578
Kramer, Gustav 1579
Kramer, Pedro 1283
Krammer, Mario 1347
Kraus, Alois 477
Krause, Fritz 1348
Krebs, Norbert 144, 879
Kretschmer, Konrad 36, 40, 145-146
Kristof, Ladis K. D. 478
Kropotkin, Peter 569
Krout, Igor V. 1660
Kühn, Arthur 696, 1349
Kühnel, Josef 1287-1288
Kugler, Ernst 1221
Kuls, Wolfgang 147

Lacoste, Yves 1545
Lagarde, Luci 39, 1671
Laget, Gustave de 1628
La Gorce, John Oliver 550
Lampe, Felix 148
Landsberg, Helmut E. 434-435
Lange, Gottfried 720, 1454
Langley, Michael 261
Larnaude, Marcel 1166, 1253
La Roncière, Charles de 262
Lauf, G. B. 341
Laussedat, Aimé 342
Lautensach, Hermann 149, 880, 1657
Lavrov (Lawrow), S. B. 150, 1148
Lawton, Richard 721
Learmonth, Andrew T. A. 1254
Lebedev, Dmitri Mikhailovitch 947
Leconte, P. 635

Lefèvre, Marguerite 963
Le Gear, Clara 363
Lehmann, Edgar 343, 1580, 1604
Lehmann, F. W. Paul 1403
Lehmann, Herbert 1516-1517
Leib, Jürgen 720
Leigh, Myee D. 570
Leighly, John 436, 771, 776, 783, 881, 1087, 1094, 1107
Lejeaux, A. 344
Lejeune, Dominique 636
LeLong, Jacques 2
Lemosof, Paul 151
Lenglet-Dufresnoy, Nicolas 2
Lentz, Eduard 1350
Lenz, Karl 571-572
Leonhardy, Hans 642
Léotard, Jacques 573-574
Leser, Hartmut 882
Leszczycki, Stanislaw 697, 1637
Lewis, Charles L. 1088
Lewis, G. Malcolm 575, 777
Lewis, Oscar 1031
Lewthwaite, Gordon 152
Lexis, Wilhelm 728
Ley, David 829
Libault, André 345
Lichtenberger, Elisabeth 883
Lincoln, W. Bruce 1633
Lindeman, M. 576
Linke, Max 1581, 1630
Linklater, Eric 501
Lipscomb, Andrew A. 1070
Liss, Carl-Christoph 964
Livingstone, David N. 81, 153, 1122-1123, 1404
Lochhead, Elspeth Nora 817-818
Löwenberg, Julius 154
Longhena, Mario 1695
Louis, Herbert 577, 699, 1504
Lowenthal, David 483, 694, 775, 1077-1080
Lüdde, Johann Gottfried 155-156
Lukermann, Fred 852
Lunn, Arnold 637
Lynam, Edward 346-347

Author Index

MacEachren, Alan M. 348
Mackinder, Halford John 157
Maclean, Kenneth 1195
Macpherson, Anne 1351
Madden, Edward H. 143
Mairet, Philip 1256
Maldonado-Koerdell, Manuel 1364
Malesani, Emilio 916
Manheim, Frank J. 1062
Mansur, Fatma 1005
Marcel, Gabriel 1445
Marcinek, Joachim 437, 1505
Marcus, Melvin G. 778
Maret, Marie-Paule 853
Margerie, Emmanuel de 854
Margot-Duclos, Jean-Luc 349
Marinelli, Giovanni 158-159
Marinelli, Olinto 438
Markham, Albert H. 1464
Markham, Clements R. 350, 578, 819, 1553
Markov, Konstantin Konstantinovitch 700
Marsden, William Edward 1260-1261
Marshall, Ann 1712
Marshall, John U. 160
Marshall-Cornwall, James 579, 1194, 1432, 1620
Martens, Robert 1473
Martin, Edwin T. 1071
Martin, Geoffrey J. 137, 439, 560-561, 779, 855, 1007, 1015-1016, 1063, 1066-1067
Martin, Lawrence 1043
Martonne, Emmanuel de 105, 161-162, 638, 653, 856-857
Maschke, Erich 1308
Mason, Kenneth 263
Masseport, Jean 1171
Mather, Cotton 772
Mathieson, John 351
Matley, Ian M. 965, 1193
Matznetter, Josef 1638
Maull, Otto 1352, 1582
May, Joseph A. 163, 1405
Maynial, Edouard 1174
McDonald, James R. 654

McGee, W J 580
McIntosh, Robert P. 440
McKay, Donald V. 858
McKinney, William M. 479
McManis, Douglas R. 480
Mead, William R. 581, 966-968, 1522, 1654-1655
Mehedinţi, Simion 1583
Meinardus, Wilhelm 441
Meinig, Donald W. 1126
Melhorn, Wilton N. 425
Melón, Amando 164, 352, 969-970, 1358
Merriman, Daniel 504
Metz, Friedrich 1308
Meurers, Joseph 1353
Meyer-Abich, Adolf 1354
Meynen, Emil 264, 884-885, 1410, 1506
Meynier, André 502, 715, 859, 1252
Middleton, Dorothy 265, 582, 1636
Migliorini, Elio 917-918
Mihailescu, Vintila 1474
Mikesell, Marvin W. 481-483, 780, 1529, 1546
Milanini Kemeny, Anna 919
Miles, Linda Jeanne 1022
Mill, Hugh Robert 165, 532, 583-584, 1480
Miller, Eldon S. 41, 108
Miller, E. Willard 585
Miller, Ronald 1496
Milne, Kathleen 1481
Minguet, Charles 1355
Mitchell, Jean B. 166
Mitra, Sevati 586
Modelska-Strzelecka, Bozena 697
Möller, Ilse 562
Moir, Donald G. 587
Mollat, Michel 270
Molyneaux, Gary A. 1
Monbeig, Pierre 1664
Mondolfo, Rodolfo 1422
Mongini, Giovanni Maria 920
Mookerjee, Sitanshu 1691
Moravia, Sergio 860
Morawetz, Sieghard 701

Morgan, L. H. 374
Mori, Attilio 159
Morison, Samuel Eliot 675
Morris, A. S. 442
Morris, Rita M. L. 702
Mücke, Erwin 703
Mühlmann, Wilhelm Emil 1262, 1289
Müller, Alice 1584
Müller, Georg 1251
Müller, Theodor 704
Müller-Wille, Christopher F. 484
Müller-Wille, Wilhelm 1224
Munzar, Jan 443
Muris, Oswald 353
Murzayev, E. M. 1161
Musset, René 861
Muthmann, Friedrich 1356

Nakano, Takamasa 997
Nardy, Jean-Pierre 1437-1438
Natali, Giovanni 921
Nedelcu, Eugen 1497
Neff, Ernst 167, 655
Nelson, Helge 168
Netter, Marie-Larence 1545
Nettlau, Max 1547-1548
Newbigin, Marion I. 588, 820
Newby, Eric 266
Nicod, Jean 1170
Nicolas-Obadia, Danielle 1585
Nicolas-Obadia, Georges 1585, 1665
Nielsen, Niels 705
Nikitin, N. P. 971
Nimigeanu, George 1177
Nischer, Ernst 354
Nitz, Bernhard 437
Nockher, Ludwig 1488
Nötlich, Kurt 1477
Nougier, Louis-Rene 270
Nyström, Johan Fredrik 267

Oberhummer, Eugen 551, 677, 886, 972, 1357
Oberlander, Hermann 169
Oelke, Eckhard 703
Oertel, Karl Otto 887
Oestreich, Karl 973

Ohlmarks, Åke 224
Oliver, John W. 1072
Olwig, Kenneth Robert 485
Ostuni, Josefina 1586-1587
Otero Pedrayo, R. 1358
Otremba, Erich 1359, 1369
Oughton, Marguerita 1640
Outhwaite, Leonard 268
Overbeck, Hermann 787, 888-889, 1530
Overton, J. D. 269
Ozouf, Marie-Vic 862

Paassen, Christiaan van 1588
Paffen, Karlheinz 706
Paisey, D. L. 289
Pallot, Judith 951
Panetta, Rinaldo 589
Papy, Louis 1140, 1188, 1191, 1554
Parias, Louis-Henri 270
Parker, Elliott S. 1651
Parker, William H. 1455
Parkins, Almon E. 781
Parry, John Trevor 992
Parsons, James J. 707, 1108
Partsch, Joseph 677, 708-709, 890
Passarge, Siegfried 710
Pastoureau, Mireille 1133
Patten, John 131, 864, 1264
Patterson, Elizabeth C. 1641-1642
Pattison, William D. 711, 782, 1047, 1103-1104
Paulukat, Inge 1365
Pauly, Philip J. 590
Payne, Melvin M. 591
Pearson, Karen S. 356
Péchoux, Pierre-Yves 1545
Pedreschi, Luigi 1297
Peel, Ronald 712
Peltier, Louis C. 444
Penck, Albrecht 592, 713, 891, 1051, 1635
Pereira da Silva, Clodomiro 170
Pergameni, Charles 593
Perejda, Andrew D. 1025
Perpillou, Aimé 863, 1219
Perthes, Bernhard 639

Author Index

Perry, Peter J. 1214
Peschel, Oscar 171-172
Pethybridge, Roger 974
Peucker, Thomas K. 1415
Pfeifer, Gottfried 783, 1109, 1308-1309, 1360-1361, 1416, 1589, 1679
Pfeiffer, Heinrich 1323, 1325, 1341, 1362, 1380
Philippson, Alfred 714
Pinchemel, Geneviève 173, 864
Pinchemel, Philippe 105, 173-176, 594, 640, 853, 864, 998, 1666
Plewe, Ernst 177, 1137, 1149-1150, 1185, 1290, 1308, 1310, 1363, 1590-1595, 1604
Plischke, Hans 271-272, 1173
Plott, Adalbert 1471, 1596
Pokshishevskiy, Vadim V. 975
Popp, Nicolae 1659
Porena, Filippo 178-179, 273, 1213
Portmann, J.-P. 976
Potter, Jefferson R. 737
Potter, S. R. 1423
Powell, Joseph Michael 1702, 1704, 1706-1707, 1709, 1713
Preuss, Helmut 1604, 1690
Prillinger, Ferdinand 180, 977
Prothero, R. Mansell 1151
Pyne, Stephen J. 399, 1052

Quaini, Massimo 486

Rado, Sándor 1435
Rahir, Maurice 598
Rahman, Shah M. H. 181
Raikov, Boris E. 1143
Raisz, Erwin J. 357-359
Rakhilin, V. K. 1241
Ramakers, Günter 445, 1245
Rassem, Mohammed 866
Ratzel, Friedrich 1528
Raudzens, G. 360
Raup, Hallock F. 446
Raup, Hugh M. 447
Ravenstein, Ernst Georg 182
Rawat, Indra Singh 274

Reclus, Elisée 1585
Reeves, Edward A. 361, 1549
Reguera Sierra, Ernesto 1701
Reich, Otto 1315
Reichman, Shalom 999
Reparaz Ruiz, Gonzalo de 199, 978
Reuther, Martin 1271
Rhein, Catherine 865
Ribeiro, Orlando 1470, 1634, 1667
Riccardi, Riccardo 922, 1462
Ricchieri, Giuseppe 183
Richards, Paul 1406
Richter, Bernhard 892
Richter, Otto 1597
Richthofen, Ferdinand von 184, 596
Rio, Maria Isabel del 979
Ristow, Walter W. 362-363, 716
Ritchie, George Stephen 364-365, 1238
Ritter, Carl 185
Robertson, Charles J. 821
Robic, Marie-Claire 487, 1668
Robinson, Adrian H. W. 366
Robinson, Arthur H. 305, 367-371
Robinson, David J. 1065
Robinson, Guy 1264
Robinson, J. Lewis 553, 992, 1000
Robson, Brian T. 1257
Robson, William S. 732
Rodrigues, Fátima 822
Roh, Do-yang 1001
Roller, David C. 750
Romanova, Maria Mikhailovna 1501
Roob, Helmut 1482
Rose, Courtice 1222
Rotberg, Robert I. 275
Rouch, J. 270
Rowley, Virginia M. 1125
Rozo M., Darío 1697
Rudmose Brown, R. N. 823
Rühl, Alfred 755
Ruge, Sophus 27, 30-31, 33, 40, 171, 186
Ruge, Walther 34, 40
Rugg, Dean S. 786
Ruiz, Ernesto A. 597
Ruiz, José Ignazio 1693

Ryabchikov, Aleksandr
  Maksimovich 717
Rzepa, Zbigniew 1433

Saarmann, Gert 353
Sáenz de la Calzada, Carlos 1364
Saey, Pieter 1456, 1531
Salmon, Pierre 598
Samuels, Marwyn S. 829
Sanchez, Francisca 980
Sanchez, Pedro C. 187
Sanderson, Marie 1643, 1714
Sandner, Gerhard 656
Sandru, Ion 599, 981
Sanke, Heinz 1365-1366
Sapper, Karl 488
Sauer, Carl O. 489-490, 787, 1119, 1510, 1532-1533, 1598
Saushkin, Yulian G. 982-983
Scargill, David Ian 718
Schamp, Heinz 600
Schick, Manfred 1624
Schiffers, Heinrich 1151-1152
Schlee, Paul 601
Schlee, Susan 503
Schmidt, Erwin 984, 1367, 1599-1600
Schmidt, Gerhard 719
Schmidt, Max Georg 1411
Schmidt, Peter Heinrich 491
Schmieder, Oskar 1368, 1625
Schmithüsen, Josef 188, 448
Schmitthenner, Heinrich 449, 1272, 1311, 1511, 1602-1603
Schneer, Cecil J. 400
Schneider-Carius, Karl 450-451
Schöller, Peter 1198-1199, 1294
Schöne, Gustav Hermann 1407
Schott, Carl 720
Schott, Gerhard 893
Schottlaender, Felix 1246
Schrader, R. 603
Schreiber, Karl-Friedrich 1416
Schröder, Karl Heinz 1275-1276
Schubert, Gustav von 1153
Schuepp, Max 452
Schulte-Althoff, Franz-Josef 894
Schultz, Hans-Dietrich 895-896

Schultze, Joachim 1333, 1338, 1359, 1369, 1373, 1379, 1385, 1604
Schulz, Heinz 1605-1607
Schulz, Wilhelm 1176
Schulze, Bruno 1608
Schumm, Stanley A. 413
Schwarz, Gabriele 189, 1305
Scurla, Herbert 1370
Sears, Mary 504
Seaver, George 1687
Semmel, Bernard 1457
Sestini, Aldo 453, 923, 1169, 1463
Seylaz, Louis 1520
Seymour, W. A. 372
Shaler, Nathaniel S. 1124
Shalowitz, Aaron L. 373
Sherwood, Morgan B. 1020, 1032
Shibanov, Fyodor A. 374
Shnitnikov, Arseniy V. 985
Siddle, D. J. 1442
Sievers, Wilhelm 276
Singer, Charles 323, 375
Sinnhuber, Karl 1371, 1609
Sion, Jules 1669
Sitwell, O. F. George 1518
Skelton, Raleigh A. 277, 289, 375-378
Slabczynski, Waclaw 1425, 1646
Slatter, John 1424
Smidt, Marc de 686
Smith, J. Russell 788
Smith, Woodruff D. 1534
Sölch, Johann 1507
Sommerlatte, Herbert W. A. 1472
Soons, Jane M. 1699
Sparn, Enrique 604
Spate, Oskar Hermann Khristian 492-493, 612, 1002, 1064
Specklin, Robert 897, 1136
Spencer, Joseph E. 789
Speth, William W. 790, 1011, 1110, 1131
Spethmann, Hans 1466
Spilhaus, Margaret Whiting 278
Spinner, Henri 1412

Author Index

Spörer, Julius   7-8
Spreitzer, Hans   605, 1508
Spreng, Alfred   606
Stäblein, Gerhard   1561
Stafford, Mary Peary   607
Stagl, Justin   866
Stams, Werner   379, 1372
Staszewski, Josef   454
Stavenhagen, Willibald   380-382
Stearn, William T.   455
Steel, Robert W.   721
Steers, James Alfred   1313
Stegner, Wallace   1099
Stein, Harry   722
Steinmetzler, Johannes   1135, 1535
Stevens, Rayfred L.   1364, 1369, 1373-1374
Stevenson, Edward L.   383
Stevenson, W. Iain   1258-1259
Stine, Gordon E.   384
Stitcher, Teresa L.   746
Stoddart, David R.   65, 79, 83, 119, 190-193, 608-609, 648, 723, 791, 824, 1101, 1183, 1215, 1279, 1388, 1421, 1540, 1682
Stoetzer, Carlos   1375
Street, John M.   682
Ström, Ernst Ture   986
Suchowa, N. G.   1376
Suizu, Ishiro   1003
Surdich, Francesco   279
Surface, George T.   1073
Sutton, Keith   1167
Swedberg, Swen   610
Sykes, Percy M.   280

Taberini, Annalena   611
Taillefer, François   1233
Takeuchi, Keiichi   1004
Tatham, George   194, 494
Taton, René   848
Taylor, Benjamin   416, 440
Taylor, Eva G. R.   385
Taylor, James A.   672
Taylor, T. Griffith   194-195, 494, 612, 620, 850, 900, 959, 962, 1715
Tazieff, Haroun   270

Teller, James T.   1192
Termer, Franz   1621-1622
Theodorides, Jean   1377
Thomale, Eckhard   495
Thoman, Richard S.   1096
Thomas, Paul   613
Thornthwaite, C. Warren   456
Thouez, Jean-Pierre   792, 1091
Thrower, Norman J.W.   386
Tichy, Franz   196
Tiessen, Ernst   1298, 1562-1563
Tiggesbäumker, Günter   1228
Tilkovszky, Lorant   1650
Tilley, Philip   457, 1431
Toni, Youssef   657
Toniolo, Antonio Renato   924
Tooley, Ronald V.   387-388
Torroja y Miret, José M.   614
Toschi, Umberto   925
Traversi, Carlo   641
Tricart, Jean   430
Trindell, Roger T.   1012
Troll, Carl   845, 898-899, 1362, 1369, 1378-1381, 1517, 1610, 1680
Tsujita, Usao   1708, 1717
Tümertekin, Erol   1005
Tulard, Jean   1382
Tulippe, Omer   698, 724, 926
Turnock, David   197
Turri, Eugenio   658
Turrill, William B.   458
Twyman, Robert W.   750
Tyrrell, Joseph Burr   1716

Ugalde, José   595
Uhlig, Harald   725-726
Ulbrich, Rolf   934
Ule, Willi   1216
Ulrich, Johannes   281
Umlauft, Friedrich   987-988
Unstead, John Frederick   1300, 1458
Urso, Tomaso   641

Valk, Jan Gerard van der   1409, 1512, 1611
Vallance, Thomas G.   1646
Vallaux, Camille   198, 1217, 1670, 1672

Valls Taberner, Fernando 199
Valverde, Orlando 1681
Vandermotten, Christian 598
Van Valkenburg, Samuel 900
Varenius, Bernhard 104
Vasovic, Milorad 1211
Vaughan, J. E. 738
Vaugondy, Robert de 200
Veggi Donati, Maria Angela 1698
Victor, Paul-Emile 270
Vietinghoff, Georg von 1365
Vilá Valentí, J. 615
Visher, Stephen S. 793-794, 1105
Vitek, John D. 444
Vivien de Saint-Martin, Louis 201
Vivó Escoto, Jorge A. 1364
Vogt, Per 1489
Vooys, Adriaan Cornelis de 1467
Vosseler, Paul 727

Wadel, C. 581
Wagner, Fritz 56
Waites, Bryan 1631
Waldbaur, Harry 890, 1499
Wallis, Helen M. 371, 390
Wanklyn, Harriet 1536
Ward, R. Gerard 1459
Warntz, William 41, 203, 795-796
Warrick, G. T. 1441
Warrington, T. C. 618
Watson, J. Wreford 659, 1496
Wayland, John W. 1089
Webb, Kempton E. 1065
Weber, Heinrich 459
Weber, Peter 1416
Wegener, Georg 1564
Wegener-Köppen, Else 1413
Wegner, Eginhard 691
Weigeldt, Paul 169
Weigend, Guido G. 1201
Weigt, Ernst 1267
Weinert, Hermann K. 1673
Weiss, J. 1392
Weller, Hugo Ewald 1513-1514
Wenk, Hans-Gunther 706, 730
West, Robert C. 1111
Wheeler, Keith 1387

Whitaker, Arthur P. 1383
Whitbeck, Ray H. 797
White, Arthur Silva 825
White, Thurman 416, 440
Whitford, Kathryn 1045
Whitford, Philip 1045
Whittemore, Katheryne T. 660
Wiche, Konrad 605
Wichmann, H. 17, 19, 21, 24, 26
Wiggers, A. J. 1230
Wilcock, Arthur A. 204, 460, 1519
Wilford, John Noble 391
Wilhelmy, Herbert 1384
Williams, Donovan 1465
Williams, Frances Leigh 1090
Williams, John 944
Williams, Juliet 731
Williams, Michael 1112
Williams, Trevor I. 323, 375
Willis, Bailey 461
Winkler, Arno 1684
Wirth, Eugen 1202
Wise, Michael J. 732, 826, 1196, 1391
Wisotzki, Emil 205
Wissmann, Hermann von 1564
Witthauer, Karl 661
Wolff, Peter 41, 203
Wolkenhauer, Wilhelm 22, 206, 392-393, 903
Wolter, John Amadeus 394-395
Wood, Herbert J. 282
Wood, J. David 160, 163, 207
Woodward, David 396
Wright, John Kirtland 208-210, 283, 505, 619-621, 827, 1018, 1054
Wrigley, Edward Anthony 211
Wrigley, Gladys 647, 662
Wroth, Lawrence C. 397
Wünsche, Ernst Alwin 1612
Wüst, George 1369, 1385

Yacher, Leon 1100
Yamato, Kasai 1565
Yli-Jokipii, Pentti 989
Yochelson, Ellis L. 1053
Yugai, R. L. 1486

Author Index

Zabelin, Igor M. 1280
Zavatti, Silvio 284
Zea, Leopoldo 1364
Zelinsky, Wilbur 496
Zemliak, Martin 1545
Zeune, Johann August 212
Zimmermann, Susanne 1494
Zittel, Karl Alfred von 462
Zögner, Lothar 398, 1613
Zondervan, H. 213
Zonneveld, Jan I. S. 990
Zuckermann, Brigitta 1566

SUBJECT INDEX

(N.B.--This index includes only key words that appear in the titles and brief annotations and is not, therefore, a complete guide to the contents of the publications.)

Abbadie, Antoine d' 1133
Aberdeen, University of 1636
Aberystwyth, University of Wales 672, 1237
Abraham 214
Accademia degli Argonauti 529
Adam of Bremen 903
Adelaide, University of 1709
Africa 527, 870, 1686; East 1314, 1481; exploration 273, 1151, 1278, 1442, 1471-1472, 1630; German East 1265-1267; Italian exploration and colonization 919; Society for Commercial Exploration in (Italy) 919
African Association 186, 513, 552, 584
Agassiz, Louis 446
Aix-Marseille, Academy of 1170
Aix-Marseille, University of 1644, 1686
Ajo, Reino 1134
Alaska 1020, 1032
Alaska Geographical Society 547
Albert I, Prince of Monaco 499
Alfred, King 584
Algeria, French expeditions to 834; geographical societies 594
Algiers, University of 1253
Aligarh Muslim University 991
Allgemeine Deutsche Versammlung von Freunden der Erdkunde. See under

Frankfurt, General German Conference of Friends of Geography.
Alpine Club 625, 637
Alpinism 636
Althusser, Louis 486
American Climatological Association 547
American Geographical Society 547, 597, 619, 621, 677, 1016, 1026, 1091, 1132
American Geographical and Statistical Society. See under American Geographical Society.
American Museum of Natural History 1131
American Philosophical Society 420
American Society for Geographical Research 585
American Society for Professional Geographers 584
Ampère, André-Marie 1208
Amundsen, Roald 216, 1135
Ancel, Jacques 1136
Andree, Karl Theodor 1137
Annales de géographie 653-654, 1229
Antarctica, exploration 256, 1235, 1713. Also see polar exploration.
Anthropogeography 463, 495, 1535
Antillón, Isidoro de 978
Antwerp Geographical Society 511
Anuchin, Dmitry Nikolaevich 934, 1138, 1193

Subject Index

Apian, Peter 70
Appalachian Mountain Club 547
Arbos, Philippe 1139
Arctic exploration 254, 256, 874, 1487-1489, 1492-1493, 1631, 1646. Also see polar exploration.
Arctic physical geography 459
Argentina 1176, 1700-1701
Aristotle 402, 448, 478
Arqué, Paul 1140
Asia 1346; Central 1372; exploration 273, 1235, 1296-1297, 1645; universities 682; vegetation 1417
Association of American Geographers 521, 558-561, 585, 760, 766, 771, 794, 1007, 1038
Atlases 310-314, 334, 362
Atwood, Wallace 665, 695, 1006
Auerbach, Bertrand 1141
Ausschuss für deutsche Geschichtsforschung 172
Australia 1702-1704, 1706-1707, 1709, 1711-1715; exploration 240-241, 273, 1238, 1646; geography 1002
Australian National University 1703
Austria 517, 977; cartography 288, 292, 354; geography 931, 933, 935, 948, 972, 988; universities 677
Austria-Hungary, geography 987

Babylonia 326, 345
Babylonians 233, 317
Bacon, Francis 73
Baer, Karl Ernst von 1142-1143
Baker, Samuel White 275
Bakker, Jan Pieter 945
Balbi, Adriano 104, 1144
Balkans 1392
Baltimore, Geographical Society of 547
Banks, Joseph 93, 218, 1145
Banse, Ewald 1146
Baranskii, Nikolai N. 983, 1147-1148
Barrett, Robert L. 1007
Barros Gomes, Bernardino 928

Barrows, Harlan H. 475
Bartels, Dieter 69
Barth, Heinrich 275, 870, 1149-1153, 1595
Bartholomew, John 825, 1154
Bartholomew, John & Son Ltd. 324
Basel Geographical and Ethnological Society 518
Basel, University of 663, 727
Bastian, Adolf 63, 1155
Bates, Henry Walter 429
Baulig, Henri 1156-1158
Beaufort, Francis 366
Beekman, Anton Albert 945
Belcher, Edward 764
Belgium 138-139, 1433; geography 875, 926, 963; Royal Belgium Geographical Society 593, 598; Society for Colonial Studies 598
Belgrade, University of 1211, 1675
Bennett, Arnold 1159
Benton, Thomas Hart 740
Berg, Lev Semenovich 1160-1161
Berghaus, Heinrich 308, 311-313, 337, 389, 1162-1164
Berghaus, Hermann 337, 389
Bering, Vitus 246
Berlin Geographical Society 513, 554, 567, 571-572, 592, 596, 642, 1557, 1567
Berlin, Handels- (Wirtschafts-) Hochschule 1684
Berlin, Technical University of 1523
Berlin, University of 667, 688, 729, 755, 1011, 1150, 1228, 1272, 1289, 1502, 1506, 1508, 1555, 1559, 1569, 1574, 1581, 1601-1602, 1609
Berlioux, Etienne-Félix 1165
Bern Geographical Society 548, 606
Bern, University of 687
Bernard, Augustin 1166-1167
Bertuch, Friedrich Justin 651
Biasutti, Renato 1168-1169
Bingham, Millicent Todd 1008

Subject Index

Biogeography 86, 101, 198, 404, 448.
  Also see Ecology and Geography
  (plant).
Biology, marine 262, 498
Birmingham, University of 1441
Bjerknes, Vilhelm 451
Blache, Jules 1170
Blanchard, Raoul 994, 1171-1172
Blodget, Lorin 1009
Blumenbach, Johann Friedrich 1173
Boas, Franz 1010-1012
Bodin, Jean 469, 486
Boeckh, August 1150
Bohr, Niels 167
Bolivia 1285, 1288
Bollaert, William 764
Bonn, University of 714, 1475-1476,
  1517, 1523, 1550, 1555, 1559, 1656,
  1679-1680
Bonpland, Aimé 1174-1176
Bordeaux Geographical Society 522
Bordeaux, University of 1140, 1188,
  1191, 1239, 1554
Bose, Nirmal Kumar 1691
Botero, Giovanni 468
Bougainville, Louis Antoine de 238
Bourne, William 803
Bowman, Isaiah 93, 439, 1013-1018
Boyle, Robert 450
Bratescu, Constantin 1177
Braun, Gustav 727
Braunschweig, Technical University
  of 704
Braunschweig, University of 1431
Brazil 1231, 1629, 1679, 1681;
  geography 993, 1065
Brazilian Historical and Geographical
  Institute 539
Bremen 903
Bremen Geographical Society 576
Breslau, University of 708
Brigham, Albert Perry 1019
Bristol, University of 712
British Arctic Expedition 528
British Association for the
  Advancement of Science 623, 634,
  1485

British Rainfall Organization
  1479-1480
British universities, geography in
  1639
Brittany 1658
Bromme, Traugott 314
Brooks, Alfred Hulse 1020
Brouillette, Benoît 994
Brown, Ralph Hall 480, 1021-1022
Brückner, Eduard 1178
Brunhes, Jean 1179-1180, 1409
Brussels, Free University of 98
Bauche, Philippe 340, 842, 1181-
  1182
Buache de Neuville, Jean-
  Nicolas 860, 1182
Buch, Leopold von 421
Buchanan, John Young 1183
Buchanan, Robert Ogilvie 826
Bucharest, University of 981,
  1177, 1474, 1659
Buckle, Henry Thomas 494
Büsching, Anton Friedrich 127,
  189, 592, 651, 1184-1185
Buffon, Georges-Louis Leclerc,
  Comte de 196, 838, 1186-1187
Bulletin de la Société de
  géographie d'Egypte 657
Bulletin of the American Bureau of
  Geography 660
Bulletin of the American
  Geographical Society 647, 662
Burgess, Ernest Watson 761
Burton, Richard 275
Byrd, Richard E. 215

Calcutta Geographical Society 586
Calcutta, University of 1691
Caldas, Francisco José de 1692-1693
California 744, 1556
California, Geographical Society
  of 537, 547
California, University of 678,
  707, 763, 790, 1028, 1031, 1054,
  1061, 1106-1109, 1712
Cambridge, University of 545, 685,
  723, 801-802, 1183, 1214, 1439

Camena d'Almeida, Pierre 1188
Cameron, Verney Lovett 275
Canada 1716; French 1172;
  geography 992, 994, 1000
Canadian Association of
  Geographers 553
The Canadian Geographer 659
Candolle, Alphonse de 402
Canterbury, University of (New
  Zealand) 1705
Cantillon, Richard 472
Cape Province, South Africa,
  exploration 242
Capot-Rey, Robert 1189
Carey, Henry 479
Carter, Clement Cyril 1190, 1220
Cartography, history of 44-45, 50,
  60, 98, 113, 146, 151, 159, 171,
  178, 182, 200-201, 277, 285-398
  passim, 731, 1212; Austrian 288,
  292, 354, 935, 940, 948, 988;
  Belgian 963; British 319,
  346-347, 377; European 315; Finnish
  966; French 319, 344, 349, 380;
  German 379, 639; Italian 915;
  Japanese 1003; Korean 996; medical
  337; polar regions 398;
  Portuguese 302; Prussian 381;
  Russian 290, 374, 382;
  Spanish 969, 978; statistical 322,
  368; Swiss 301, 327; thematic 340,
  348, 367, 371; United States 291,
  321-322, 336, 358
Cassini, Jean Dominique 328
Catastrophism 405
Caucasus region 1524
Cavaillès, Henri 1191
Cayley, Arthur 203
Cecchi, Antonio 908
Celtis, Conrad 328
Central America 1621
Central place theory 468, 487,
  1198-1202
Cernauti, University of 1177
Challenger Expedition 501, 1183
Chamonix 1571
Champollion, Jean-François 240

Charleston, South Carolina 403
Charpentier, Jean de 1192
Chekhov, Anton 1193
Chesney, Francis Rawdon 1194,
  1432, 1620
Chicago, Geographic Society of 547
Chicago, University of 674, 711,
  1023, 1096, 1103-1105, 1113,
  1115, 1119, 1712-1715
Chichester, Francis 214
China 878, 1316, 1556, 1558, 1561;
  geography 997, 1003
Chisholm, George Goudie 821,
  1195-1196
Cholley, André 1197
Christaller, Walter 472, 484, 487,
  1198-1202
Cities, mapping of 317
Clark, William 243, 757
Clark University 665, 695, 716,
  1006, 1113, 1115, 1119
Claval, Paul 486
Clermont-Ferrand, University
  of 1139, 1644
Climatic classification 409, 452,
  456
Climatology 421, 434-436, 450,
  452, 1413, 1674-1675
Club Alpin Français. See under
  French Alpine Club.
Cluj, University of 1659
Coast Survey, United States 373,
  1028, 1030-1031
Codazzi, Agustín 1694-1698
Coello, Francisco 1203
Coimbra, University of 928
Colby, Charles Carlyle 1023
Cole, Grenville Arthur James 1204
Colgate University 1019
Collège de France 1179-1180, 1437
Colombia 1692-1698
Colonialism 279, 858, 908
Colonization, African 510
Columbia University 1011, 1125
Columbus, Christopher 185, 240,
  249, 1355
Comte, Auguste 81, 1208

Subject Index

Constantinople 1571
Continental drift 1683
Contours 333
Cook, James 148, 218-219, 240, 246, 376, 459, 1205, 1245, 1247-1248
Cook, Orator Fuller 1024
Cooley, William Desborough 520
Copenhagen, University of 705, 1443
Coral reefs, theories of origin 412, 426
Cornish, Vaughan 1206-1207
Cortambert, Eugène 1208-1209
Cosmographer (the term) 204
Cotton, Charles A. 1699
Cournot, Antoine-Augustin 852, 1208
Cowles, Henry 761
Cracow, University of 670, 697, 1521, 1637
Credner, Rudolf 691
Credner, Wilhelm 878
Cressey, George Babcock 1025
Curzon, George Nathaniel 1210
Cvijić, Jovan 112, 1211
Czechoslovakia 517; geography 962

Dagenais, Pierre 994
Dainville, François de 1212
Dalla Vedova, Giuseppe 905, 1213
Daly, Charles Patrick 1026
Dana, James Dwight 426
Darby, Henry Clifford 802, 1214
Darby, William 1027
Darwin, Charles 81, 404, 412, 426, 429, 791, 1215-1217, 1672
Darwinism 1678; social 765
David, Mihai 1218
Davidson, George 1028-1033
Davis, William Morris 203, 239, 407, 413, 425, 430, 433, 439, 457, 461, 521, 671, 702, 739, 755, 761-762, 779, 1034-1043, 1575
De Brahm, William Gerard 1044
De Geer, Sten 112, 946
Delisle, Guillaume 352
Demangeon, Albert 189, 1219
Demolins, Edmond 494

Denmark, geography 927, 937
Desmarest, Nicolas 1511
Determinism, environmental 492-494
DeWitt, Benjamin 446
Dickinson, Basil Bentham 1190, 1220
Dickinson, Emily 1008
Dictionaries, geographical 502
Dietrich, Philippe-Frédéric de 1221
Dilthey, Wilhelm 1222
Dimitrescu-Aldem, Alexandre 1223
Dörries, Hans 1224
Dokuchaev, Vasily Vasilyevich 934, 1225-1226, 1340
Dove, Heinrich Wilhelm 441, 1399
Drapeyron, Ludovic 831, 1227
Drygalski, Erich von 1228
Dubois, Marcel 1229
Dumont d'Urville, Jules Sébastien César 238
Durkheim, Emile 829, 865
Dwight, Timothy 1045

Ecology 416, 440, 761
Economics 491
Edelman, C. H. 1230
Eden, Richard 514, 578
Edinburgh 799
Edinburgh, Royal Scottish Geographical Society 541, 583, 587-588, 818, 820, 825, 1154
Edinburgh, University of 1154, 1195-1196, 1240, 1254, 1260-1261, 1496
Egyptian Geographical Society 600, 657, 1630
Egyptians, ancient 185, 201
Emin Pasha 1528
England, teaching 736, 738
Enlightenment 860
Environmentalism 492-494
Erde 642
Ergänzungshefte zu Petermanns Geographische Mitteilungen 661
Erlangen, University of 1202, 1275-1276

Eschwege, Wilhelm Ludwig von 1231
Ethiopia 1133
Europe, cartography 315; geography 1125; geomorphology 444
Evans, Lewis 756-757, 1046
Everest, George 1232
Exploration 214-284, 519-520, 524, 565, 591, 629; Africa 527; Arctic 528; Austrian 948, 988; British 819; German 872, 874, 886; history of 45, 60, 75, 90, 94, 96, 98, 106-107, 110, 117, 144-145, 151, 154-155, 166, 171, 178, 180, 185-187, 199-201, 206, 214-284, 918; Italian 907, 909; polar 566; Russian 934, 942, 947, 951; Scandinavian 581; Spanish 969; Viennese researches 931

Fabri, Johan Ernst 722
Fairbanks, Harold W. 1047
Farrington, Anthony 536
Faucher, Daniel 1233
Fawcett, P. H. 249
Filchner, Wilhelm 1235
Finland 581; geography 967, 989; mapping 966
Fischer, Theobald 706, 730
Fleure, Herbert John 523, 672, 1236-1237
Flinders, Matthew 1238
Florence, Military Geographical Institute 641
Florence, University of 1463
Fockema Andreae, Sybrandus Johannes 945
Foncin, Pierre 1239
Forbes, James David 1240
Formozov, Alexander Nikolayevich 1241
Forster, Georg 887, 1242-1246, 1248
Forster, Johann Reinhold 887, 1242, 1247-1248
Foucault, Michel 85, 486
Fourier, Joseph 1249
France 139; cartography 319, 344, 349, 380; Collège de 839, 1180; connection with W. M. Davis 671, 1037; geographical societies 594; geography 828-865, 986, 1461; geomorphology 415, 1156; in German geographical literature 897; topographic mapping 285, 1386; universities 677; urban geography 864
Frankfurt, General German Conference of Friends of Geography 630
Frankfurt Geographical Society 516, 538, 617
Franklin, Benjamin 461
Franz, Johann Michael 78, 1250
Frémont, John C. 740
French Alpine Club 627, 636
Fröbel, Julius 764, 868, 881, 1251

Galletti, Johann 902
Gallois, Lucien 1252
Galton, Francis 112, 203
Gama, Vasco da 185
Gambi, Lucio 486
Gandhi, Mahatma 1691
García, Pedro Andrés 1700-1701
Gautier, Emile-Félix 1253
Geddes, Arthur 1254
Geddes, Patrick 100, 624, 821, 823, 1254-1259
Geikie, Archibald 1260-1261
Geikie, James 1260-1261
Geneva Geographical Society 531, 546
Geodesy 323, 351
Geographical Association. See under Great Britain, Geographical Association.
Geographical Club 579
Geographical Magazine 648
Geographical Review 647, 662
Geographical Society of India. See under India, Geographical Society of.
Geographische Gesellschaft in Bern. See under Bern Geographical Society.

Subject Index

Geographische Gesellschaft in Bremen.
See under Bremen Geographical
Society.
Geographische Gesellschaft in
Hamburg. See under Hamburg
Geographical Society.
Geographische Gesellschaft in
München. See under Munich
Geographical Society.
Geographische Gesellschaft in Wien.
See under Vienna Geographical
Society.
Geographische Gesellschaft zu
Hannover. See under Hannover
Geographical Society.
Geographische-Ethnologisch
Gesellschaft Basel. See under
Basel Geographical and Ethnological
Society.
Geographische Zeitschrift 645, 656, 1307
Geography, American 80, 100, 131, 140; applied 500; British 100, 104, 111, 131, 140, 157, 165; cultural 471, 483, 489-490, 785; economic 113, 477, 488, 1435; European 138-139; French 80, 99, 104, 107, 654, 671; German 53, 56, 99, 109, 120, 126, 128, 147-149, 157, 186, 188, 194, 221-222; historical 198, 475, 480, 485, 799, 802, 1214; human 86, 101, 130, 140, 160, 463-496, 761-762, 785, 1002, 1005, 1668; mathematical 293; medical 1091; physical 86, 101, 113, 399-462 passim, 776, 778, 990, 1394-1395, 1398, 1427, 1434, 1440, 1446; plant 414, 440, 445, 447, 454-455, 458, 1380; political 113, 478, 1530, 1658; regional 101, 113, 783; Russian 131; social 113, 198, 466, 470, 495; statistical 497; Swedish 168; theoretical 1415-1416, 1537, 1652; urban 484, 864, 1066
Geology 198, 405, 461-462, 805

Geomorphology 399, 406, 411, 413, 415, 418, 430-431, 433, 439, 442, 444, 449, 461, 803, 805, 869, 882, 992, 1156, 1511
Geopolitics 478, 1291-1294, 1453, 1455-1456, 1534
Gerland, Georg 881, 1095, 1262
Germany 139, 517, 656; cartography 379, 639; East, publications on history of geography 958; geography 866-903, 986, 1246, 1573; geography in schools in 18th century 735; geomorphology 882; universities 677, 724, 728
Gesellschaft für Erdkunde zu Berlin. See under Berlin Geographical Society.
Giessen, University of 725-726, 1431
Gilbert, Edmund William 1263-1264
Gilbert, Grove Karl 399, 430, 461, 755, 1048-1053
Gillman, Clement 1265-1267
Gilman, Daniel Coit 1054
Gioia, Melchiorre 925, 1213
Giraud-Soulavie, Jean-Louis 445
Glaciation 401, 446, 1192
Globes 296-297, 299, 306, 316, 353, 383
Gmelin, Johann Georg 887
Goethe, Johann Wolfgang von 188, 1231, 1268-1272, 1578, 1602
Göttingen 901
Göttingen, University of 696, 1250, 1440
Gogol, Nikolai 1273
Gordon, Patrick 827
Gotha 344, 389, 902, 1509
Gotha Geographical Institute 632-633, 639
Gothenburg Geographical Society 610
Gourou, Pierre 1274
Goyder, George Woodroofe 1702
Gradmann, Robert 483, 1202, 1275-1276
Granö, Johannes Gabriel 189, 1277

Grant, James 1278
Graz, University of 701
Great Britain 51, 139-140;
 cartography 319, 346; Geographical
 Association 540, 575, 618, 1190,
 1220; geography 798-827, 986;
 geology 405; Institute of British
 Geographers 608; mapping 287;
 universities 677, 680
Great Siberian Expedition 515
Greece, French expeditions to 834
Greeks, ancient 45, 78, 116, 135,
 163, 188, 202, 427
Greenwood, George 1279
Gregory, John Walter 821
Greifswald, University of 691, 1431
Grenoble, University of 839, 1170-1171
Gribaudi, Pietro 914
Grigoryev, Andrei Alexandrovich 942, 1280
Guthrie, William 827, 1281
Guyot, Arnold Henri 203, 757, 1055-1060, 1575

Haack, Hermann 632
Haeckel, Ernst 402, 761, 1282, 1528
Hägerstrand, Torsten 84
Haenke, Thaddaeus 1283-1288
Hahn, Eduard 476, 483, 1289-1290
Hakluyt, Richard 584, 810
Hakluyt Society 520, 628-629, 1485
Halle, University of 703, 1410-1411, 1623-1624
Halley, Edmund 203, 364
Hamburg 870
Hamburg Geographical Society 543, 562, 601, 872
Hamburg, University of 710
Hannover Geographical Society 508
Hanoverian England 385
Hariot, Thomas 416
Harness, Henry Drury 368
Hartshorne, Richard 207
Harvard University 671, 675, 702, 768, 774-775, 1008, 1036, 1038, 1042-1043, 1100, 1120-1124, 1127, 1130

Hassenstein, Bruno 1528
Hatshepsut, Pharaoh 61
Hauber, Eberhard David 78
Haushofer, Karl 1291-1294
Haviland, Alfred 1295
Hebrews, ancient 185
Hedin, Sven 223, 253, 968, 1296-1298, 1562
Heidelberg, University of 1306-1307, 1309-1310, 1682
Heisenberg, Werner 167
Helsinki, University of 1277, 1654-1655
Henry, Prince 391
Herbertson, Andrew John 806, 823, 1299-1301, 1436
Herder, Johann Gottfried von 103, 196, 1302-1305, 1535, 1612
Herodotus 168, 290, 465, 488
Hesiod 143, 451
Hettner, Alfred 124, 458, 645, 656, 881, 1095, 1306-1311, 1361, 1368, 1409, 1467, 1589
Hilgard, Eugene Woldemar 1061
Hillard, George Stillman 1062
Himalayas 263
Himly, Louis-Auguste 1312
Hinds, Richard Brinsley 409
Hinks, Arthur Robert 1313
Hippocrates 448, 492
Historicism 82
History, connections with geography 175
Hitler, Adolf 899
Ho, Robert 1703
Höhnel, Ludwig von 1314
Hoff, Karl Ernst Adolf von 1315
Holmes, James Macdonald 1704
Homer 43, 261, 451
Hooker, Joseph Dalton 458
Horton, Robert 203
Hüttner, Johann Christian 1316
Humboldt, Alexander von 47, 49, 63, 70, 73, 82, 93, 99, 101, 104, 122, 137, 143, 145-146, 148-149, 154, 161, 167, 171, 184, 188, 196, 199, 211, 213, 240, 276, 308, 314, 333, 371, 400-401, 410,

Subject Index

Humboldt, Alexander von (cont'd)
  414, 417, 423, 438, 441, 443, 445,
  450, 452, 454-455, 466-467, 490,
  494, 646, 887, 891-892, 1174-1176,
  1226, 1231, 1247, 1283-1284, 1286,
  1317-1385, 1399, 1402, 1583, 1589,
  1599, 1610, 1694
Humboldt Society for the History of
  Geography and Cartography 555, 616
Hungarian Geographical Society 563
Hungary, geography 563
Huntington, Ellsworth 112, 493,
  1063-1064
Hurault, Louis 1386
Hutchings, Geoffrey Edward 1387
Huxley, Thomas Henry 803, 1388
Hydrographic mapping 364
Hydrography 305, 342, 364-366, 372,
  375-376

Iasi University (Romania) 1218
Imperialism 133, 894
India 1389; Anthropological Survey
  1681; Geographical Society of 586;
  geography 991, 1003; Survey of 274,
  294, 330, 1232, 1552
Indiana 793
Indians, American 777
Industrial Revolution 423
Institut géographique national. See
  under Paris, National Geographical
  Institute.
Institute of British Geographers 544,
  644
Instituto Histórico e Geográfico
  Brasileiro. See under Brazilian
  Historical and Geographical
  Institute.
Instituto Histórico e Geográfico de
  São Paulo. See under São Paulo
  Historical and Geographical
  Institute.
International Geographical Congresses
  622, 635, 638, 640
International Geographical Union 635,
  638

Ireland, Geographical Society of
  536; Geological Survey of 1204;
  geomorphology 411; mapping 286;
  Ordnance Survey 1429; Royal
  College of Science 1204
Irish Board of Works 1429
Irish Geographical Association 536
Isarithms 333
Isobaths 305, 333, 369
Isogons 333
Isometric lines 369
Isopleths 369
Isotherms 333, 371, 441, 443, 454
Israel, geography 999
Istituto Geografico Militare. See
  under Florence, Military
  Geographical Institute.
Italian Geographical Congresses 908
Italian Geographical Society 506-
  507, 527, 535, 589, 611, 908
Italian Science Congresses 920-921
Italy 1213; geography 904-925,
  1169

Jackson, Julian 1389
James, Preston 1065
Japan, geography 995, 997-998,
  1003-1004; German research on 898
Jefferson, Mark S. W. 1066-1067
Jefferson, Thomas 434, 740, 787,
  1068-1073, 1338
Jena, University of 722
Jobberns, George 1705
Johannesburg 557
Johns Hopkins University 1016, 1054
Johnson, Douglas Wilson 1074
Johnston, Harry H. 1390
Johnston, W. & A. K., Ltd. 355, 825
Jones, Llewellyn Rodwell 1391, 1498
Journal of Geography 660
Journal of School Geography 660
Julius Caesar 584

Kaempfer, Engelbert 898
Kanitz, Felix 1392

Kant, Immanuel 78, 99, 103, 108, 124, 197, 213, 1303, 1351, 1393-1407, 1409, 1446
Kapp, Ernst 764, 1408-1409
Keltie, John Scott 1424
Kennedy, Joseph C. G. 322
Kentucky 743
Keuning, Hendrik Jacob 945
Kharkov 1418
Kiel, University of 706, 730, 1679
Kiepert, Heinrich 729
Kircher, Athanasius 404
Kirchhoff, Alfred 703, 1410-1411
Kjellén, Rudolf 968, 1293
Klöden, Gustav Adolf von 868
Klute, Fritz 725
Knapp, Charles 1412
Köhler, Günther 878
Königsberg, University of 1395, 1398, 1403, 1405
Köppen, Wladimir 452, 456, 460, 1413
Kohl, Johann Georg 148, 216, 1414-1416
Kolb, Peter 887
Kolosovsky, Nikolai Nikolaevich 983
Komarov, Vladimir Leontyevitch 1417
Koninklijk Aardrijkskundig Genootschap van Antwerpen. See under Antwerp, Royal Geographical Society.
Koninklijk Nederlandsch Aardrijkskundig Genootschap. See under Netherlands, Royal Dutch Geographical Society.
Korea, geography 996, 1001
Kosmographische Gesellschaft in Nürnberg. See under Nuremberg Cosmographical Society.
Kosmos (Humboldt) 1321, 1341-1342, 1347-1348, 1353, 1384
Krasnov, Andrey Nikolaevich 1418
Krause, Carl Christian Friedrich 1419
Kroeber, Alfred L. 471
Kropotkin, Peter 1420-1424
Kubary, Jan Stanislaw 1425
Kümmerly, Gottfried 301
Kümmerly and Frey 301

Küttner, Carl Gottlob 1426
Kuhn, Thomas 193, 474, 896
Kyoto, University of 1708

Lalanne, Léon-Louis-Chrétien 369, 468, 864
Lamarck, Jean-Baptiste 81
La Métherie, Jean-Claude de 422, 1427
Landscape 123, 481, 895, 1341, 1384
Landscape history 180
La Noë, O. de 830
Lapparent, Albert de 830, 1428
Larcom, Thomas Aiskew 1429
Latin America 1065, 1111, 1334, 1350, 1382-1383
Laurie, James 799
Lautensach, Hermann 1430-1431
Laval University 994
Leake, William Martin 1194, 1432, 1620
Leeds, University of 1391
Leibniz, Gottfried Wilhelm 134
Leipzig, University of 664, 679, 709, 1499, 1526, 1529, 1555, 1559
Lelewel, Joachim 929, 1433
Lencewicz, Stanislaw 1434
Lenin, Vladimir 1435
Leningrad, University of 700, 1161, 1280, 1417
Le Play, Frédéric 495, 624, 1436
Le Play Society 624
Lesley, John Peter 430
Levasseur, Emile 467, 1437-1438
Lewis, Meriwether 243, 757
Lewis, William Vaughan 1439
Leyden, University of 1446
Libraries, geographical 395
Lichtenberg, Georg Christoph 1440
Liège, University of 698
Lille Geographical Society 613
Lille, University of 1644
Limagne (France) 862
Linton, David Leslie 1441
Lisbon Geographical Society 928

Subject Index 379

Lisbon, University of 928, 1634
Liverpool, University of 721, 801-802, 1214, 1617
Livingstone, David 201, 216, 275, 1442
Location theory 1682. Also see Central Place Theory and Geography (theoretical).
Löffler, Ernst 1443
Lösch, August 472
Löwenberg, Julius 59
Lomonosov, Mikhail 442, 1444
London Institute of Geography 526
London, Royal Geographical Society 514, 519-520, 524, 526, 528, 532-534, 542, 545, 552, 563-564, 569, 578-579, 581-582, 584, 609, 815, 824, 1210, 1278, 1313, 1389, 1424, 1464, 1479-1480, 1485, 1549, 1651, 1687
London, University of (Bedford College) 1263; Birkbeck College 1490, 1648; King's College 732; London School of Economics 732, 1391, 1452, 1498; University College 680, 801-802, 1214, 1234, 1459
López, Tómas 1445
Louisiana 1542
Louisiana State University 1102
Lowie, Robert 471
Lüdde, Johann Gottfried 59
Lulofs, Johann 1446
Lwow, University of 1616
Lyell, Charles 430, 1315
Lyon, University of 839, 1165

Machiavelli, Niccolo 468
Mackinder, Halford John 93, 100, 732, 1293, 1447-1458, 1685
Maclure, William 461
Maconochie, Alexander 680, 1459
Madrid, Royal Geographical Society 509, 614-615, 1203
Magazin für Erd- und Völkerkunde 646
Magazin für Historie und Geographie 1184

Magazin für die neue Historie und Geographie 651
Magellan, Ferdinand 219
Mahan, Alfred 1293
Malaspina expedition 1283, 1285-1288
Malaya, University of 1703
Malte-Brun, Conrad 104, 968, 1460-1461, 1519
Manchester Geographical Society 523, 541, 570
Manchester, University of 523, 681, 1237
Marburg, University of 720
Marchand, Etienne 238
Margerie, Emmanuel de 830
Marinelli, Giovanni 905
Marinelli, Olinto 905, 1462-1463
Marine charts 364-366
Marine science 498-499
Marine survey 376
Markham, Clements R. 1464-1465
Marseille Society of Geography and Colonial Studies 573-574
Marsh, George Perkins 196, 483, 485, 1075-1080
Marthe, Friedrich 1466-1467
Martonne, Emmanuel de 715, 1156, 1468-1470
Marxism 82, 960, 965
Massif Central 840
Mauch, Karl 1471-1472
Maury, Matthew Fontaine 438, 1081-1090, 1575
Max, King of Bavaria 172
Maxwell, James Clerk 1473
May, Jacques 1091
McFarlane, John 821
Mehedinți, Simion 1474
Meitzen, August 483
Melanesia 230
Mendelssohn, Georg Benjamin 1475-1476
Mentelle, Edme 860
Mercantilism 469, 491
Mercator 203, 285
Merriam, Clinton Hart 416

Merz, Alfred 1477
Meteorological Society of London 424
Meteorology 198, 403, 419-421, 424, 428, 432-433, 450-451. Also see Climatology.
Methodology, geographical 126, 155-156
Mexican Geographical and Statistical Society 595
Mexico 1111, 1357, 1374; French expeditions 834; geography 595; National Autonomous University 1710
Michelet, Jules 1478
Michigan State Normal College, Ypsilanti 1067
Micronesia 230
Middle East 1686
Mill, Hugh Robert 1479-1480
Milne, Geoffrey 1481
Minard, Charles-Joseph 370
Minnesota, University of 1022
Minoans, ancient 215
Mitchell, Thomas Livingstone 1706
Möller, Johann Heinrich 1482
Mongolian People's Republic 943
Mont Eagle University 676
Montesquieu 492, 838, 1612
Morgan, Lewis Henry 476
Morse, Jedidiah 741, 767, 827, 1092-1093
Morton, John 1483
Moscow, University of 700, 717, 1138, 1147, 1241, 1280, 1501, 1660
Moses 201
Mountaineering 220, 257, 262, 625, 627, 631, 637
Mountain measurement 505
Mountains, study of 836
Mueller, Ferdinand Jakob Heinrich von 1707
Muir, John 1094
Munich Geographical Society 551, 577
Munich, Technical University of 689
Munich, University of 699, 1228, 1292
Murchison, Roderick Impey 1484-1485
Murray, John 823
Mushketov, Ivan Vasylievitch 1486

Nancy, University of 1141, 1170
Nansen, Fridtjof 216, 968, 1487-1489
Nares, George 528
National Council for Geographic Education 512, 660
National Council of Geography Teachers 512
National Geographic Magazine 549, 658
National Geographic Society 547, 549-550, 580, 590-591
Naumann, Carl Friedrich 679
Nautical atlases 313
Nautical charting 332, 373
Navigation 282, 318, 323, 350, 376, 385, 397
Nelson, Helge 946
Neo-Lamarckism (or Neo-Lamarckianism) 81
Neopositivism 82
Netherlands, geography 686, 938, 945, 956, 973, 990
Netherlands, Royal Dutch Geographical Society 568, 602, 956
Neuchâtel Geographical Society 1412
Neuchâtel, geography 976
Neumann, C. 148
Newbigin, Marion Isabel 821, 1490, 1648
New Brunswick, University of 1000
New England Meteorological Society 433
Newfoundland 376
New South Wales Geographical Society 612
Newton, Isaac 134
New Zealand 269, 1699, 1705
Noé, Heinrich 1528
Nordenskiöld, Adolf Erik 968, 1491-1493
Norske Geografiske Selskab. See under Norwegian Geographical Society.
North America 140; exploration 245, 273
Northeast Passage 278

Subject Index

North Pole 257
Northwest Passage 278
Norway, geography 927, 939
Norwegian Geographical Society 566
Nouvelle géographie universelle
 (Reclus) 1541
Nuremberg Cosmographical Society 513

Oberhummer, Eugen 1494
Obruchev, Vladimir 1495
Oceania 1425
Oceanography 198, 262, 281, 375-376,
 498-499, 501, 503-504, 893, 1385,
 1658
O'Dell, Andrew 821
Ogawa, Takuji 1708
Ogilvie, Alan Grant 112, 821, 1496
Oldham, H. Yule 681
Ontography 779
Ordnance Survey of Great Britain 286-
 287, 325, 335, 351, 360, 372, 377
Orghidan, Nicolai 1497
Ormsby, Hilda 1498
Oxford, University of 526, 545, 684-
 685, 718, 731, 798-799, 810, 1263-
 1264, 1300-1301, 1449-1450, 1452

Pacific Basin, exploration 219, 222,
 230, 238, 244, 272
Pacific, Geographical Society of the
 537
Pacific Railroad Survey 1126
Panzer, Wolfgang 878
Paris, Ecole Normale Supérieure 1662
Paris Geographical Society 513, 530,
 603, 636, 854, 858, 1460
Paris, National Geographical
 Institute 349, 626
Paris, University of 839, 851, 856,
 1136, 1166-1167, 1197, 1219, 1229,
 1252, 1312, 1468-1470, 1614, 1644,
 1662, 1666, 1670
Park, Robert E. 761, 1095
Parker, Francis 1575
Partsch, Joseph 148, 890, 1499
Passarge, Siegfried 1500

Pavlov, Aleksei Petrovich 1501
Penck, Albrecht 148, 439, 688,
 1502-1508
Penck, Walther 431, 439
Pennsylvania, University of 1125
Perthes, Justus 632-633, 639,
 1509; Anstalt 389, 639, 902
Peru 1285-1286, 1288, 1464
Peschel, Oscar 122, 679, 709, 868,
 1409, 1510-1512, 1528, 1583, 1611
Peter the Great 942, 950-951
Petermann, August 379, 389, 1513-
 1514
Petermanns Geographische
 Mitteilungen 379, 643, 652, 655, 661
Philadelphia, Geographical Society
 of 547, 607
Philippson, Alfred 1515-1517
Phoenicians 499
Physiography 418
Pinkerton, John 827, 1518-1519
Platt, Robert S. 1096
Playfair, John 431, 1520
Pleistocene studies 437
Pöppig, Eduard 1528
Pol, Wincenty 697, 1521
Poland, geography 929-930, 941,
 957, 959, 961
Polar exploration 252, 256, 258,
 273, 277
Polar regions, cartography 398
Polo, Marco 216, 247, 277
Polybius 473
Polynesia 230
Popper, Karl 160, 896
Porthan, Henrik Gabriel 1522
Portugal, cartography 302;
 geography 928
Positivism 82, 1535
Possibilism 492, 494
Powell, John Wesley 461, 739-740,
 755, 1097-1099
Price, Archibald Grenfell 1709
Princeton University 1055-1056,
 1058-1059
Projections, map 320

Prussia, cartography 381
Ptolemy, Claudius 203, 327, 392
Pundits 263, 274, 1552
Punt, Land of 61
Pytheas 214, 223, 254, 258

Quebec Geographical Society 525, 547, 994
Queen's University of Belfast 673
Quelle, Otto 1523

Radde, Gustav 1524
Raisz, Erwin 1100
Raleigh Club 579, 584
Ramaer, Johan Christoffel 945
Ramond de Carbonnières, Louis-François-Elisabeth 838
Ramsay, Andrew 430
Ranke, Leopold von 1525
Ratzel, Friedrich 49, 63, 93, 127, 161, 177, 216, 463-464, 467, 473, 483, 495, 679, 889, 1217, 1293, 1409, 1526-1536, 1672, 1678
Ravenstein, Ernst Georg 1537-1538
Ravn, Nils F. 369
Rawlinson, Henry 545
Reading, University of 1263
Real Sociedad Geográfica. See under Madrid, Royal Geographical Society.
Reclus, Elisée 63, 138, 306, 485, 1079, 1539-1548, 1661
Reeves, Edward A. 1549
Regel, Fritz 722
Rein, Johannes Justus 720, 898, 1550
Renaissance 44, 75, 98
Rennell, James 294, 330, 1551-1553
Rennes, University of 715
Revert, Eugène 1554
Revue de géographie 858, 1227
Reyher, Samuel 730
Reynaud, Jean 487
Richmond, Duke of 377
Richthofen, Ferdinand von 63, 93, 143, 449, 667, 679, 688, 729, 755, 881-882, 1298, 1466-1467, 1555-1566
Riehl, Wilhelm Heinrich 889
Rio de Janeiro 539

Ritter, Carl 47, 49, 63, 70, 82, 93, 99, 101, 104, 122, 137, 145-146, 148-149, 154, 161, 169, 171, 184, 188, 196, 199, 211-213, 309-311, 343, 410, 466-467, 490, 494, 688, 728-729, 868, 881, 889, 891, 1062, 1149-1150, 1272, 1361, 1367, 1381, 1402, 1409, 1512, 1528, 1535, 1567-1613
Robequain, Charles 1614
Rohlfs, Gerhard 275, 1528
Romania, geography 981
Romanian Geographical Society 599
Romer, Eugeniusz 1615-1616
Rosler, Robert 701
Ross, James 227
Rostock, University of 719
Rousseau, Jean-Jacques 486, 838
Roxby, Percy Maude 112, 721, 1617
Roy, William 325, 377
Royal Dutch Geographical Society. See under Netherlands, Royal Dutch Geographical Society.
Royal Geographical Society. See under London, Royal Geographical Society.
Royal Scottish Geographical Society. See under Edinburgh, Royal Scottish Geographical Society.
Royce, Josiah 1101
Rühl, Alfred 1618
Russell, Richard Joel 1102
Russia 1389; cartography 290, 374, 382; contacts with non-Russian geographers 975; geographical observations of British visitors in 19th century 974; geography 934, 942, 944, 947, 950-952, 971; German travelers in 984. Also see Soviet Union.
Russian Geographical Society 515, 569

Sabine, Edward 1194, 1432, 1619-1620
Sahara 1189
St. Andrews, University of 1240

Subject Index

St. Petersburg Academy of Sciences 1444, 1660
St. Petersburg Mining Institute 1486
St. Petersburg, University of 1417, 1674
Salisbury, Rollin 1103-1105
Salmon, Thomas 827
Salzburg 977
San Francisco 537
Santarém, Viscount of 302
São Paulo Historical and Geographical Institute 556
Sapper, Karl 1621-1622
Sauer, Carl Ortwin 189, 471, 473, 481, 483, 707, 743, 761-763, 790, 1106-1112, 1712
Saussure, Horace-Bénédict de 437
Scandinavia 581; geography 927, 968
Schelling, Friedrich Wilhelm Joseph von 1606
Schimper, Andreas Franz Wilhelm 402
Schlagintweit brothers (Hermann, Adolf, and Robert) 1346
Schlagintweit, Robert von 725-726, 1346
Schlüter, Otto 1409, 1623-1624
Schmieder, Oskar 1625
Schmitthenner, Heinrich 656
Schott, Charles Anthony 434
Schouw, Joachim Frederik 485
Schrader, Franz 1626-1628
Schultz, Woldemar 1629
Schurtz, Heinrich 1528
Schweinfurth, Georg 1630
Schweizer Alpen-Club. See under Swiss Alpine Club.
Scientific Revolution 423
Sclater, Philip Lutley 429
Scoresby, William 1631
Scotland 818, 820-821, 825
Scott, Robert F. 1711
Scottish Geographical Magazine 649-650, 1490, 1648
Scylax of Caryanda 376
Semenov Tian-Shanskii, Petr 1193, 1632-1633

Semple, Ellen Churchill 473, 494, 743, 762, 1113-1119
Sequent occupance 482
Shaler, Nathaniel Southgate 702, 743, 1120-1124
Shaw, William Napier 451
Sheffield, University of 1441
Siberia 1328, 1343, 1524
Siebold, Philipp Franz von 898
Sierra Club 547
Sierra Nevada, California 1094
Sievers, Wilhelm 725-726
Silva Telles, Francisco Xavier da 1634
Simony, Friedrich 517, 668-669, 713, 935, 1635
Singh, Kishen 263, 274
Singh, Nain 274
Slicher van Bath, Bernard Hendrik 945
Smith, Adam 469
Smith, George Adam 1636
Smith, John 416
Smith, Joseph Russell 1125
Smithsonian Institution 1098
Smolenski, Jerzy 1637
Sociedad Mexicana de Geografía y Estadística. See under Mexican Geographical and Statistical Society.
Società Geografica Italiana. See under Italian Geographical Society.
Societas Meteorologica Palatina 432
Société de géographie de Bordeaux. See under Bordeaux Geographical Society.
Société de géographie d'Egypte. See under Egyptian Geographical Society.
Société de géographie et d'études coloniales de Marseille. See under Marseille Society of Geography and Colonial Studies.
Société de géographie de Genève. See under Geneva Geographical Society.

Société de géographie de Lille. See under Lille Geographical Society.
Société de géographie de Paris. See under Paris Geographical Society.
Société de géographie de Québec. See under Quebec Geographical Society.
Société roumaine de géographie. See under Romanian Geographical Society.
Société royale belge de géographie. See under Belgium, Royal Belgian Geographical Society.
Societies, geographical 604, 620; Great Britain 811
Society for the History of Discoveries 629
Sociografie 956
Sociclogy 495
Soil science 1061, 1225, 1230, 1241, 1483
Sölch, Johann 1638
Somerville, Mary 410, 1639-1643
Sorbonne. See under Paris, University of.
Sorre, Maximilien 483
South Africa, cartography 341
South African Geographical Society 557
South America 1174-1176, 1283, 1285-1288, 1335, 1350; exploration 249, 273, 276
South Australia 1702
South Pole 228, 257
Southeast Asia, geography 997
Soviet Union, geography 942-944, 947, 950-955, 960, 965, 982-983, 985; publications on history of geography 958. Also see Russia.
Space, outer 235-236, 266
Spain, geography 932, 964, 969, 978-980; topographic mapping 1203
Speke, John Hanning 275, 519
Speleology 220
Spencer, Herbert 479
Spinoza, Benedictus de 1303
Stamp, Laurence Dudley 732, 826
Stanley, Henry Morton 216, 275
Stein, Marc Aurel 1645

Steller, Georg Wilhelm 887
Stevens, Alexander 821
Stevens, Isaac 1126
Stewart, John Q. 203
Stieler, Adolf 334, 389
Strabo 99, 300, 1612
Strasbourg (Strassburg), University of 1156-1157, 1262
Strzelecki, Pawel Edmund 1646
Stuttgart, University of 1431
Suess, Eduard 406
Sumerians 280
Surveying 293, 319, 321, 323, 325, 329-330, 335, 341, 350, 360, 372, 374-377, 385, 396
Swansea, University College of 666
Sweden, geography 927, 936, 946, 949, 986
Swedish Society for Anthropology and Geography 949
Swiss Alpine Club 631
Swiss Foundation for Alpine Research 637
Switzerland, cartography 301, 327; geography 976; universities 663, 683, 687, 727
Sydney, University of 1704, 1712-1715
Sydow, Emil von 369, 1647
Syracuse University 1025, 1065

Tamayo, Jorge Leonides 1710
Tanganyika 1265-1267
Tartu, University of 1277
Taylor, Eva Germaine Rimington 1490, 1648
Taylor, Thomas Griffith 494, 1711-1715
Teilhard de Chardin, Pierre 1649
Teleki, Paul 1650
Teleology 1568, 1597
Texas 764, 1408
Thales 361
Theophrastus 402, 447-448
Thompson, David 1716
Thomson, John 1651
Thomson, Joseph 275

Subject Index 385

Thucydides 465
Thünen, Johan Heinrich von 203, 468-469, 1652
Tibet 263
Tiflis Caucasian Museum 1524
Tillo, Alexey Andreyevich 1653
Tokyo, University of 1717
Topelius, Zachris 967-968, 1654-1655
Topographic mapping 285, 329, 344, 1203
Torell, Otto 437
Toronto, University of 1000, 1712-1715
Toulouse, University of 1233
Tournefort, Joseph Pitton de 417
Town(e)ley, Richard 450
Toynbee, Arnold 493
Transactions of the Institute of British Geographers 644
Troll, Carl 1656-1657
Tübingen, University of 1276
Türst, Conrad 301
Turkey, geography 1005
Turku, University of 1277, 1522
Turner, Frederick Jackson 1127-1128

Ule, Wille 719
Ullman, Edward 1129
Ulysses 214
Uniformitarianism 405
United States 1679; Army Medical Department 428; cartography 291, 321, 336, 358; Coast and Geodetic Survey 1028, 1030-1031; Department of Agriculture 1024; geographers 794; geography 739-797, 952, 986, 1125; Geological Survey 1020, 1050, 1052, 1098; geology 461; geomorphology 425, 444; meteorology 403, 420; teaching 733; universities 677, 693-694; western exploration 245
Universities, French, geography in 847, 851, 856; German 876; German, geography in 867; Great Britain 801, 808, 815-817; Russian, geography in 953; Spanish 980;

United States 746, 749, 753, 757-758, 775, 788-789, 796
Utrecht, University of 686

Vallaux, Camille 1658
Valsan, George 1659
Varenius, Bernhard 122, 164, 177, 486, 795, 938, 970
Vauban, Sébastien Le Prestre de 486
Vega expedition 1493
Vegetation classification 409
Venezuela 1359, 1695, 1697-1698
Verein für Geographie und Statistik zu Frankfurt am Main. See under Frankfurt Geographical Society.
Vernadsky, Vladimir Ivanovich 1660
Verne, Jules 1661
Vico, Giambattista 908
Vidal de la Blache, Paul 93, 112, 164, 189, 211, 467, 828-829, 837, 841, 843, 845, 850, 852-853, 855, 865, 970, 1229, 1409, 1662-1670
Vienna Geographical Society 517, 605
Vienna, Military Geographical Institute 940
Vienna, researches on history of geography and exploration 931
Vienna, University of 517, 668-670, 713, 935, 1178, 1494, 1502, 1506, 1508, 1635, 1638
Virginia, University of 677
Vivien de Saint-Martin, Louis 1671
Vogel, Eduard 1528
Volney, Constantin Francois Chasse-Boeuf, Comte de 467, 838, 860, 1217, 1672
Voltaire, Francois-Marie Arouet de 1673
Voyeikov, Alexander Ivanovitch 934, 1193, 1674
Vujević, Pavle 1675
Vulcanology 1221

Wageningen, State Agricultural University of 1230
Wagner, Hermann 148, 555, 881, 1676

Wagner, Moritz 49, 1528, 1677-1678
Waibel, Leo 1679-1681
Wallace, Alfred Russel 429
Ward, Robert DeCourcy 1130
Warming, Eugenius 402
Warsaw, University of 1434
Washington, George 760
Washington, University of 1129
Weber, Alfred 1682
Wegener, Alfred 1683
Wegener, Georg 1684
Weimar Geographical Institute 876
Wells, Herbert George 1685
Werner, Abraham Gottlob 431
Weulersse, Jacques 1686
Whitney, Josiah Dwight 744, 755
Whittlesey, Derwent S. 482, 761
Wilkes, Charles 227
Wilkes Exploring Expedition 426
Windham, William 625
Wissler, Clark 1131
Wissmann, Hermann von 878
Wolff, Christian 78
Women travelers 217, 260, 265
Wooldridge, Sidney william 826
Wright, John Kirtland 1132
Würzburg, University of 690
Wuttke, Heinrich 679

Yale University 1045, 1054
Yamasaki, Naomasa 1717
Younghusband, Francis 1687

Zach, Franz Xaver von 651, 1688
Zannoni, Giovanni Antonio Rizzi 1689
Zeitschrift der Gesellschaft
   für Erdkunde zu Berlin 642
Zeune, Johann August 688, 729, 1690
Zoogeography 429
Zürich, Eidgenossischen
   Technischen Hochschule 683

Ref Z 6001 .D86 1985
Dunbar, Gary S.
The history of modern
 geography